Universitext

Universitext

Universitext is a series of textbooks that presents material from a wide variety of mathematical disciplines at master's level and beyond. The books, often well class-tested by their author, may have an informal, personal even experimental approach to their subject matter. Some of the most successful and established books in the series have evolved through several editions, always following the evolution of teaching curricula, to very polished texts.

Thus as research topics trickle down into graduate-level teaching, first textbooks written for new, cutting-edge courses may make their way into *Universitext*.

More information about this series at http://www.springer.com/series/223

Qingkai Kong

A Short Course in Ordinary Differential Equations

 Springer

Qingkai Kong
Department of Mathematical Sciences
Northern Illinois University
DeKalb, IL, USA

ISSN 0172-5939   ISSN 2191-6675 (electronic)
ISBN 978-3-319-35426-2   ISBN 978-3-319-11239-8 (eBook)
DOI 10.1007/978-3-319-11239-8
Springer Cham Heidelberg New York Dordrecht London

Mathematics Subject Classifications (2010): 34-01, 34Axx, 34Bxx, 34Cxx, 34Dxx, 37-01, 37Cxx

Printed on acid-free paper

Springer is part of Springer Science+Business Media (www.springer.com)

Contents

Preface

Ordinary differential equations (ODEs) is a well developed field in mathematics. Due to its broad applications in many areas of mathematics, science, and technology, its modern branches are still actively developing. Courses on ODEs are offered in most schools either as a required course or as an important elective for graduate students in various mathematical disciplines such as analysis, algebra, and mathematics education. Such courses provide students with a basic training in theoretical and applied analysis. Currently, many standard graduate textbooks for ODEs are excellent for research purposes. They are especially useful for the Ph.D. students whose dissertation areas are related to ODEs. However, most of these texts cover too much to teach; especially for a general one semester ODEs course. Some of them require knowledge from higher level courses such as real, complex, and functional analyses. Very often, instructors who teach such a course have to use their own lecture notes as textbooks or must select and reorganize materials from several existing textbooks in an effort to fit the course targets and to meet the needs of the students. This practice is inconvenient and students are often left without a text.

This text is not intended to serve as a "complete textbook" for ODEs as most existing textbooks do. Rather, it is intended to be used mainly for general knowledge courses on the qualitative theory of ODEs for beginning graduate students in mathematics. For this purpose, the book covers only the subjects that are of fundamental importance, and the prerequisites are strictly limited to elementary differential equations, linear algebra, and advanced calculus. Here it should be pointed out that the Jordan canonical form for square matrices, which are used in Chaps. 2–5, are usually covered, not in a first, but in a second undergraduate linear algebra course. To avoid adding the second linear algebra course as another prerequisite, I present the Jordan canonical form clearly in Sect. 2.4, and this is all that is needed in this book.

I would also like to point out that this book contains more than what can be covered in one semester. It is possible that it can be used for up to two semesters of teaching. However, it is designed specifically for a one semester ODEs course. More material gives the instructors options to choose which topics they will teach. In fact, this book is organized to allow instructors flexibility in choosing chapters and sections based on their teaching hours and preferences. The first three chapters form the basic part of the book, but Chaps. 4–6 are less dependent on one another. Chapter 5 must be taught after Sects. 4.1, 4.3, and 4.4, and 4.6 is used in Sect. 5.4. For a one semester course with three credit hours, I would suggest the following three possible selections of the course materials:

1. Chapters 1–3 and Sects. 4.1–4.6 with the proof of the Poincaré–Bendixson theorem optional.
2. Chapters 1–3; Sects. 4.1, 4.3, and 4.4; Sects. 5.1–5.3 and a brief sketch of Sect. 5.4.
3. Chapters 1–3, Sects. 6.1–6.4, and either Sects. 6.5 or 6.6.

In all three plans, Sects. 3.3 and 3.6 are optional, and the differentiable dependence in Sect. 1.5 need not be covered unless the complete proof of Theorem 6.6.4 for Sturm–Liouville problems with periodic boundary conditions will be given. Unlike most textbooks for ODEs, this book can be used as a "true text" in the sense that once a plan is chosen, it can be taught in the section-by-section manner, and no further adjustment of materials is necessary.

This book is written in an easy-to-understand style and at a level that is appropriate for the targeted audience. The presentation is rigorous. Results are stated clearly and accurately with proofs given for most of them. On the other hand, a few details of proofs are purposely left as exercises. In this way, the student gains knowledge of ODEs and further develops his abilities in analysis and logical thinking. Some existing results, which are usually given in a general or abstract setting, are stated and proved in elementary ways; and many nontrivial mathematical details, which were taken for granted in classical textbooks, are elaborated upon to assist in the students' understanding. Statements and proofs of many existing results have been modified/simplified. A multitude of examples and figures are given for illustration purposes, and carefully selected exercises, together with some answers and hints, help the reader to understand the course materials more deeply and with better understanding.

Although this book is mainly geared toward entry-level graduate teaching, it can also serve as a textbook for the basic ODEs courses for those pursuing a Ph.D. degree in differential equations. On the other hand, since no further training from analysis is required, if appropriate chapters and sections are selected, the text is suitable for senior undergraduates as well. At the same time, every effort has been made to keep this text concise and to cover the most significant results, avoiding lengthy and less significant discussion.

Here I briefly summarize the topics covered in this book.

Chapter 1: Initial Value Problems. This chapter discusses the basic results on initial value problems for systems of differential equations. First, several preliminaries from analysis are introduced: Lipschitz continuity, Gronwall inequality, and properties of uniformly convergent sequences and series of functions including the Arzelà–Ascoli Lemma. Next, the existence-uniqueness theorem by Picard and the existence theorem by Cauchy–Peano are presented. Results on the continuation of solutions and continuous and differentiable dependence of solutions on parameters and initial conditions are given with detailed proofs. Kneser's theorem on the number of non-unique solutions is stated without proof.

Chapter 2. Linear Differential Equations. In this chapter, we first study the general structure of solution spaces for homogeneous and nonhomogeneous linear equations and derive formulas for the solutions in terms of fundamental matrix solutions. We further discuss the behavior of the homogeneous linear equations with constant coefficients and with periodic coefficients, respectively. For the former, the matrix exponential form of fundamental matrix solutions is introduced, whose calculations are provided for both the diagonalizable matrix case and the general case. Here, a variation of the Putzer algorithm is given to avoid the involvement of the generalized eigenvectors for matrices. For the latter, the Floquet theory, including characteristic multipliers/exponents and the structure of fundamental matrix solutions, is presented.

Chapter 3. Lyapunov Stability Theory. We study the Lyapunov stability in this chapter. After the introduction of the basic concepts, we discuss the stability of linear equations. Necessary and sufficient conditions for stabilities are given in terms of the fundamental matrix solutions. Explicit conditions for the stability of homogeneous linear equations with constant coefficients and with periodic coefficients are derived using the eigenvalues of the coefficient matrix and the characteristic multipliers/exponents, respectively. We also include in this chapter the Lozinskii measure method for the stability of linear equations with variable coefficients presented in Coppel [12] and Vidyasagar [43], which is useful but rarely covered by the standard textbooks for basic ODEs. The latter part of this chapter deals with nonlinear stability. Several criteria by linearization are given, and theorems on the Lyapunov function method are proved for both autonomous and nonautonomous equations. Examples are given to demonstrate their applications. We comment that the nonlinear stability studied here is restricted to local stability only.

Chapter 4. Dynamic Systems and Planar Autonomous Equations. This chapter covers topics that may be contained in several chapters in many other textbooks. Because they are closely related in nature, we combined these topics into one chapter. We begin with a general discussion on autonomous equations, followed by an introduction to dynamic systems including flows, orbits, limit sets, and invariant sets. Then we turn our attention to planar autonomous equations and study the topological structures of different types of equilibria for linear equations and their preservation under nonlinear perturbations. The well known Poincaré–Bendixson theorem and its generalization are proved. The applications of the Poincaré–Bendixson theorem and the Lyapunov function method to the existence and nonexistence of periodic solutions and limit cycles are discussed. Orbital stability and its determination are also discussed. Indices of equilibria are introduced and used as additional means to determine the nonexistence of periodic solutions. The last section of this chapter is devoted to the further investigation of flows of n-dimensional autonomous equations. The existence and characterization of invariant subspaces for linear equations and invariant manifolds for nonlinear equations are studied. The center manifold theorem is applied to reduce the dimensions of the systems in order to determine the stability of the equilibria. Due to the lengths of the proofs, some theorems in the last section are stated without proofs.

Chapter 5. Introduction of Bifurcation Theory. This chapter introduces some basic concepts and theorems on one-dimensional bifurcations and Hopf bifurcations. Results on various one-dimensional bifurcations for scalar equations are developed. The Lyapunov–Schmidt method for one-dimensional bifurcations for planar systems is introduced. Results on the Hopf bifurcations for planar systems are derived using the Lyapunov function method and the Friedrich method. The normal form approach is avoided in this book. The implicit function theorem plays a critical role in the proofs. Many examples are given for intuitive explanations.

Chapter 6. Second-Order Linear Equations. Second-order linear differential equations have been studied for almost two centuries and have broad applications in science, engineering, and many other fields. This is the motivation for us to discuss second-order linear equations and related problems in this chapter. We first introduce the Prüfer transformation and present the Sturm comparison and separation theorems. Based on these theorems, we investigate the properties of nonoscillatory equations and derive several results on oscillation including the celebrated classical Fite–Wintner criterion, Wintner criterion, Kamenev criterion, Hartman criterion, and the newly obtained interval oscillation criterion. Green's functions are presented for boundary value problems with separated boundary condition and with periodic boundary condition in the general setting. This sets up a foundation for the further study of nonlinear boundary value problems. Sturm–Liouville problems with separated boundary conditions are studied using the Prüfer angle approach. Sturm–Liouville problems with periodic boundary conditions are also included in this chapter. However, they are studied by an elementary approach rather than by the spectral theory approach as given in some other books. Dependence of eigenvalues on the equation and on the boundary condition is discussed.

Finally, I would like to mention that this book has been developed from my teaching and research experience on ODEs for more than 20 years. I would like to take this opportunity to express my sincere appreciation to many of my students for their helpful suggestions for corrections and improvements to this manuscript.

DeKalb, IL, USA   Qingkai Kong

Notation and Abbreviations

Notation

$\Longleftrightarrow$	Necessary and sufficient condition
$\Longrightarrow$	Necessary condition
$\Longleftarrow$	Sufficient condition
0	Zero scalar or zero vector
$\mathbf{0}$	Zero matrix
$A\ [B]$ holds	A [respectively B] holds
$\|A\| = \sum_{i,j=1}^{n}\|a_{ij}\|$	Norm of matrix $A = \{a_{ij}\}$
$\|A\|_i,\ i = 1, 2, \infty$	Induced i-norm of matrix A
$\|a\|$	Absolute value of a if a is a scalar or norm of a if a is a vector
$\|a\|_1 := \sum_{i=1}^{n}\|a_i\|$	1-norm of $a = (a_1, \ldots, a_n)^T$
$\|a\|_2 := (\sum_{i=1}^{n}\|a_i\|^2)^{1/2}$	2-norm of $a = (a_1, \ldots, a_n)^T$
$\|a\|_\infty := \max_{1\le i\le n}\|a_i\|$	∞-norm of $a = (a_1, \ldots, a_n)^T$
$C(A, B)$	Space of continuous functions from A to B
$C^k(A, B)$	Space of k-th continuously differentiable functions from A to B
$\mathbb{C}$	Set of all complex numbers
$d(A, B)$	Distance between sets A and B
$\det A$	Determinant of matrix A
∇f	Gradient of function f
$\mathbf{i}$	Imaginary-part unit
$\mu_i(A),\ i = 1, 2, \infty$	i-Lozinskii measure of matrix A
$\nu_i(A),\ i = 1, 2, \infty$	$-\mu_i(-A)$
$\mathbb{N} := \{1, 2, \ldots\}$	Set of natural numbers
$\mathbb{N}_0 := \{0, 1, 2, \ldots\}$	Set of nonnegative integers
o-notation	$f(t) = o(g(t))$ as $t \to a$ if $\lim_{t\to a} f(t)/g(t) = 0$
O-notation	$f(t) = O(g(t))$ as $t \to a$ if $\|f(t)/g(t)\| \le M$ with $M > 0$ in a neighborhood of a
$\dfrac{\partial f}{\partial x} := \dfrac{\partial(f_1, \ldots, f_n)}{\partial(x_1, \ldots, x_n)}$	Jacobian matrix of $f = (f_1, \ldots, f_n)^T$ with respect to $x = (x_1, \ldots, x_n)^T$

$\operatorname{Re}\lambda$ Real-part of the complex number λ

$\mathbb{R} := (-\infty, \infty)$ Set of all real numbers

$\mathbb{R}_+ := [0, \infty)$ Set of all nonnegative numbers

$\mathbb{R}^n := (-\infty, \infty)^n$ Cartesian product of n $\mathbb{R}$s

$\mathbb{Z} := \{0, \pm 1, \pm 2, \dots\}$ Set of integers

Abbreviations

(A-N)	Left-asymptotically-stable and right-unstable
BVP	Boundary value problem
BC	Boundary condition
IC	Initial condition
IVP	Initial value problem
(N-A)	Left-unstable and right-asymptotically-stable
SLP	Sturm–Liouville problem

CHAPTER 1

Initial Value Problems

1.1. Introduction

As you have learned from an elementary differential equations course, a differential equation, more precisely an ordinary differential equation, is an equation for an unknown function of a single variable which involves certain derivatives of the unknown function. The order of the highest derivative of the unknown function appearing in the equation is called the order of the differential equation. Based on this, nth-order scalar differential equations have the general form

$$(1.1.1) \qquad F(t, x, x', \ldots, x^{(n)}) = 0,$$

where F is a function of $n+2$ variables. If F satisfies certain conditions, then Eq. (1.1.1) can be written into the standard form

$$(1.1.2) \qquad x^{(n)} = f(t, x, x', \ldots, x^{(n-1)}),$$

where f is a function of $n+1$ variables. A solution of Eq. (1.1.1) or Eq. (1.1.2) is a scalar function $x(t)$ such that $x'(t), \ldots, x^{(n)}(t)$ exist and the equation will become an identity on the domain of the argument t when x and $x', \ldots, x^{(n)}$ are replaced by $x(t)$ and $x'(t), \ldots, x^{(n)}(t)$, respectively. A differential equation may or may not have a solution; and if it does, most likely it has infinitely many solutions. The general solution of Eq. (1.1.1) or Eq. (1.1.2), if it exists, is a solution that contains n arbitrary constants $c_1, \ldots, c_n$. Under certain assumptions on the function F or f, the values of $c_1, \ldots, c_n$ can be uniquely determined for a particular solution by n initial conditions (ICs) defined at a single point (initial point) t_0:

$$(1.1.3) \qquad x(t_0) = z_1, \; x'(t_0) = z_2, \; \ldots, \; x^{(n-1)}(t_0) = z_n;$$

where $z_1, \ldots, z_n \in \mathbb{R}$. The problem consisting of Eq. (1.1.2) and IC (1.1.3) is called an initial value problem (IVP).

The above concepts for scalar differential equations and IVPs can be extended to vector-valued differential equations and IVPs. For instance, the standard form of first-order systems of differential equations is

© Springer International Publishing Switzerland 2014
Q. Kong, *A Short Course in Ordinary Differential Equations*, Universitext,
DOI 10.1007/978-3-319-11239-8_1

$$(1.1.4) \qquad \begin{cases} x_1' = f_1(t, x_1, \ldots, x_n) \\ x_2' = f_2(t, x_1, \ldots, x_n) \\ \qquad \cdots \cdots \\ x_n' = f_n(t, x_1, \ldots, x_n) \end{cases} \qquad \text{or} \qquad x' = f(t, x),$$

where $x = (x_1, \ldots, x_n)^T$ and $f = (f_1, \ldots, f_n)^T$. A solution of system (1.1.4) is an n-dimensional vector-valued function $x(t)$ such that $x'(t)$ exists and (1.1.4) will become an identity on the domain of the independent variable t when x and x' are replaced by $x(t)$ and $x'(t)$, respectively. The general solution of system (1.1.4), if it exists, is a solution that contains n arbitrary constants $c_1, \ldots, c_n$. Under certain assumptions on the function f, the values of $c_1, \ldots, c_n$ can be uniquely determined for a particular solution by n ICs

$$(1.1.5) \qquad x_1(t_0) = z_1, \ x_2(t_0) = z_2, \ \ldots, \ x_n(t_0) = z_n; \quad \text{or} \quad x(t_0) = z;$$

where $z_1, \ldots, z_n \in \mathbb{R}$ and $z = (z_1, \ldots, z_n)^T$. The problem consisting of system (1.1.4) and ICs (1.1.5) is an IVP.

We comment that any standard scalar or vector-valued differential equation or IVP can be changed to a first-order system of differential equations or IVP. For example, if we let

$$x_1 = x, \ x_2 = x', \ldots, \ x_n = x^{(n-1)},$$

then Eq. (1.1.2) becomes the system of differential equations

$$\begin{cases} x_1' = x_2 \\ x_2' = x_3 \\ \qquad \cdots \cdots \\ x_{n-1}' = x_n \\ x_n' = f(t, x_1, \ldots, x_n) \end{cases}$$

and IC (1.1.3) becomes (1.1.5). Therefore, the existence and uniqueness of any scalar or vector-valued IVP can be represented by those of IVP (1.1.4), (1.1.5).

Under what conditions does IVP (1.1.4), (1.1.5) have a unique solution? Before answering this question, let's look at the following examples of scalar IVPs.

EXAMPLE 1.1.1. Consider the IVP

$$(1.1.6) \qquad\qquad x' = tx, \quad x(0) = x_0.$$

By solving the problem we see that the solution of IVP (1.1.6) is $x = x_0 e^{t^2/2}$ which exists on the whole real number line $\mathbb{R}$ and is unique.

EXAMPLE 1.1.2. Consider the IVP

$$(1.1.7) \qquad\qquad x' = x^2, \quad x(0) = x_0.$$

By solving the problem we see that when $x_0 \neq 0$, IVP (1.1.7) has a unique solution $x = x_0/(1 - tx_0)$. This solution exists on $(-\infty, 1/x_0)$ when $x_0 > 0$ and on $(1/x_0, \infty)$ when $x_0 < 0$. We note that this solution is never zero. When $x_0 = 0$, we observe that $x \equiv 0$ is a solution which exists on the whole real number line $\mathbb{R}$. We claim that it is the only solution. Otherwise, IVP (1.1.7) has a solution $x_1(t)$ satisfying $x_1(t_1) = x_1^* \neq 0$ for some $t_1 \in \mathbb{R}$. This means that $x_1(t)$ is a solution of the IVP $x' = x^2$, $x(t_1) = x_1^*$. However, by solving this IVP, we see that $x_1(t)$ can never be zero. This contradicts the assumption that $x_1(0) = 0$. Therefore, for all $x_0 \in \mathbb{R}$, IVP (1.1.7) has a unique solution, but solutions for different x_0 may be defined on different intervals. See Fig. 1.1.

EXAMPLE 1.1.3. Consider the IVP

$$(1.1.8) \qquad\qquad x' = |x|^{1/2}, \quad x(0) = 0.$$

Clearly, $x = 0$ is a solution of IVP (1.1.8) which exists on the whole real number line $\mathbb{R}$. By solving the equation, we see that for any $c > 0$, $x = (t-c)^2/4$ is a solution of the equation in (1.1.8) for $t \in [c, \infty)$. By combining this solution with the zero solution, we have that for any $c > 0$

$$x = \begin{cases} (t-c)^2/4, & t \geq c \\ 0, & t < c \end{cases}$$

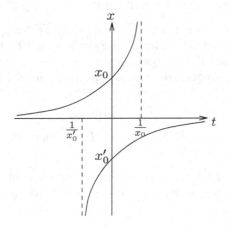

FIGURE 1.1. Solution may not exist on the whole line

is also a solution of IVP (1.1.8) which exists on $\mathbb{R}$. Similarly, for any $c < 0$

$$x = \begin{cases} -(t-c)^2/4, & t \leq c \\ 0, & t > c \end{cases}$$

is also a solution of IVP (1.1.8) which exists on $\mathbb{R}$. This means that IVP (1.1.8) has an infinite number of solutions on $\mathbb{R}$. See Fig. 1.2.

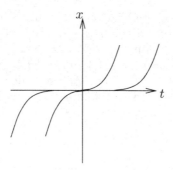

FIGURE 1.2. Nonunique solutions

EXAMPLE 1.1.4. Consider the IVP

$$(1.1.9) \qquad x' = \operatorname{sgn} t := \begin{cases} 1, & t > 0 \\ 0, & t = 0 \\ -1, & t < 0 \end{cases}, \qquad x(0) = 0.$$

We claim that IVP (1.1.9) has no solution at all. In fact, if it had a solution, then the solution would be $x = |t|$. However, this function is not differentiable at $t = 0$. We have reached a contradiction.

From the above examples we observe that an IVP may or may not have a solution, and if it does, the solution may or may not be unique. Moreover, the interval where the solution exists depends on the equation and the IC. To study the existence, uniqueness, and further problems for IVPs, we consider IVPs in the form

$$(V) \qquad\qquad x' = f(t, x), \quad x(t_0) = x_0,$$

where $f = (f_1, \ldots, f_n)^T : D \to \mathbb{R}^n$ with D an open subset of $\mathbb{R} \times \mathbb{R}^n$ and $(t_0, x_0) \in D$.

DEFINITION 1.1.1. A function $\phi(t)$ is said to be a solution of IVP (V) if $\phi(t)$ is differentiable in an interval J containing t_0 as an interior point such that $(t, \phi(t)) \in D$ and $\phi'(t) \equiv f(t, \phi(t))$ on J, and $\phi(t_0) = x_0$.

To establish existence and uniqueness results for IVPs, we will need the following equivalence between IVP (V) and an integral equation.

LEMMA 1.1.1. *Assume* $f \in C(D, \mathbb{R}^n)$. *Then* $x(t)$ *is a solution of the IVP* (V) $\iff$ $x(t)$ *is a solution of the integral equation*

$$(I) \qquad\qquad x(t) = x_0 + \int_{t_0}^t f(s, x(s)) \, ds,$$

where

$$\int_{t_0}^t f \, ds = \left(\int_{t_0}^t f_1 \, ds, \ldots, \int_{t_0}^t f_n \, ds \right)^T.$$

PROOF. Assume $x(t)$ is a solution of IVP (V). Then

$$x'(t) \equiv f(t, x(t)), \quad x(t_0) = x_0.$$

It follows that

$$x(t) - x_0 = \int_{t_0}^{t} f(s, x(s)) \, ds.$$

Hence $x(t)$ is a solution of Eq. (I).

Assume $x(t)$ is a solution of Eq. (I). Then $f(s, x(s))$ is integrable from t_0 to t for any t in the domain. Hence $\int_{t_0}^{t} f(s, x(s)) \, ds$ is continuous in t. It follows from (I) that $x(t)$ is continuous. Moreover, since $f \in C(D, \mathbb{R}^n)$, $f(t, x(t))$ is continuous, from which it follows that the right-hand side of (I) is differentiable, and so is $x(t)$. Differentiating both sides of (I) we have $x'(t) \equiv f(t, x(t))$. Note that $x(t_0) = x_0$. Hence $x(t)$ is a solution of IVP (V).   □

There are three commonly used norms on $\mathbb{R}^n$, $|\cdot|_1$, $|\cdot|_2$, and $|\cdot|_\infty$, which are defined as follows: for any $x = (x_1, \ldots, x_n) \in \mathbb{R}^n$

$$|x|_1 := \sum_{i=1}^{n} |x_i|, \quad |x|_2 := \left(\sum_{i=1}^{n} |x_i|^2 \right)^{1/2}, \quad \text{and} \quad |x|_\infty := \max_{1 \le i \le n} |x_i|.$$

It is well known that the three norms are equivalent, i.e., for any $j, k \in \{1, 2, \infty\}$, there exist $c_1, c_2 > 0$ such that the inequalities $c_1 |x|_k \le |x|_j \le c_2 |x|_k$ hold for all $x \in \mathbb{R}^n$. However, for convenience, we will use the $|\cdot|_1$-norm in $\mathbb{R}^n$ throughout this book and simply denote it by $|\cdot|$, unless indicated specifically as in Sects. 3.3 and 3.6. Without confusion we denote by $0 = (0, \ldots, 0)^T$ the zero vector in $\mathbb{R}^n$. Recall that the norm $|\cdot|$ satisfies the following properties:

(i) $|x| \ge 0$, and $|x| = 0 \iff x = 0$;
(ii) $|cx| = |c||x|$ for any $c \in \mathbb{R}$ and $x \in \mathbb{R}^n$;
(iii) $|x + y| \le |x| + |y|$ for any $x, y \in \mathbb{R}^n$.

Note that this norm also satisfies

(iv) $|\int_a^b f(t) \, dt| \le \int_a^b |f(t)| \, dt$ if $a \le b$ and $f \in L([a, b], \mathbb{R}^n)$.

As is traditional, we let $\mathbb{R}_+ = [0, \infty)$.

1.2. Preliminaries

In this section, we will introduce several preliminaries from analysis which are important in dealing with differential equations. The first one is on the Lipschitz continuity of functions which plays a key role in establishing existence and uniqueness of solutions of IVPs.

I. Lipschitz continuity.

DEFINITION 1.2.1. Let $f : E \subset \mathbb{R}^n \to \mathbb{R}^n$. Then f is Lipschitz on E, or f satisfies a Lipschitz condition on E, if there exists a constant $k > 0$ such that

$$|f(x_1) - f(x_2)| \le k|x_1 - x_2|$$

for any $x_1, x_2 \in E$; f is locally Lipschitz on E, or f satisfies a local Lipschitz condition on E, if for any $x^* \in E$, there are a neighborhood $\mathcal{N}^*$ of x^* in E and a constant $k^* > 0$ such that

$$|f(x_1) - f(x_2)| \le k^*|x_1 - x_2|$$

for any $x_1, x_2 \in \mathcal{N}^*$.

REMARK 1.2.1. From the definition we have the following observations:

(i) If f is Lipschitz on a set E, then f is uniformly continuous on E, and hence is continuous on E. However, it is easy to see from advanced calculus that not every continuous function or even uniformly continuous function on E is Lipschitz on E.

(ii) Assume f is differentiable on a convex set E (i.e., for any $x_1, x_2 \in E$, the line segment between x_1 and x_2 is contained in E). Then the Jacobian matrix $\dfrac{\partial f}{\partial x} := \dfrac{\partial(f_1, \ldots, f_n)}{\partial(x_1, \ldots, x_n)}$ of f is bounded on E implies that f is Lipschitz on E, and $\partial f/\partial x$ is continuous on E implies that f is locally Lipschitz on E.

The student is expected to figure out the details of the above for the case $n = 1$ as shown in Exercises 1.2 and 1.3. The proof for the general case should utilize the mean value theorem in $\mathbb{R}^n$.

DEFINITION 1.2.2. Let $f : D \subset \mathbb{R} \times \mathbb{R}^n \to \mathbb{R}^n$. Then f is Lipschitz in x on D, or f satisfies a Lipschitz condition in x on D, if there exists a constant $k > 0$ such that

$$|f(t, x_1) - f(t, x_2)| \le k|x_1 - x_2|$$

for any $(t, x_1), (t, x_2) \in D$; f is locally Lipschitz in x on D, or f satisfies a local Lipschitz condition in x on D, if for any $(t^*, x^*) \in D$, there are a neighborhood $\mathcal{N}^*$ of (t^*, x^*) in D and a constant $k^* > 0$ such that

$$|f(t, x_1) - f(t, x_2)| \le k^*|x_1 - x_2|$$

for any $(t, x_1), (t, x_2) \in \mathcal{N}^*$.

REMARK 1.2.2. Assume $f(t, x)$ is continuous and has partial derivative (Jacobian matrix) $\partial f/\partial x$ with respect to x on a convex set $D \subset \mathbb{R} \times \mathbb{R}^n$. Then, as in Remark 1.2.1, $\partial f/\partial x$ is bounded on D implies that f is Lipschitz in x on D, and $\partial f/\partial x$ is continuous on D implies that f is locally Lipschitz in x on D.

II. Gronwall inequality.

The well known Gronwall inequality, due to T. H. Gronwall, gives estimates for upper bounds of solutions of certain linear integral inequalities.

LEMMA 1.2.1 (Gronwall Inequality). *Assume that $M \in \mathbb{R}$, $h \in C([t_0, \infty)$, $\mathbb{R}_+)$ for some $t_0 \in \mathbb{R}$, and $u(t)$ is a continuous solution of the inequality*

$$(1.2.1) \qquad u(t) \leq M + \int_{t_0}^{t} h(s)u(s)\, ds, \quad t \geq t_0.$$

Then

$$(1.2.2) \qquad u(t) \leq M e^{\int_{t_0}^{t} h(s)\, ds}, \quad t \geq t_0.$$

PROOF. Let

$$r(t) = M + \int_{t_0}^{t} h(s)u(s)\, ds.$$

Then for $t \geq t_0$

$$r'(t) = h(t)u(t) \leq h(t)r(t),$$

i.e., $r'(t) - h(t)r(t) \leq 0$. Multiplying both sides by $e^{-\int_{t_0}^{t} h(s)\, ds}$ we find that $(r(t)e^{-\int_{t_0}^{t} h(s)\, ds})' \leq 0$ and hence $r(t)e^{-\int_{t_0}^{t} h(s)ds}$ is decreasing. Thus for $t \geq t_0$

$$r(t)e^{-\int_{t_0}^{t} h(s)ds} \leq r(t_0) = M.$$

This means that

$$r(t) \leq M e^{\int_{t_0}^{t} h(s)\, ds}.$$

Then the conclusion follows since $u(t) \leq r(t)$. $\qquad \square$

The following result is a direct consequence of Lemma 1.2.1 with $M = 0$.

COROLLARY 1.2.1. *Assume that $h \in C([t_0, \infty), \mathbb{R}_+)$ for some $t_0 \in \mathbb{R}$, and $u(t) \geq 0$ is a continuous solution of the inequality*

$$u(t) \leq \int_{t_0}^{t} h(s)u(s)\, ds, \quad t \geq t_0.$$

Then $u(t) \equiv 0$, $\quad t \geq t_0$.

The original Gronwall inequality given in Lemma 1.2.1 can be extended to the case where the constant M is replaced by a function of t.

LEMMA 1.2.2 (Generalized Gronwall Inequality). *Assume that $f \in C([t_0, \infty), \mathbb{R})$ and $h \in C([t_0, \infty), \mathbb{R}_+)$ for some $t_0 \in \mathbb{R}$, and $u(t)$ is a continuous solution of the inequality*

$$(1.2.3) \qquad u(t) \leq f(t) + \int_{t_0}^{t} h(s)u(s)\, ds, \quad t \geq t_0.$$

Then

$$(1.2.4) \qquad u(t) \leq f(t) + \int_{t_0}^{t} f(s)h(s)e^{\int_{s}^{t} h(\tau)\, d\tau}\, ds, \quad t \geq t_0.$$

The proof is left as an exercise, see Exercise 1.5.

REMARK 1.2.3. (i) The upper bound for $u(t)$ given in Lemma 1.2.1 [Lemma 1.2.2] is the best in the sense that when the equality in (1.2.1) [(1.2.3)] holds, then the equality in (1.2.2) [(1.2.4)] holds.

(ii) In Lemma 1.2.2, when $f(t) \equiv M$ for some $M \in \mathbb{R}$, then from (1.2.4) we have

$$
\begin{aligned}
u(t) \quad &\leq \quad M + M \int_{t_0}^t h(s) e^{\int_s^t h(\tau) \, d\tau} \, ds \\
&= \quad M - M(1 - e^{\int_{t_0}^t h(\tau) \, d\tau}) = M e^{\int_{t_0}^t h(\tau) \, d\tau}.
\end{aligned}
$$

This shows that Lemma 1.2.2 covers Lemma 1.2.1 as a special case.

(iii) The Gronwall inequalities given in Lemmas 1.2.1 and 1.2.2 can be extended to the case when $t \leq t_0$. This is left as an exercise, see Exercise 1.6.

III. Sequences and series of functions.

This last part helps the student to recall and to further develop some knowledge of sequences and series of functions learned in advanced calculus.

DEFINITION 1.2.3. Let $f_m : [a, b] \to \mathbb{R}^n$, $m = 1, 2, \ldots$. Then the sequence of functions $\{f_m(t)\}_{m=1}^\infty$ is uniformly convergent on $[a, b]$ if there exists a function $f : [a, b] \to \mathbb{R}^n$ satisfying that for any $\epsilon > 0$, there exists an $N \in \mathbb{N}$ such that for any $m \geq N$ and any $t \in [a, b]$ we have that $|f_m(t) - f(t)| < \epsilon$. In this case, we say that $\{f_m(t)\}_{m=1}^\infty$ is uniformly convergent to $f(t)$ on $[a, b]$, or $f(t)$ is the uniform limit of $\{f_m(t)\}_{m=1}^\infty$ on $[a, b]$.

Uniformly convergent sequences satisfy the following properties:

LEMMA 1.2.3. (a) Assume $f_m \in C([a, b], \mathbb{R}^n)$ for $m = 1, 2, \ldots$, and the sequence of functions $\{f_m(t)\}_{m=1}^\infty$ is uniformly convergent to $f(t)$ on $[a, b]$. Then $f \in C([a, b], \mathbb{R}^n)$, and

$$
\lim_{m \to \infty} \int_a^b f_m(t) \, dt = \int_a^b f(t) \, dt.
$$

(b) Assume $f_m \in C([a, b], G)$ for $m = 1, 2, \ldots$, where G is a compact subset of $\mathbb{R}^n$, and the sequence of functions $\{f_m(t)\}_{m=1}^\infty$ is uniformly convergent to $f(t)$ on $[a, b]$. Then for any $g \in C(G, \mathbb{R}^n)$, the sequence of functions $\{g(f_m(t))\}_{m=1}^\infty$ is uniformly convergent to $g(f(t))$ on $[a, b]$.

PROOF. Part (a) is a basic result in advanced calculus, see [17, Theorems 7.3 and 7.5] for the proof. Now we prove Part (b).

By the assumption, $f \in C([a, b], G)$. Note that g is uniformly continuous on G. Hence for any $\epsilon > 0$, there exists a $\delta > 0$ such that for any $u_1, u_2 \in G$ with $|u_1 - u_2| < \delta$, $|g(u_1) - g(u_2)| < \epsilon$. Since $\{f_m(t)\}_{m=1}^\infty$ is uniformly convergent to f on $[a, b]$, there exists an $N \in \mathbb{N}$ such that for all $m \geq N$ and $t \in [a, b]$ we have $|f_m(t) - f(t)| < \delta$. Combining the above, we obtain that for any $\epsilon > 0$, there exists an $N \in \mathbb{N}$ such that for all $m \geq N$ and $t \in [a, b]$ we have $|g(f_m(t)) - g(f(t))| < \epsilon$. This verifies the uniform convergence of $g(f_m(t))$. □

REMARK 1.2.4. Lemma 1.2.3, Part (b) can be extended with a similar proof to the following: For $g \in C([a, b] \times G, \mathbb{R}^n)$, the sequence of functions $\{g(t, f_m(t))\}_{m=1}^{\infty}$ is uniformly convergent to $g(t, f(t))$ on $[a, b]$.

DEFINITION 1.2.4. Let $g_m : [a, b] \to \mathbb{R}^n$, $m = 1, 2, \ldots$. Then the series of functions $\sum_{m=1}^{\infty} g_m(t)$ is uniformly convergent on $[a, b]$ if the partial sum sequence

$$\{f_m(t)\}_{m=1}^{\infty} := \left\{ \sum_{i=1}^{m} g_i(t) \right\}_{m=1}^{\infty}$$

is uniformly convergent on $[a, b]$.

LEMMA 1.2.4 (Weierstrass M-Test). Assume there exist constants M_m, $m = 1, 2, \ldots$, such that

(a) $|g_m(t)| \leq M_m$ for all $t \in [a, b]$ and $m = 1, 2, \ldots$, and

(b) $\sum_{m=1}^{\infty} M_m$ is convergent.

Then the series of functions $\sum_{m=1}^{\infty} g_m(t)$ is uniformly convergent on $[a, b]$.

DEFINITION 1.2.5. Let $\mathfrak{F}$ be a collection of functions defined on a compact interval J. Then

(a) $\mathfrak{F}$ is uniformly bounded on J if there exists an $M > 0$ such that $|f(t)| \leq M$ on J for all $f \in \mathfrak{F}$;

(b) $\mathfrak{F}$ is equicontinuous on J if for any $\epsilon > 0$, there exists a $\delta > 0$ such that for any $f \in \mathfrak{F}$ and any $t_1, t_2 \in J$ satisfying $|t_1 - t_2| < \delta$, we have $|f(t_1) - f(t_2)| < \epsilon$.

EXAMPLE 1.2.1. Let $0 \leq k_1 < k_2 \leq \infty$ and

$$\mathfrak{F} = \{f_k(t) = t^k : k_1 < k < k_2, t \in [0, 1]\}.$$

Clearly, $\mathfrak{F}$ is uniformly bounded on J with $M = 1$.

Now we show that it is equicontinuous on J if $1 \leq k_1$ and $k_2 < \infty$. In fact, for any $k_1 \leq k \leq k_2$ and $t_1, t_2 \in [0, 1]$, by the mean value theorem, there exists a c between t_1 and t_2 such that

$$
\begin{aligned}
|f(t_1) - f(t_2)| &= |t_1^k - t_2^k| = |kc^{k-1}(t_1 - t_2)| \\
&\leq k|t_1 - t_2| < k_2|t_1 - t_2|.
\end{aligned}
$$

This implies that the definition of equicontinuity is satisfied.

It is easy to see that $\mathfrak{F}$ is not equicontinuous on J if $k_1 < 1$ or $k_2 = \infty$.

LEMMA 1.2.5 (Arzelà–Ascoli). Suppose that a collection of functions $\mathfrak{F}$ is uniformly bounded and equicontinuous on a compact interval J. Then every sequence $\{f_n\} \subset \mathfrak{F}$ has a subsequence which is uniformly convergent on J.

See Walter [**44**, P74] or Coddington and Levinson [**11**, P5] for the proof.

1.3. Existence and Uniqueness Theorems

Now we go back to IVP (V). We first introduce the results on the existence and uniqueness of the solution of IVP (V) due to Charles Émile Picard.

LEMMA 1.3.1. *Let*

$$(1.3.1) \qquad G = \{(t, x) : |t - t_0| \le a, |x - x_0| \le b\}.$$

Suppose that $f \in C(G, \mathbb{R}^n)$ and is Lipschitz in x on G. Then IVP (V) has a unique solution which exists for $|t - t_0| \le \gamma$, where $\gamma = \min\{a, b/M\}$ with $M = \max_{(t,x) \in G} |f(t, x)|$.

This result is illustrated in Fig. 1.3.

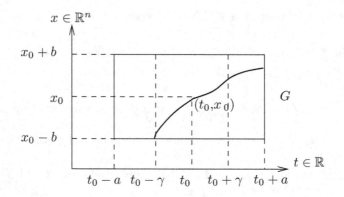

FIGURE 1.3. Rectangular domain

PROOF. Note that IVP (V) is equivalent to the integral equation (I).

(i) We first prove uniqueness.

Assume IVP (V) has two solutions $x_1(t)$ and $x_2(t)$. By Lemma 1.1.1, they are solutions of the integral equation (I), i.e.,

$$x_i(t) = x_0 + \int_{t_0}^t f(s, x_i(s))\, ds, \quad i = 1, 2.$$

Subtracting the equations with $i = 1, 2$ and using the Lipschitz condition we have that for $t \ge t_0$

$$|x_1(t) - x_2(t)| \quad = \quad \left| \int_{t_0}^t (f(s, x_1(s)) - f(s, x_2(s)))\, ds \right|$$

$$\le \quad \int_{t_0}^t |f(s, x_1(s)) - f(s, x_2(s))|\, ds \le \int_{t_0}^t k|x_1(s) - x_2(s)|\, ds,$$

where k is the Lipschitz constant. Let $u(t) = |x_1(t) - x_2(t)|$. Then $u(t) \geq 0$ and $u(t) \leq \int_{t_0}^{t} ku(s)\,ds$ for $t \geq t_0$. By the Gronwall inequality given in Corollary 1.2.1, $u(t) \equiv 0$ for $t \geq t_0$. The case for $t < t_0$ can be proved similarly using the left-hand Gronwall inequality shown in Exercise 1.6.

(ii) Then we prove existence.

Define a sequence of functions $\{x_m(t)\}_{m=0}^{\infty}$ by recurrence as follows:

$$x_0(t) = x_0,$$

$$(1.3.2) \quad x_{m+1}(t) = x_0 + \int_{t_0}^{t} f(s, x_m(s))\,ds, \quad m \in \mathbb{N}_0 := \{0, 1, 2, \dots\}.$$

We claim that $x_m(t)$ is defined on the interval $J := [t_0 - \gamma, t_0 + \gamma]$ and

$$(1.3.3) \qquad\qquad |x_m(t) - x_0| \leq b \quad \text{for } m \in \mathbb{N}_0,$$

and hence $(t, x_m(t)) \in G$ for $t \in J$. Clearly, the claim holds for $m = 0$. Assume the claim holds for some $m \in \mathbb{N}_0$. Then $(t, x_m(t)) \in G$ for $t \in J$. By the assumption, $|f(t, x_m(t))| \leq M$ for $t \in J$. From (1.3.2), $x_{m+1}(t)$ exists on J and

$$|x_{m+1}(t) - x_0| = \left| \int_{t_0}^{t} f(s, x_m(s))\,ds \right| \leq M|t - t_0| \leq M\gamma \leq b,$$

i.e., $(t, x_{m+1}(t)) \in G$ for $t \in J$.

We then show that $\{x_m(t)\}_{m=0}^{\infty}$ is uniformly convergent on J. To do so, rewrite $x_m(t)$ as

$$x_m(t) = x_0 + \sum_{i=0}^{m-1} (x_{i+1}(t) - x_i(t)), \quad m = 1, 2, \dots.$$

By induction, we have that for $t \in J$ and $i = 0, \dots, m - 1$

$$(1.3.4) \qquad |x_{i+1} - x_i(t)| \leq \frac{Mk^i |t - t_0|^{i+1}}{(i+1)!} \leq \frac{Mk^i \gamma^{i+1}}{(i+1)!}.$$

Clearly, (1.3.4) holds for $i = 0$. Assume (1.3.4) holds for some $i \in \mathbb{N}$. Then from (1.3.2) and using the Lipschitz condition we see that for $t \in J$ and $t \geq t_0$

$$|x_{i+2}(t) - x_{i+1}(t)| \leq \int_{t_0}^{t} |f(s, x_{i+1}(s)) - f(s, x_i(s))|\,ds$$

$$\leq k \int_{t_0}^{t} |x_{i+1}(s) - x_i(s)|\,ds \leq k \int_{t_0}^{t} \frac{Mk^i (s - t_0)^{i+1}}{(i+1)!}\,ds$$

$$= \frac{Mk^{i+1}(t - t_0)^{i+2}}{(i+2)!} \leq \frac{Mk^{i+1}\gamma^{i+2}}{(i+2)!}.$$

Similarly for $t \in J$ and $t < t_0$. Thus (1.3.4) holds for $i + 1$. Since the constant series $\sum_{i=0}^{\infty} \dfrac{Mk^i \gamma^{i+1}}{(i+1)!}$ is convergent, by the Weierstrass M-test, we see that the series of functions $\sum_{i=1}^{\infty}(x_i(t) - x_{i-1}(t))$ is uniformly convergent on J. This means that the sequence of functions $\{x_m(t)\}_{m=0}^{\infty}$ is uniformly convergent on J. Let $x(t)$ be the uniform limit of $\{x_m(t)\}_{m=0}^{\infty}$. By Lemma 1.2.3 and Remark 1.2.4, $x(t)$ is continuous on J and $f(t, x_m(t))$ is uniformly convergent to $f(t, x(t))$ on J. Thus $f(t, x(t))$ is continuous on J. Taking limits as $m \to \infty$ in (1.3.2) we obtain that

$$x(t) = x_0 + \int_{t_0}^{t} f(s, x(s))\, ds \quad \text{for } t \in J,$$

i.e., $x(t)$ is a solution of Eq. (I) and hence a solution of IVP (V) on J.

$\square$

The result in Lemma 1.3.1 can be extended to a general domain as shown in the next theorem.

THEOREM 1.3.1 (Picard Existence-Uniqueness Theorem). *Let D be an open subset of $\mathbb{R} \times \mathbb{R}^n$ and $(t_0, x_0) \in D$. Assume $f \in C(D, \mathbb{R}^n)$ and is locally Lipschitz in x on D. Then there exists a $\gamma > 0$ such that IVP (V) has a unique solution which exists for $|t - t_0| \leq \gamma$.*

PROOF. Since D is an open set and $(t_0, x_0) \in D$, there exist $a, b > 0$ such that the set G defined by (1.3.1) is contained in D, see Fig. 1.4. Since f is locally Lipschitz in x on D, we may choose a and b sufficiently small so that a Lipschitz condition with a fixed Lipschitz constant $k > 0$ holds on G. Then the conclusion follows from Lemma 1.3.1. $\square$

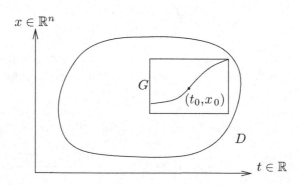

FIGURE 1.4. Any open domain

The next set of results are due to Cauchy and Peano and are for the existence of solutions of IVP (V) only. The proof we offer here is from Hartman [**22**, p. 10]. A different proof can also be found in Coddington and Levinson [**11**, p. 6] and Birkhoff and Rota [**3**, p. 166].

LEMMA 1.3.2. *Let G be defined by (1.3.1) and assume $f \in C(G, \mathbb{R}^n)$. Then IVP (V) has at least one solution which exists for $|t - t_0| \le \gamma$, where $\gamma = \min\{a, b/M\}$ with $M = \max_{(t,x) \in G} |f(t, x)|$.*

PROOF. We only show the existence of a solution of IVP (V) on $[t_0, t_0 + \gamma]$. A similar argument works on $[t_0 - \gamma, t_0)$. For $0 < r < \gamma$ we define

$$(1.3.5) \qquad x_r(t) = \begin{cases} x_0, & t_0 - r < t \le t_0; \\ x_0 + \displaystyle\int_{t_0}^t f(s, x_r(s - r))\, ds, & t_0 \le t \le t_0 + \gamma. \end{cases}$$

It is easy to see that $x_r(t)$ is well defined on $[t_0, t_0 + \gamma]$ by iteration. Let

$$\mathfrak{F} = \{x_r(t) : t \in [t_0, t_0 + \gamma], 0 < r < \gamma\}.$$

We first show that $\mathfrak{F}$ is uniformly bounded. More specifically, we show that for $t \in [t_0, t_0 + \gamma]$,

$$(1.3.6) \qquad |x_r(t) - x_0| \le b \quad \text{for } 0 < r < \gamma$$

by induction. In fact, for $t \in [t_0, t_0 + r]$,

$$\begin{aligned} |x_r(t) - x_0| &= \left| \int_{t_0}^t f(s, x_r(s - r))\, ds \right| \le \int_{t_0}^t |f(s, x_r(s - r))|\, ds \\ &= \int_{t_0}^t |f(s, x_0)|\, ds \le M(t - t_0) \le M\alpha \le b. \end{aligned}$$

Hence (1.3.6) holds for $t \in [t_0, t_0 + r]$. Assume (1.3.6) holds for $t \in [t_0 + (k-1)r, t_0 + kr] \cap [t_0, t_0 + \gamma]$. Then $(t, x_r(t - r)) \in G$ and thus $|f(t, x_r(t - r))| \le M$ for $t \in [t_0 + (k - 1)r, t_0 + kr] \cap [t_0, t_0 + \gamma]$. When $t \in [t_0 + kr, t_0 + (k + 1)r] \cap [t_0, t_0 + \gamma]$, we have

$$\begin{aligned} |x_r(t) - x_0| &= \left| \int_{t_0}^t f(s, x_r(s - r))\, ds \right| \le \int_{t_0}^t |f(s, x_r(s - r))|\, ds \\ &\le M(t - t_0) \le M\gamma \le b. \end{aligned}$$

Hence (1.3.6) holds for $t \in [t_0 + kr, t_0 + (k + 1)r] \cap [t_0, t_0 + \gamma]$.

Next we show that $\mathfrak{F}$ is equicontinuous. For $\epsilon > 0$ let $\delta = \epsilon/M$. Let $0 < r < \gamma$ and $t_1, t_2 \in [t_0, t_0 + \gamma]$ such that $|t_1 - t_2| < \delta$. Note that for $i = 1, 2$,

$$x_r(t_i) = x_0 + \int_{t_0}^{t_i} f(s, x_r(s - r))\, ds.$$

Then

$$|x_r(t_1) - x_r(t_2)| = \left| \int_{t_1}^{t_2} f(s, x_r(s-r)) \, ds \right| \leq \left| \int_{t_1}^{t_2} |f(s, x_r(s-r))| \, ds \right|$$
$$\leq M|t_2 - t_1| < M\delta = \epsilon.$$

This verifies the equicontinuity of $\mathfrak{F}$.

By Lemma 1.2.5, there exists a sequence $r_n \to 0+$ such that the corresponding sequence of functions $x_{r_n}(t)$ is uniformly convergent to a continuous function $x(t)$ on $[t_0, t_0 + \gamma]$. As shown in the proof of Lemma 1.3.1, by taking limits on both sides of (1.3.5) as $n \to \infty$, we obtain that for $t \in [t_0, t_0 + \gamma]$

$$x(t) = x_0 + \int_{t_0}^{t} f(s, x(s)) \, ds.$$

This means that $x(t)$ is a solution of the integral equation (I) and hence a solution of the IVP (V) on $[t_0, t_0 + \gamma]$. In the same way, we can show that IVP (V) has a solution on $[t_0 - \gamma, t_0]$. $\qquad \square$

Similar to Theorem 1.3.1 we obtain the following theorem from Lemma 1.3.2.

THEOREM 1.3.2 (Cauchy–Peano Existence Theorem). *Let D be an open subset of $\mathbb{R} \times \mathbb{R}^n$ and $(t_0, x_0) \in D$. Assume $f \in C(D, \mathbb{R}^n)$. Then there exists a $\gamma > 0$ such that IVP (V) has at least one solution which exists for $|t - t_0| \leq \gamma$.*

EXAMPLE 1.3.1. Consider the IVP

(1.3.7) $$x' = 1 + t^2 x^2, \quad x(t_0) = x_0.$$

Clearly, the domain $D = \mathbb{R}^2$ and hence $(t_0, x_0) \in D$ for any $t_0, x_0 \in \mathbb{R}$. Although the equation looks simple, it belongs to the class of Riccati equations and hence cannot be solved for solutions in closed form.

Let $f(t, x) = 1 + t^2 x^2$. Then $f \in C(D, \mathbb{R})$. Since $f_x(t, x) = 2t^2 x$ is continuous on D, we see that f is locally Lipschitz in x on D. By Theorem 1.3.1, IVP (1.3.7) has a unique solution in a neighborhood of t_0 for any $t_0, x_0 \in \mathbb{R}$.

EXAMPLE 1.3.2.

(1.3.8) $$x' = |x|^{1/2}, \quad x(0) = 0.$$

Clearly, the domain $D = \mathbb{R}^2$ and hence $(0, 0) \in D$. Let $f(t, x) = |x|^{1/2}$. Then $f \in C(D, \mathbb{R})$. By Theorem 1.3.2, IVP (1.3.8) has at least one solution in a neighborhood of 0. However, as shown in Exercise 1.1, Part (a), f is not Lipschitz in x in any neighborhood of $(0, 0)$. Hence Theorem 1.3.1 fails to apply for the uniqueness of the solution. In fact, we have already shown in Example 1.1.3 that IVP (1.3.8) has an infinite number of solutions.

REMARK 1.3.1. We comment that the conditions in Theorem 1.3.1 are sufficient but not necessary for IVP (V) to have a unique solution. To see this, let us look at the example below.

EXAMPLE 1.3.3. Consider the IVP

(1.3.9) $x' = f(x) := \begin{cases} x \ln|x|, & x \neq 0, \\ 0, & x = 0; \end{cases}$ $x(0) = x_0$

for any $x_0 \in \mathbb{R}$. We leave it to the reader to show that the right-hand side function f of the equation is continuous on $\mathbb{R}$ but not Lipschitz in any neighborhood of 0, see Exercise 1.1, Part (b). Hence Theorem 1.3.1 fails to apply to this problem when $x_0 = 0$.

Nevertheless, solving the equation by separating variables we find that the general solution $x(t)$ satisfies $|x(t)| = c^{e^t}$ for $c > 0$. When $x_0 \neq 0$, using the IC $x(0) = x_0$ we obtain a unique solution of the IVP, $x = |x_0|^{e^t} \operatorname{sgn} x_0$, which exists on the whole real number line. However, due to the way that the general solution is obtained, it does not include the solution for the IC $x(0) = 0$. In this case, since $x \equiv 0$ is a solution and the solution of IVP (1.3.9) can never be zero if $x_0 \neq 0$, we have that $x \equiv 0$ is the only solution of the IVP. Therefore, IVP (1.3.9) has a unique solution for any $x_0 \in \mathbb{R}$ including $x_0 = 0$.

Now, we apply Theorems 1.3.1 and 1.3.2 to derive results on the existence and uniqueness of solutions of the n-th order scalar IVP
(1.3.10)
$$y^{(n)} = g(t, y, y', \ldots, y^{(n-1)}), \quad y(t_0) = a_1, \, y'(t_0) = a_2, \, \ldots, \, y^{(n-1)}(t_0) = a_n.$$

THEOREM 1.3.3. Let D be an open subset of $\mathbb{R} \times \mathbb{R}^n$ and $(t_0, a_1, a_2 \ldots, a_n) \in D$.

(a) Assume $g \in C(D, \mathbb{R}^n)$. Then there exists a $\gamma > 0$ such that IVP (1.3.10) has at least one solution which exists for $|t - t_0| \leq \gamma$.

(b) Assume $g \in C(D, \mathbb{R}^n)$, and as a function of $(t, y_1, y_2, \ldots, y_n)$, g is locally Lipschitz in $(y_1, y_2, \ldots, y_n)$ on D. Then there exists a $\gamma > 0$ such that IVP (1.3.10) has a unique solution which exists for $|t - t_0| \leq \gamma$.

PROOF. Let
$$y_1 = y, \; y_2 = y', \; \ldots, \; y_n = y^{(n-1)}.$$
Then the n-th order scalar IVP (1.3.10) becomes the first-order vector-valued IVP
$$Y' = f(t, Y), \quad Y(t_0) = Y_0;$$
where $Y = (y_1, y_2, \ldots, y_n)^T$, $f(t, Y) = (y_2, \ldots, y_n, g(t, y_1, y_2, \ldots, y_n))^T$, and $Y_0 = (a_1, a_2, \ldots, a_n)^T$. Note that f is continuous on D if and only if g is continuous on D, and f is locally Lipschitz in Y if and only if g is locally Lipschitz in $(y_1, y_2, \ldots, y_n)$. Then the conclusions follow from Theorems 1.3.1 and 1.3.2. $\square$

EXAMPLE 1.3.4. Consider the IVP

(1.3.11) $y'' = \dfrac{1}{t}(y')^{2/3} - y^{3/2} + e^t, \quad y(t_0) = a_1, \, y'(t_0) = a_2.$

Let $g(t, y_1, y_2) = \frac{1}{t}y_2^{2/3} - y_1^{3/2} + e^t$. Then g is continuous on the open set $D := \{(t, y_1, y_2) \in \mathbb{R}^3 : t \neq 0, y_1 > 0\}$. Since

$$g_{y_1}(t, y_1, y_2) = -\frac{3}{2}y_1^{1/2} \quad \text{and} \quad g_{y_2}(t, y_1, y_2) = \frac{2}{3t}y_2^{-1/3},$$

g_{y_1} and g_{y_2} are continuous on D whenever $y_2 \neq 0$. Therefore, if $t_0 \neq 0$ and $a_1 > 0$, then IVP (1.3.11) has at least one solution in a neighborhood of t_0. Furthermore, if $a_2 \neq 0$, then the solution is unique.

The theorem below provides a result on the number of solutions of IVP (V) when uniqueness is not guaranteed. See [**22**, Theorem 4.1 in P15] for the proof.

THEOREM 1.3.4 (Kneser). *Let G be defined by (1.3.1). Assume $f \in C(G, \mathbb{R}^n)$ and let $\gamma = \min\{a, b/M\}$ with $M = \max_{(t,x)\in G} |f(t,x)|$. Then for any $c \in [t_0 - \gamma, t_0 + \gamma]$, the set $S_c = \{\phi(c) : \phi(t) \text{ is a solution of (V)}\}$ is closed and connected.*

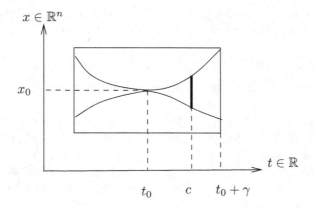

FIGURE 1.5. Nonuniqueness

Theorem 1.3.4 shows that IVP (V) has either a unique solution or an uncountably infinite number of solutions. The highlighted line segment in Fig. 1.5 represents the set S_c for the case $n = 2$ when the solutions are not unique.

1.4. Maximal Interval of Existence

Let D be an open subset of $\mathbb{R} \times \mathbb{R}^n$ and $(t_0, x_0) \in D$. Assume $f \in (D, \mathbb{R}^n)$. By Theorem 1.3.2, there exists a solution $x = \phi_0(t)$ of IVP (V) on an interval $[a_0, b_0]$ such that $t_0 \in (a_0, b_0)$. Note that $(a_0, \phi_0(a_0)) \in D$ since D is open. By Theorem 1.3.2, there exists a solution $x = \phi_-(t)$ of the IVP

$$x' = f(t, x), \quad x(a_0) = \phi_0(a_0)$$

on $[a_1, a_0)$ for some $a_1 < a_0$. Similarly, since $(b_0, \phi_0(b_0)) \in D$, by Theorem 1.3.2, there exists a solution $x = \phi_+(t)$ of the IVP

$$x' = f(t, x), \quad x(b_0) = \phi_0(b_0)$$

on $(b_0, b_1]$ for some $b_1 > b_0$. Define

$$\phi_1(t) = \begin{cases} \phi_-(t), & t \in [a_1, a_0) \\ \phi_0(t), & t \in [a_0, b_0] \\ \phi_+(t), & t \in (b_0, b_1]. \end{cases}$$

Then $x = \phi_1(t)$ is a solution of IVP (V) defined on $[a_1, b_1]$. We call $\phi_1(t)$ an extension of $\phi(t)$ from $[a_0, b_0]$ to $[a_1, b_1]$. This extension is illustrated in Fig. 1.6.

Clearly, $\phi_1(t)$ can be further extended. In general, let ϕ_{m+1} be an extension of ϕ_m from $[a_m, b_m]$ to $[a_{m+1}, b_{m+1}]$ for $m \in \mathbb{N}$. Then $\{a_m\}_{m=1}^\infty$ is strictly decreasing and $\{b_m\}_{m=1}^\infty$ is strictly increasing. Let

$$\alpha = \lim_{m \to \infty} a_m \quad \text{and} \quad \beta = \lim_{m \to \infty} b_m.$$

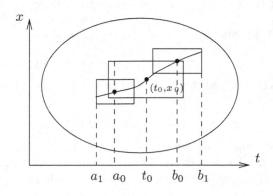

FIGURE 1.6. Extension of a solution

For ease of notation, we denote by $\phi(t)$ the extension of $\phi_0(t)$ on (α, β). Then we have the following result.

LEMMA 1.4.1. (a) Assume the set $\{(a_m, \phi(a_m))\}_{m=1}^\infty$ has an accumulation point (α, x_α) in D. Then $\lim_{t \to \alpha+} \phi(t) = x_\alpha$. In this case, $\phi(t)$ can be further extended to $t = \alpha$.

(b) Assume the set $\{(b_m, \phi(b_m))\}_{m=1}^\infty$ has an accumulation point (β, x_β) in D. Then $\lim_{t \to \beta-} \phi(t) = x_\beta$. In this case, $\phi(t)$ can be further extended to $t = \beta$.

PROOF. Without loss of generality we only give the proof for Part (a). The proof for Part (b) is essentially the same.

Let $\epsilon > 0$ be so small that

$$G_\epsilon := \{(t, x) : |t - \alpha| \le \epsilon, |x - x_\alpha| \le \epsilon\} \subset D.$$

Let $M_\epsilon = \max\{\max_{(t,x)\in G_\epsilon} |f(t,x)|, 1\}$. Then there exists an $N \in \mathbb{N}$ such that

(1.4.1) $\qquad\qquad a_N - \alpha < \dfrac{\epsilon}{2M_\epsilon}$ and $\quad |\phi(a_N) - x_\alpha| < \dfrac{\epsilon}{2}.$

We claim that

(1.4.2) $\qquad\quad |\phi(t) - \phi(a_N)| < M_\epsilon(a_N - \alpha) \quad$ for $\alpha < t \le a_N.$

Assume the contrary. Since (1.4.2) holds for $t = a_N$, there exists a $t^* \in (\alpha, a_N)$ such that

$$|\phi(t) - \phi(a_N)| < M_\epsilon(a_N - \alpha) < \frac{\epsilon}{2} \quad \text{for } t^* < t \le a_N$$

and

(1.4.3) $\qquad\qquad\qquad |\phi(t^*) - \phi(a_N)| = M_\epsilon(a_N - \alpha).$

Thus for $t^* < t \le a_N$

$$|\phi(t) - x_\alpha| \le |\phi(t) - \phi(a_N)| + |\phi(a_N) - x_\alpha| < \epsilon.$$

This means that $(t, \phi(t)) \in G_\epsilon$ for $t^* < t \le a_N$, and hence

$$\begin{aligned}
|\phi(t^*) - \phi(a_N)| &= \left| \int_{t^*}^{a_N} f(s, \phi(s))\, ds \right| \\
&\le M_\epsilon(a_N - t^*) < M_\epsilon(a_N - \alpha).
\end{aligned}$$

This contradicts (1.4.3) and therefore confirms (1.4.2). By (1.4.1) and (1.4.2) we have that for $\alpha < t \le a_N$

$$|\phi(t) - x_\alpha| \le |\phi(t) - \phi(a_N)| + |\phi(a_N) - x_\alpha| < \epsilon.$$

This implies that $\lim_{t \to \alpha+} \phi(t) = x_\alpha$.

From the equivalence between IVP (V) and the integral equation (I) given by Lemma 1.1.1, we see that $\phi(t)$ can be extended to $t = \alpha$. $\qquad \square$

Lemma 1.4.1 implies that if $(\alpha, \phi(\alpha))$ or $(\beta, \phi(\beta))$ is an accumulation point of the solution curve of (V) in D, then the solution can be extended further to a larger interval.

DEFINITION 1.4.1. An interval J is called the maximal interval of existence of a solution $\phi(t)$ of IVP (V) if $\phi(t)$ exists on J and there does not exist an interval $J_1 \supsetneq J$ such that $\phi(t)$ has an extension from J to J_1.

From the definition we see that when the domain D is an open set, then the maximal interval of existence of a solution must be an open interval. Otherwise, it can be extended further.

THEOREM 1.4.1. *Let D be an open subset of $\mathbb{R} \times \mathbb{R}^n$, $f \in C(D, \mathbb{R}^n)$, and $(t_0, x_0) \in D$. Assume $x = \phi(t)$ is a solution of IVP (V). Then $\phi(t)$ can be extended to a maximal interval of existence (α, β) in a countably infinite number of steps; and $(t, \phi(t)) \to \partial D$, the boundary of D, as $t \to \alpha+$ or $t \to \beta-$.*

Furthermore, if f is locally Lipschitz in x on D, then the extension is unique.

REMARK 1.4.1. In Theorem 1.4.1, by $(t, \phi(t)) \to \partial D$ as $t \to \alpha+$ or $t \to \beta-$ we mean that for any compact set $G \subset D$, $(t, \phi(t)) \notin G$ when t is sufficiently close to α or β.

PROOF OF THEOREM 1.4.1. For $k = 1, 2, \ldots$, define

$$D_k = \left\{ (t, x) \in D : |t - t_0| < k, |x - x_0| < k, \text{ and } d((t, x), \partial D) > \frac{1}{k} \right\}$$

and $\bar{D}_k$ is the closure of D_k, where $d((t, x), \partial D)$ means the distance from the point (t, x) to ∂D. Clearly,

$$\bar{D}_k \subset D_{k+1} \text{ for } k \in \mathbb{N} \text{ and } \cup_{k=1}^{\infty} D_k = D.$$

Let $k \in \mathbb{N}$. It is easy to see that for any $(t^*, x^*) \in D_k$, the set

$$G_k := \{ (t, x) \in D : |t - t^*| \le \delta_k \text{ and } |x - x^*| \le \delta_k \},$$

where $\delta_k = \frac{1}{\sqrt{2}} \left(\frac{1}{k} - \frac{1}{k+1} \right)$, is contained in D_{k+1}. Let

$$M_k = \max_{(t,x) \in \bar{D}_{k+1}} \{ |f(t, x)| \} \text{ and } \gamma_k = \min \{ \delta_k, \delta_k / M_k \}.$$

By Lemma 1.3.2, the IVP

$$x' = f(t, x), \quad x(t^*) = x^*$$

has a solution which exists for $|t - t^*| \le \gamma_k$. This implies that we can make extensions of the solution $\phi_0(t)$ of IVP (V) with the extension size at least γ_k, i.e., the mth extension extends the solution from $[a_{m-1}, b_{m-1}]$ to $[a_m, b_m]$ with $a_m - a_{m-1} \le -\gamma_k$ and/or $b_m - b_{m-1} \ge \gamma_k$, as long as the endpoint $(a_m, \phi(a_m))$ and/or $(b_m, \phi(b_m))$ is in D_k. Without loss of generality, we only consider the right-extensions. Since $\bar{D}_k$ is compact, by a finite number of steps of such right-extensions, the endpoint $(b_m, \phi(b_m))$ will go out of $\bar{D}_k$, i.e., there exists an $m = m_k$ such that $(b_{m_k}, \phi(b_{m_k})) \notin \bar{D}_k$.

In this way, we obtain a sequence $\{(b_{m_k}, \phi(b_{m_k}))\}_{k=1}^{\infty} \subset D$ with $(b_{m_k}, \phi(b_{m_k})) \notin \bar{D}_k$ for $k = 1, 2, \ldots$. Therefore, $(b_{m_k}, \phi(b_{m_k})) \to \partial D$ as $k \to \infty$. Let $\beta = \lim_{k \to \infty} b_{m_k}$. Then the extension $\phi(t)$ of $\phi_0(t)$ exists on $[t_0, \beta)$. We need to show further that $(t, \phi(t)) \to \partial D$ as $t \to \beta-$. Assume the contrary. Then by Remark 1.4.1, there exist a compact set $G \subset D$ and a sequence $\{t_j\}_{j=1}^{\infty}$ in $[t_0, \beta)$ such that $t_1 < t_2 < \ldots, t_j \to \beta$, and $(t_j, \phi(t_j)) \in G$. Hence the set $\{(t_j, \phi(t_j))\}_{j=1}^{\infty}$ has an accumulation point (β, ϕ_β) in G. On the other hand, for $j = 1, 2, \ldots$, we may treat $(t_j, \phi(t_j))$ as the endpoint of the j-th extension of $\phi_0(t)$ for a new set of extensions. Then by Lemma 1.4.1, Part (b), we have that $\lim_{t \to \beta-} \phi(t) = \phi_\beta$ for some $\phi_\beta \in \mathbb{R}^n$. Note that $G \subset D_k$ for some $k \in \mathbb{N}$. Thus $\lim_{t \to \beta-} \phi(t) = \phi_\beta$ with $(\beta, \phi_\beta) \in D_k$. This contradicts the fact that $(b_{m_k}, \phi(b_{m_k})) \notin \bar{D}_k$ for sufficiently large k.

If f is locally Lipschitz in x on D, then the uniqueness of the extension is guaranteed by Theorem 1.3.1. □

COROLLARY 1.4.1. *Let $D = (a, b) \times \mathbb{R}^n$ with $-\infty \le a < b \le \infty$, $t_0 \in (a, b)$. Assume $f \in C(D, \mathbb{R}^n)$ and $(\alpha, \beta) \subset (a, b)$ is the maximal interval of existence of a solution $\phi(t)$ of* (V). *Then we have*

$$\alpha = a \quad or \quad \alpha > a \text{ and } \lim_{t \to \alpha+} |\phi(t)| = \infty, \quad and$$

$$\beta = b \quad or \quad \beta < b \text{ and } \lim_{t \to \beta-} |\phi(t)| = \infty.$$

Moreover, if f is bounded on D, then every solution of (V) *exists on (a, b).*

PROOF. The first part follows from Theorem 1.4.1 directly. To prove the second part by contradiction, let $(\alpha, \beta) \subsetneq (a, b)$. Without loss of generality assume $\beta < b$. From the first part, $\lim_{t \to \beta-} |\phi(t)| = \infty$. On the other hand, since

$$\phi(t) = x_0 + \int_{t_0}^t f(s, \phi(s)) \, ds,$$

we have that for $t \ge t_0$

$$|\phi(t)| \le |x_0| + \int_{t_0}^t |f(s, \phi(s))| \, ds \le |x_0| + M(t - t_0),$$

where M is an upper bound of $|f|$ on D. This shows that

$$|\phi(t)| \le |x_0| + M(\beta - t_0) < \infty.$$

We have reached a contradiction. □

EXAMPLE 1.4.1. Consider the scalar equation

(1.4.4) $x' = t + \sin(x^2 - t).$

Let $f(t, x) = t + \sin(x^2 - t)$. Then f is defined on $D = \mathbb{R}^2$. It is easy to see that f and f_x are continuous on D. Hence the IVP consisting of Eq. (1.4.4) and any IC $x(t_0) = x_0$ has a unique solution $\phi(t)$. Let T be any positive number such that $|t_0| < T$ and let $D_T = (-T, T) \times \mathbb{R}$. Then f is bounded on D_T. By Corollary 1.4.1, $\phi(t)$ exists on $(-T, T)$ and hence exists on $\mathbb{R}$ due to the arbitrariness of $T > 0$.

1.5. Dependence of Solutions on Parameters and Initial Conditions

Differential equation models for real world problems may contain parameters. For instance, the second-order IVP reflecting the basic mechanical vibrations is

$$mx'' + cx' + kx = \omega(t), \quad x(t_0) = x_0, x'(t_0) = v_0,$$

where m is the mass of the vibrator; c and k are the damping constant and spring constant, respectively; $\omega(t)$ is the external force; and x_0 and v_0 are the initial position and velocity of the vibrator, respectively. In this model, we are often concerned about how the motion of the vibrator is affected by the parameters m, c, k and the IC. The same concern arises in mathematical model building. To set up an IVP for a practical problem, one must measure the magnitudes of the ICs and parameters. However, there are always errors in measurements. A natural question is: How sensitive is the solution of the IVP to the errors? Recall that most IVPs cannot be solved explicitly. Therefore, to address such concerns, we need to investigate the dependence of solutions of IVPs on parameters and IC without solving the problem.

I. Continuous dependence

First, we consider the IVP

$$(V[\mu]) \qquad\qquad x' = f(t, x; \mu), \quad x(t_0) = x_0,$$

where $f : D \to \mathbb{R}^n$ with D an open subset of $\mathbb{R} \times \mathbb{R}^n \times \mathbb{R}^k$ and $(t_0, x_0; \mu) \in D$. Here we assume that $\mu \in \mathbb{R}^k$ is a k-dimensional parameter and consider the dependence of the solution of IVP $(V[\mu])$ on μ.

THEOREM 1.5.1. *Assume that* $f \in C(D, \mathbb{R}^n)$ *and IVP* $(V[\mu])$ *has a unique solution* $x = x(t; \mu)$. *Then* $x(t; \mu)$ *is continuous in* (t, μ) *in its domain.*

PROOF. Let $\mu_0 \in \mathbb{R}^k$ such that $(t_0, x_0, \mu_0) \in D$. Choose $a, b, c > 0$ such that the set

$$G = \{(t, x, \mu) : |t - t_0| \le a, |x - x_0| \le b, |\mu - \mu_0| \le c\}$$

is contained in D. Denote $\gamma = \min\{a, b/M\}$ with $M = \max_{(t,x,\mu) \in G} |f(t, x; \mu)|$. Without loss of generality, we only prove the result for $t_0 \le t \le t_0 + \gamma$ and $|\mu - \mu_0| \le c$. The general conclusion for $t \ge t_0$ can be derived by the continuous extensions of the solution, and a similar argument works for $t < t_0$.

For $0 < r < \gamma$ define

$$(1.5.1) \qquad x_r(t; \mu) = \begin{cases} x_0, & t_0 - r \le t < t_0; \\ x_0 + \displaystyle\int_{t_0}^t f(s, x_r(s - r); \mu)\, ds, & t_0 \le t \le t_0 + \gamma. \end{cases}$$

As in the proof of Lemma 1.3.2 we can show that there exists a sequence $r_n \to 0^+$ such that the corresponding sequence of functions $x_{r_n}(t; \mu)$ is uniformly convergent to a continuous function $x(t; \mu)$ for $t_0 \le t \le t_0 + \gamma$ and $|\mu - \mu_0| \le c$. Taking limits on both sides of (1.5.1) as $n \to \infty$ we obtain that for $t_0 \le t \le t_0 + \gamma$ and $|\mu - \mu_0| \le c$,

$$x(t; \mu) = x_0 + \int_{t_0}^t f(s, x(s; \mu); \mu)\, ds.$$

This shows that $x(t; \mu)$ is the unique solution of IVP (V) which is continuous in (t, μ) for $t_0 \le t \le t_0 + \gamma$ and $|\mu - \mu_0| \le c$. $\qquad\square$

Based on Theorem 1.5.1 we may discuss the dependence of solutions of differential equations on parameters and ICs.

Consider the IVP

$$(V[t_0, x_0, \mu]) \qquad\qquad x' = f(t, x; \mu), \quad x(t_0) = x_0,$$

where $f : D \to \mathbb{R}^n$ with D an open subset of $\mathbb{R} \times \mathbb{R}^n \times \mathbb{R}^k$ and $(t_0, x_0; \mu) \in D$. Here we assume that $\mu \in \mathbb{R}^k$ is a k-dimensional parameter and t_0, x_0 in the IC may vary. We consider the dependence of the solution of IVP ($V[t_0, x_0, \mu]$) on t_0, x_0, and μ.

THEOREM 1.5.2. *Let $f \in C(D, \mathbb{R}^n)$ and assume IVP ($V[t_0, x_0, \mu]$) has a unique solution $x = x(t; t_0, x_0, \mu)$. Then $x(t; t_0, x_0, \mu)$ is continuous in (t, t_0, x_0, μ) in its domain.*

PROOF. Let

(1.5.2) $s = t - t_0, \quad y(s) = x(t) - x_0.$

Then the IVP ($V[t_0, x_0, \mu]$) with a parameter μ and a variable IC becomes the IVP with parameters t_0, x_0, μ and a fixed IC

($V^*[t_0, x_0, \mu]$) $\dfrac{dy}{ds} = g(s, y; t_0, x_0, \mu), \quad y(0) = 0,$

where $g(s, y; t_0, x_0, \mu) = f(s + t_0, y + x_0; \mu)$. Obviously, g is continuous when $(s + t_0, y + x_0; \mu) \in D$. We observe that $x = x(t; t_0, x_0, \mu)$ is the unique solution of IVP ($V[t_0, x_0, \mu]$) implied that $y = y(s; t_0, x_0, \mu) := x(t; t_0, x_0, \mu) - x_0$ is the unique solution of IVP ($V^*[t_0, x_0, \mu]$). By Theorem 1.5.1, $y(s; t_0, x_0, \mu)$ is continuous in $(s; t_0, x_0, \mu)$ and hence $x(t; t_0, x_0, \mu) = y(t - t_0; t_0, x_0, \mu) + x_0$ is continuous in $(t; t_0, x_0, \mu)$.   □

EXAMPLE 1.5.1. Consider the scalar IVP

(1.4.5-$[\mu]$) $x' = \cos t + e^{-(\mu x)^2}, \quad x(\mu) = 0,$

where $\mu \in \mathbb{R}$ is a parameter. Similar to Example 1.4.1 we can show that the solution $x(t; \mu)$ of IVP (1.4.5-$[\mu]$) exists on the whole real number line $\mathbb{R}$ and is unique. We note that for $\mu \neq 0$, $x(t; \mu)$ cannot be solved explicitly. However, for $\mu = 0$, IVP (1.4.5-$[\mu]$) becomes

$$x' = \cos t + 1, \quad x(0) = 0$$

which can be solved easily to obtain the solution $x(t; 0) = \sin t + t$. By Theorem 1.5.2, we have that $\lim_{\mu \to 0} x(t; \mu) = x(t; 0) = \sin t + t$ for any $t \in \mathbb{R}$. This convergence should be uniform for t in any compact interval. Therefore, we can use $x(t; 0)$ to approximate $x(t; \mu)$ for t in a compact interval and μ sufficiently close to 0.

II. Differentiable dependence

In addition to the continuous dependence result given by Theorem 1.5.2, we are also interested in finding conditions for the function $f(t, x; \mu)$ which guarantee that the unique solution $x(t; t_0, x_0, \mu)$ of the IVP ($V[t_0, x_0, \mu]$) depends on t_0, x_0, and μ in a smooth way. The answer is given by the following theorem.

THEOREM 1.5.3. *Assume that $f \in C(D, \mathbb{R}^n)$, $\partial f/\partial x \in C(D, \mathbb{R}^{n \times n})$, and $\partial f/\partial \mu \in C(D, \mathbb{R}^{n \times k})$. Then IVP ($V[t_0, x_0, \mu]$) has a unique solution $x(t; t_0, x_0, \mu)$ which is C^1 in t_0, x_0, and μ in its domain.*

Furthermore, let

$$J(t; t_0, x_0, \mu) := \frac{\partial f}{\partial x}(t, x(t; t_0, x_0, \mu); \mu).$$

Then

(a) $(\partial x/\partial \mu)(t; t_0, x_0, \mu)$ is the solution of the IVP

(1.5.3) $$z' = J(t; t_0, x_0, \mu)z + \frac{\partial f}{\partial \mu}(t, x(t; t_0, x_0, \mu); \mu), \quad z(t_0) = \mathbf{0};$$

(b) $(\partial x/\partial x_0)(t; t_0, x_0, \mu)$ is the solution of the IVP

(1.5.4) $$z' = J(t; t_0, x_0, \mu)z, \quad z(t_0) = I;$$

(c) $(\partial x/\partial t_0)(t; t_0, x_0, \mu)$ is the solution of the IVP

(1.5.5) $$z' = J(t; t_0, x_0, \mu)z, \quad z(t_0) = -f(t_0, x_0; \mu).$$

Here I stands for the $n \times n$ identity matrix.

PROOF. Note that $\partial f/\partial x \in C(D, \mathbb{R}^{n \times n})$ implies that f is locally Lipschitz in x. Hence by Theorem 1.3.1, IVP ($V[t_0, x_0, \mu]$) has a unique solution $x(t; t_0, x_0, \mu)$.

(a) We first discuss the differentiable dependence of $x(t; t_0, x_0, \mu)$ on μ for fixed t_0 and x_0. To ease the notation, we denote $x(t; t_0, x_0, \mu)$ by $x(t; \mu)$ and consider the case when $k = 1$, i.e., μ is a scalar parameter. Given a fixed μ, let T be any point in the maximal interval of existence (α, β) of $x(t; \mu)$ and without loss of generality assume $T \in (t_0, \beta)$. Since the solution curve for $x(t; \mu)$, $t \in [t_0, T]$, is a compact set, there exists a $l > 0$ so small that

$$\mathcal{N} := \{(t, y(t), \mu^*) : t \in [t_0, T], |y(t) - x(t, \mu)| \leq l, |\mu^* - \mu| \leq l\} \subset D.$$

Define

$$\mathcal{N}_1 := \{(t, \mu^*) : t \in [t_0, T], |\mu^* - \mu| \leq l\}.$$

Note that $\mathcal{N}$ is a "square tube" centered at the solution curve for $x(t; \mu)$ and $\mathcal{N}_1$ is the projection of $\mathcal{N}$ into the $t\mu$-plane, see Fig. 1.7.

Let $t \in [t_0, T]$. For any small $\Delta\mu > 0$, let $\Delta x(t; \mu) := x(t; \mu + \Delta\mu) - x(t; \mu)$. By Lemma 1.1.1 we have

$$x(t; \mu) = x_0 + \int_{t_0}^{t} f(s, x(s; \mu); \mu)\, ds$$

and

$$x(t; \mu + \Delta\mu) = x_0 + \int_{t_0}^{t} f(s, x(s; \mu + \Delta\mu); \mu + \Delta\mu)\, ds.$$

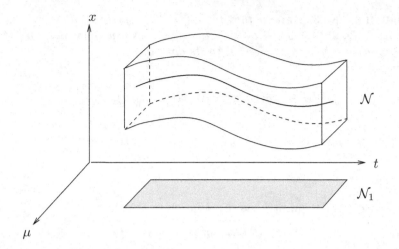

FIGURE 1.7. Sets $\mathcal{N}$ and $\mathcal{N}_1$

For $\Delta\mu$ small enough we have that $(t, x(t; \mu + \Delta\mu), \mu + \Delta\mu) \in \mathcal{N}$. Thus by linearization,

$$(1.5.6) \qquad \Delta x(t; \mu) = \int_{t_0}^{t} [f(s, x(s; \mu + \Delta\mu); \mu + \Delta\mu) - f(s, x(s; \mu); \mu)] \, ds$$

$$= \int_{t_0}^{t} \left\{ [\frac{\partial f}{\partial x}(s, x(s; \mu); \mu) + \epsilon_1(s; \mu, \Delta\mu)] \Delta x(s; \mu) \right.$$

$$\left. + [\frac{\partial f}{\partial \mu}(s, x(s; \mu); \mu) + \epsilon_2(s; \mu, \Delta\mu)] \Delta\mu \right\} ds.$$

Since $(\partial f/\partial x)(t, x(t; \mu); \mu)$ and $(\partial f/\partial \mu)(t, x(t; \mu); \mu)$ are continuous on $\mathcal{N}_1$, they are uniformly continuous on $\mathcal{N}_1$. It follows that for $i = 1, 2$, $|\epsilon_i(t; \mu, \Delta\mu)| \to 0$ as $\Delta\mu \to 0$ uniformly on $\mathcal{N}_1$.

On the other hand, the solution $y(t; \mu)$ of IVP (1.5.3) satisfies

$$(1.5.7) \qquad y(t; \mu) = \int_{t_0}^{t} \left[\frac{\partial f}{\partial x}(s, x(s; \mu); \mu) y(s; \mu) + \frac{\partial f}{\partial \mu}(s, x(s; \mu); \mu) \right] ds.$$

We show $\Delta x(t; \mu)/\Delta\mu \to y(t; \mu)$ as $\Delta\mu \to 0$ uniformly on $\mathcal{N}_1$. In fact, from (1.5.6) and (1.5.7) we see that

$$\frac{\Delta x(t; \mu)}{\Delta\mu} - y(t; \mu) = \int_{t_0}^{t} \left[(\frac{\partial f}{\partial x}(s, x(s; \mu); \mu) + \epsilon_1(s; \mu, \Delta\mu)) \right.$$

$$\left. \left(\frac{\Delta x(s; \mu)}{\Delta\mu} - y(s; \mu) \right) + (\epsilon_1(s; \mu, \Delta\mu) y(s; \mu) + \epsilon_2(s; \mu, \Delta\mu)) \right] ds.$$

Let $u(t; \mu, \Delta\mu) = |\Delta x(t; \mu)/\Delta\mu - y(t; \mu)|$. We observe that there exists a $k > 0$ such that $|y(t; \mu)| \le k$ and $|\partial f/\partial x(s, x(s; \mu); \mu)| \le k$ on $\mathcal{N}_1$; and for any $\epsilon > 0$, there exists a $\delta > 0$ such that

$|\epsilon_i(s; \mu, \Delta\mu)| < \epsilon$ for $i = 1, 2$ if $|\Delta\mu| < \delta$. Thus for $t \in [t_0, t_0 + T]$ and $|\Delta\mu| < \delta$,

$$u(t; \mu, \Delta\mu) \le (k + 1)\epsilon T + \int_{t_0}^t (k + \epsilon)u(s; \mu, \Delta\mu)\, ds.$$

By the Gronwall inequality,

$$u(t; \mu, \Delta\mu) \le (k + 1)\epsilon T e^{(k+\epsilon)(t-t_0)} \le (k + 1)\epsilon T e^{(k+\epsilon)T}.$$

This implies that $u(t; \mu, \Delta\mu) \to 0$ as $\Delta\mu \to 0$. Therefore, $\partial x/\partial\mu = y$ and hence is continuous on $\mathcal{N}_1$ and satisfies (1.5.3) for $t \in [t_0, T]$. Note that $T \in (\alpha, \beta)$ is chosen arbitrarily, so the conclusion holds for any $t \in (\alpha, \beta)$.

(b) Then we consider the differentiable dependence of $x(t; t_0, x_0, \mu)$ on x_0 for fixed t_0 and μ. For this purpose, we denote $x(t; t_0, x_0, \mu)$ by $x(t; x_0)$. By way of the transformation (1.5.2), Eq. (V$[t_0, x_0, \mu]$) becomes Eq. (V*$[t_0, x_0, \mu]$), where x_0 is a parameter and t_0 and μ are fixed. Hence the right-hand side function of Eq. (V*$[t_0, x_0, \mu]$) can be compressed as $g(s, y; x_0)$. Using the result in Part (a), where μ is replaced by x_0, we see that the solution $y(s; x_0)$ of Eq. (V*$[t_0, x_0, \mu]$) is C^1 in x_0 and hence $x(t; x_0) = y(t - t_0; x_0) + x_0$ is C^1 in x_0; furthermore,

$$\left(\frac{\partial y}{\partial x_0}\right)'(s; x_0) = \frac{\partial g}{\partial y}(s, y(s; x_0); x_0)\frac{\partial y}{\partial x_0}(s; x_0) + \frac{\partial g}{\partial x_0}(s, y(s; x_0); x_0),$$

$$\frac{\partial y}{\partial x_0}(0) = \mathbf{0}.$$

From (1.5.2) we see that $(\partial y/\partial x_0)(s; x_0) = (\partial x/\partial x_0)(t; x_0) - I$. Thus

$$\begin{aligned}
\left(\frac{\partial x}{\partial x_0}\right)'(t; x_0) &= \frac{\partial f}{\partial x}(t, x(t; x_0); \mu)\left(\frac{\partial x}{\partial x_0}(t; x_0) - I\right) + \frac{\partial f}{\partial x}(t, x(t; x_0); \mu) \\
&= \frac{\partial f}{\partial x}(t, x(t; x_0); \mu)\frac{\partial x}{\partial x_0}(t; x_0).
\end{aligned}$$

and $(\partial x/\partial x_0)(t_0) = I$. Therefore, $(\partial x/\partial x_0)(t, x_0)$ satisfies (1.5.4).

(c) The method in Part (b) works also for the case when t_0 varies provided f is C^1 in t. However, it fails to work for the general case due to the observation that the function g in Eq. (V*$[t_0, x_0, \mu]$) may not be C^1 in t_0. However, the differentiable dependence of $x(t; t_0, x_0, \mu)$ on t_0 and (1.5.5) can be established essentially the same way as in Part (a) for μ. We leave it as an exercise, see Exercise 1.25.

$\square$

Exercises

1.1. Use the definition to show that the following functions are uniformly continuous on any compact interval but not Lipschitz in any neighborhood of 0:

(a) $f(x) = \sqrt{|x|}$,   (b) $f(x) = \begin{cases} x \ln|x|, & x \neq 0 \\ 0, & x = 0 \end{cases}$.

1.2. Assume f is a function defined on an interval $I \subset \mathbb{R}$. Show that f is Lipschitz on I implies that f is uniformly continuous on I, but the converse is not true in general.

1.3. Assume $f : (a, b) \to \mathbb{R}$ is differentiable with $-\infty \leq a < b \leq \infty$. Show that

(a) f' is bounded on (a, b) if and only if f is Lipschitz on (a, b);
(b) if f' is continuous on (a, b), then f is locally Lipschitz on (a, b).

1.4. Let $f(t, x) = t + x^2$. Show that f satisfies a local Lipschitz condition in x but does not satisfy any Lipschitz condition in x on $\mathbb{R}^2$.

1.5. (a) Suppose that $u(t)$ is a continuous solution of the inequality

$$u(t) \leq f(t) + \int_{t_0}^t h(s)u(s)\,ds, \quad t \geq t_0,$$

where f, h are continuous and $h(t) \geq 0$ for $t \geq t_0$. Show that

$$u(t) \leq f(t) + \int_{t_0}^t f(s)h(s)e^{\int_s^t h(\sigma)\,d\sigma}\,ds, \quad t \geq t_0.$$

(b) Moreover, if $f(t)$ is increasing for $t \geq t_0$, show that

$$u(t) \leq f(t)e^{\int_{t_0}^t h(s)\,ds}, \quad t \geq t_0.$$

Note that when $f(t) \equiv M$, it reduces to the original Gronwall inequality.

1.6. Prove the following left-hand side Gronwall inequality: Assume that $M \in \mathbb{R}$, $h \in C([t_0, \infty), \mathbb{R}_+)$ for some $t_0 \in \mathbb{R}$, and $u(t)$ is a continuous solution of the inequality

$$u(t) \leq M + \int_t^{t_0} h(s)u(s)\,ds, \quad t \leq t_0.$$

Then

$$u(t) \leq M e^{\int_t^{t_0} h(s)\,ds}, \quad t \leq t_0.$$

1.7. Discuss the existence and uniqueness of the following IVP's:

(a) $x' = \sin(tx) + (1 - x^2)^3$, $x(0) = 1/2$.
(b) $x' + p(t)\sqrt{|x|} = q(t)$, $x(t_0) = x_0$,   where $p, q \in C(a, b)$, $t_0 \in (a, b)$ and $x_0 \in \mathbb{R}$.
(c) $x' = x^{1/3}$, $x(0) = 0$.

1.8. Consider the IVP $x' = t + \dfrac{(\mu x)^2}{1 + (\mu x)^2}$, $x(0) = 0$, where $\mu \in \mathbb{R}$ is a parameter.

(a) Show that for any $\mu \in \mathbb{R}$, the solution $x_\mu(t)$ of the equation exists on $(-\infty, \infty)$.

(b) Let $\mathcal{F} = \{x_\mu(t) : \mu \in \mathbb{R}\}$ be the set of all solutions of the equation with $\mu \in \mathbb{R}$. Show that $\mathcal{F}$ is equicontinuous on $[a, b]$ with $-\infty < a < b < \infty$.

1.9. Consider the IVP

$$\begin{cases} x_1' = x_1 + \dfrac{1}{t - 1}x_2^2, & x_1(t_0) = a_1 \\ x_2' = t - x_1 x_2^{1/3}, & x_2(t_0) = a_2. \end{cases}$$

Based on the existence and uniqueness theorems, what can you say about the local existence and uniqueness of the solutions of the IVP for the following values of t_0, a_1, and a_2? Justify your answer.

(a) $t_0 = 2$, $a_1 = 1$, $a_2 = -1$;

(b) $t_0 = 2$, $a_1 = 1$, $a_2 = 0$;

(c) $t_0 = 1$, $a_1 = 1$, $a_2 = -1$.

1.10. For which values of t_0, a_1, a_2 and a_3, does the IVP

$$y''' = (\sin t)e^y + \frac{(y'')^{5/3}}{\cos t} + (y')^{1/3}, \quad y(t_0) = a_1, y'(t_0) = a_2, y''(t_0) = a_3,$$

have a solution and have a unique solution, respectively? Justify your answers.

1.11. Let $f(t, x) \in C(\mathbb{R} \times \mathbb{R}^n, \mathbb{R}^n)$ and satisfy a local Lipschitz condition in x. Assume $f(t, 0) = 0$. If $x(t)$ is a solution of the equation $x' = f(t, x)$ such that $x(0) \neq 0$, show that $x(t) \neq 0$ for any $t \in \mathbb{R}$.

1.12. Consider the equation $y'' + p(t)y' + q(t)y^3 = 0$ with $p, q \in C(\mathbb{R}, \mathbb{R})$.

(a) Is it possible for the function $y = e^t - t^2/2 - t - 1$ to be a solution of this equation?

(b) Is it possible for the functions $y_1 = t$ and $y_2 = e^t - 1$ to be solutions of this equation?

Justify your answers.

1.13. (a) Assume $f(t, x)$ is continuous on the rectangle

$$R := \{(t, x) : t_0 \leq t \leq t_0 + a, \; |x - x_0| \leq b\} \subset \mathbb{R}^2$$

and is nonincreasing in x, i.e.,

$$[f(t, x_1) - f(t, x_2)](x_1 - x_2) \leq 0$$

for all $(t, x_1), (t, x_2) \in R$. Show that the IVP

$$x' = f(t, x), \quad x(t_0) = x_0$$

has a unique solution in a right-neighborhood of t_0.

(b) State a parallel result for the IVP in Part (a) to have a unique solution in a left-neighborhood of t_0.

(c) Can you extend the above results to a system IVP?

1.14. Show that for $\alpha \in (0, 1]$, the IVP

$$x' = \begin{cases} x \ln^\alpha |x|, & x \neq 0, \\ 0, & x = 0; \end{cases} \qquad x(0) = 0$$

has a unique solution, but Theorem 1.3.1 fails to apply to this problem.

1.15. Determine the maximal intervals of existence of the solutions of the equation $\dfrac{dx}{dt} = \dfrac{x^2 - 1}{2}$ through $(0, 0)$ and $(\ln 2, -3)$, respectively, by finding the solutions. In this problem, you are expected to see that different solutions of the same equation may have different maximal intervals of existence.

1.16. Discuss the existence, uniqueness, and the maximal interval of existence of the solution of the IVP

$$x' = \ln t + \frac{x}{x^2 + 1}, \qquad x(1) = 0.$$

1.17. For each of the following IVPs

(a) $x' = \dfrac{t^2 x}{1 + e^x}$, $x(t_0) = x_0$,

(b) $x' = \sin(tx)x + t$, $x(t_0) = x_0$,

(c) $x' = e^t \dfrac{x^3}{1 + x^2}$, $x(t_0) = x_0$;

show that the solution exists on the interval $[t_0, \infty)$ by changing it to an integral equation and using a Gronwall inequality.

1.18. Show that for $\alpha \in [0, 1/2]$, all solutions of the equation $x' = (x^2 + e^t)^\alpha$ exist on the whole interval $[0, \infty)$.

1.19. Let $-\infty \leq a < b \leq \infty$, $t_0 \in (a, b)$, and $x_0 \in \mathbb{R}^n$.

(a) Assume $f \in C((a, b) \times \mathbb{R}^n, \mathbb{R}^n)$ such that

$$|f(t, x)| \leq p(t)|x| + q(t), \qquad (t, x) \in (a, b) \times \mathbb{R}^n,$$

where p, q are continuous on (a, b). Show that every solution of the IVP

$$x' = f(t, x), \qquad x(t_0) = x_0$$

exists on the whole interval (a, b).

(b) What can you say about the solutions of the linear equation $x' = A(t)x + g(t)$ where $A \in C((a, b), \mathbb{R}^{n \times n})$ and $g \in C((a, b), \mathbb{R}^n)$?

1.20. Let $D = (a, b) \times \mathbb{R}^n$, $-\infty \leq a < b \leq \infty$ and $(t_0, x_0) \in D$. Suppose that $f \in C(D, \mathbb{R}^n)$ and satisfies a Lipschitz condition in x on D. (In particular, the Jacobian matrix $\partial f / \partial x$ is bounded on D.) Show that the IVP

$$x' = f(t, x), \qquad x(t_0) = x_0$$

has a unique solution which exists on the whole interval (a, b).

1.21. Consider the IVP

$$x' = h(t)g(x), \quad x(t_0) = x_0,$$

where $h \in C([a, \infty), [0, \infty))$, $g \in C([0, \infty), (0, \infty))$, $t_0 \in [a, \infty)$, and $x_0 \in [0, \infty)$. Moreover, $\int_{x_0}^{\infty} \frac{dx}{g(x)} = \infty$. Show that every solution of the IVP exists on $[t_0, \infty)$.

1.22. Consider the IVP

$$x' = f(x), \quad x(t_0) = x_0,$$

where $f \in C(\mathbb{R}, \mathbb{R})$ such that $xf(x) < 0$ for $x \neq 0$ and $(t_0, x_0) \in \mathbb{R}^2$. Show that every solution $x(t)$ of the IVP exists and is monotone on $[t_0, \infty)$, and satisfies $\lim_{t \to \infty} x(t) = 0$.

1.23. Prove the following result on the continuous dependence of solutions of IVPs on the equation and IC: Let $x(t)$ and $y(t)$ be solutions of the two IVPs

$$x' = f(t, x), \ x(t_0) = x_0, \quad \text{and} \quad y' = g(t, y), \ y(t_0) = y_0,$$

respectively, where $f, g \in C([t_0, t_0 + T] \times \mathbb{R}, \mathbb{R})$ with $T > 0$ such that

(i) f satisfies the Lipschitz condition in x with a Lipschitz constant k,
(ii) $|f(t, z) - g(t, z)| \leq \epsilon$ for some $\epsilon > 0$ and $(t, z) \in [t_0, t_0 + T] \times \mathbb{R}$.

Show that $x(t)$ and $y(t)$ exist on $[t_0, t_0 + T]$ and satisfy that

$$|x(t) - y(t)| \leq (|x_0 - y_0| + \epsilon T)e^{Tk}, \quad t \in [t_0, t_0 + T].$$

1.24. Consider the IVP (E_μ): $x' = \dfrac{t}{1 + e^{\mu x^2}}$, $x(0) = 1$, where $\mu \in \mathbb{R}$ is a parameter.

(a) Show that for each $\mu \in \mathbb{R}$, the solution $x_\mu(t)$ of (E_μ) exists on $\mathbb{R}$ and is unique.
(b) For the solution $x_\mu(t)$ of (E_μ), find $\lim_{\mu \to 0} x_\mu(t)$ and justify your answer.

1.25. Prove Theorem 1.5.3, Part (c).

CHAPTER 2

Linear Differential Equations

2.1. Introduction

Consider the system of first-order linear differential equations

$$\begin{cases} x_1' = a_{11}(t)x_1 + a_{12}(t)x_2 + \cdots + a_{1n}(t)x_n + f_1(t) \\ x_2' = a_{21}(t)x_1 + a_{22}(t)x_2 + \cdots + a_{2n}(t)x_n + f_2(t) \\ \qquad\qquad \cdots\cdots\cdots\cdots \\ x_n' = a_{n1}(t)x_1 + a_{n2}(t)x_2 + \cdots + a_{nn}(t)x_n + f_n(t). \end{cases}$$

Let

$$x = \begin{bmatrix} x_1 \\ x_2 \\ \vdots \\ x_n \end{bmatrix}, \quad A(t) = \begin{bmatrix} a_{11} & a_{12} & \cdots & a_{1n} \\ a_{21} & a_{22} & \cdots & a_{2n} \\ \cdots & \cdots & \cdots & \cdots \\ a_{n1} & a_{n2} & \cdots & a_{nn} \end{bmatrix}(t), \quad \text{and} \quad f(t) = \begin{bmatrix} f_1 \\ f_2 \\ \vdots \\ f_n \end{bmatrix}(t).$$

Then the system of linear equations can be expressed in the vector form

(NH) $$x' = A(t)x + f(t),$$

where for any fixed t in the domain, $x(t), f(t) \in \mathbb{R}^n$ and $A(t) \in \mathbb{R}^{n \times n}$. If $f(t) \equiv 0$, then Eq. (NH) becomes

(H) $$x' = A(t)x.$$

We call Eq. (H) a homogeneous linear equation and Eq. (NH) with $f(t) \not\equiv 0$ a nonhomogeneous linear equation.

Unless specified otherwise such as in Sect. 3.3, we will adopt the notation below.

Notation.

(i) For $A \in \mathbb{R}^{n \times n}$, we use the matrix norm $|A| = \sum_{i,j=1}^{n} |a_{ij}|$. We recall that for $x \in \mathbb{R}^n$, $|x| = \sum_{i=1}^{n} |x_i|$. It is known that the following properties are satisfied by these norms:
  (a) $|A| \geq 0$ for any $A \in \mathbb{R}^{n \times n}$, and $|A| = 0 \iff A = 0$;
  (b) $|kA| = |k||A|$ for any $A \in \mathbb{R}^{n \times n}$ and $k \in \mathbb{R}$;
  (c) $|A + B| \leq |A| + |B|$ and $|AB| \leq |A||B|$ for any $A, B \in \mathbb{R}^{n \times n}$;
  (d) $|Ax| \leq |A||x|$ for any $A \in \mathbb{R}^{n \times n}$ and $x \in \mathbb{R}^n$.

© Springer International Publishing Switzerland 2014
Q. Kong, *A Short Course in Ordinary Differential Equations*, Universitext,
DOI 10.1007/978-3-319-11239-8_2

(ii) Let $k \in \mathbb{N}_0$. By $A \in C^k((a,b), \mathbb{R}^{n \times n})$ we mean that $a_{ij} \in C^k((a,b),$ $\mathbb{R})$, $i, j = 1, \ldots, n$; and we denote $A^{(k)}(t) = \left[a_{ij}^{(k)}(t) \right]_{n \times n}$.

The following theorem provides the existence and uniqueness result for IVPs associated with linear equations.

THEOREM 2.1.1. *Let $-\infty \leq a < b \leq \infty$. Assume that $A \in C((a,b),$ $\mathbb{R}^{n \times n})$, $f \in C((a,b), \mathbb{R}^n)$, $t_0 \in (a,b)$, and $x_0 \in \mathbb{R}^n$. Then the IVP consisting of Eq. (NH) and the IC $x(t_0) = x_0$ has a unique solution which exists on the whole interval (a,b).*

PROOF. Let $F(t,x) = A(t)x + f(t)$. By the assumption, $F(t,x)$ is continuous on its domain $D = (a,b) \times \mathbb{R}^n$. It is easy to check that

$$\frac{\partial F}{\partial x}(t,x) := \frac{\partial(F_1, \ldots, F_n)}{\partial(x_1, \ldots, x_n)}(t,x) = A(t).$$

Thus $\partial F/\partial x$ is continuous on D, which implies that F is locally Lipschitz in x on D. By Theorem 1.3.1, the IVP has a unique local solution. Note that for $(t,x) \in D$ we have

$$|F(t,x)| \leq |A(t)||x| + |f(t)|.$$

Since $p(t) := |A(t)|$ and $q(t) := |f(t)|$ are continuous on (a,b), by Exercise 1.19, the solution exists on (a,b). $\qquad\square$

2.2. General Theory for Homogeneous Linear Equations

Consider the homogeneous linear equation

(H) $x' = A(t)x,$

where $A \in C((a,b), \mathbb{R}^{n \times n})$ with $-\infty \leq a < b \leq \infty$. Clearly, $x(t) \equiv 0$ is a solution of Eq. (H) satisfying $x(t_0) = 0$ for any $t_0 \in (a,b)$. We call it the zero solution or the trivial solution of Eq. (H). Furthermore, the solutions of Eq. (H) satisfy the following important property which can be easily verified.

THEOREM 2.2.1. *(Superposition Principle) If $x_1(t)$ and $x_2(t)$ are solutions of Eq. (H), then for any $\alpha, \beta \in \mathbb{R}$, $x(t) := \alpha x_1(t) + \beta x_2(t)$ is a solution of Eq. (H).*

Theorem 2.2.1 shows that the solutions of Eq. (H) form a vector space. To determine the dimension of this space, we need the following definitions and lemmas on linear dependence and independence.

DEFINITION 2.2.1. *Let $x_i = [x_{i1}, \cdots, x_{in}]^T : (a,b) \to \mathbb{R}^n$, $i = 1, \ldots, m$. Then $x_1, \ldots, x_m$ are said to be linearly dependent on (a,b) if there exist $c_1, \ldots, c_m \in \mathbb{R}$, not all zero, such that*

(2.2.1) $c_1 x_1(t) + c_2 x_2(t) + \cdots + c_m x_m(t) \equiv 0$ *on (a,b).*

Otherwise, $x_1, \ldots, x_m$ are said to be linearly independent on (a,b).

REMARK 2.2.1. It is easy to see that

(i) $x_1, \ldots, x_m$ are linearly dependent on (a, b) means that one of $x_1(t)$, $\ldots, x_m(t)$ can be expressed as a linear combination of the others with the same combination coefficients for all $t \in (a, b)$;

(ii) $x_1, \ldots, x_m$ are linearly independent on (a, b) means that if (2.2.1) holds for all $t \in (a, b)$, then $c_1 = \cdots = c_m = 0$.

EXAMPLE 2.2.1. (i) Let $x_1(t) = \begin{bmatrix} \cos^2 t \\ t \end{bmatrix}$ and $x_2(t) = \begin{bmatrix} \sin^2 t - 1 \\ -t \end{bmatrix}$. Then x_1 and x_2 are linearly dependent on $\mathbb{R}$ since $c_1 x_1(t) + c_2 x_2(t) \equiv \begin{bmatrix} 0 \\ 0 \end{bmatrix}$ holds with $c_1 = c_2 = 1$.

(ii) Let $x_1(t) = \begin{bmatrix} 1 \\ 0 \end{bmatrix}$ and $x_2(t) = \begin{bmatrix} t \\ 0 \end{bmatrix}$. Then x_1 and x_2 are linearly independent on $\mathbb{R}$ since

$$c_1 x_1(t) + c_2 x_2(t) = \begin{bmatrix} c_1 + c_2 t \\ 0 \end{bmatrix} = \begin{bmatrix} 0 \\ 0 \end{bmatrix}$$

for all $t \in \mathbb{R}$ implies that $c_1 = c_2 = 0$.

The Wronskian of n functions of n components as defined below is often used to discuss the linear dependence and linear independence of the functions.

DEFINITION 2.2.2. Let $x_i = [x_{i1}, \ldots, x_{in}]^T : (a, b) \to \mathbb{R}^n$, $i = 1, \ldots, n$. Then the Wronskian of $x_1, \ldots, x_n$ is defined by

$$W(t) = W[x_1, \ldots, x_n](t) := \det \begin{bmatrix} x_{11} & \cdots & x_{n1} \\ \cdots & \cdots & \cdots \\ x_{1n} & \cdots & x_{nn} \end{bmatrix} (t).$$

LEMMA 2.2.1. Let $x_i : (a, b) \to \mathbb{R}^n$, $i = 1, \ldots, n$, and $W(t)$ their Wronskian. If $x_1, \ldots, x_n$ are linearly dependent on (a, b), then $W(t) \equiv 0$ on (a, b).

PROOF. Since $x_1, \ldots, x_n$ are linearly dependent on (a, b), there exist $c_1, \ldots, c_n$, not all zero, such that

(2.2.2) $c_1 x_1(t) + c_2 x_2(t) + \cdots + c_n x_n(t) \equiv 0$ on (a, b).

Thus, for any $t \in (a, b)$, (2.2.2) as a homogeneous linear equation for $c_1, \ldots, c_n$ has a nonzero solution. By linear algebra, the coefficient determinant $W(t) = 0$ for any $t \in (a, b)$.   □

REMARK 2.2.2. Let $x_i : (a, b) \to \mathbb{R}^n$, $i = 1, \ldots, n$, and $W(t)$ their Wronskian. By Lemma 2.2.1, if $W(t) \not\equiv 0$ on (a, b), then $x_1, \ldots, x_n$ are linearly independent on (a, b). However, $W(t) \equiv 0$ on (a, b) does not guarantee that $x_1, \ldots, x_n$ are linearly dependent on (a, b). To see this, let $x_1(t) = \begin{bmatrix} 1 \\ 0 \end{bmatrix}$ and

$x_2(t) = \begin{bmatrix} t \\ 0 \end{bmatrix}$. From Example 2.2.1, (b), x_1 and x_2 are linearly independent on $\mathbb{R}$. On the other hand, their Wronskian $W(t) = \det \begin{bmatrix} 1 & t \\ 0 & 0 \end{bmatrix} \equiv 0$.

However, the Wronskian of n solutions of the homogeneous linear equation (H) satisfies a better property.

LEMMA 2.2.2. *Let $x_i(t)$, $i = 1, \ldots, n$, be solutions of Eq. (H) and $W(t)$ their Wronskian. Then $x_1, \ldots, x_n$ are linearly independent on (a, b) $\Longleftrightarrow$ $W(t) \neq 0$ for any $t \in (a, b)$.*

PROOF. ($\Longleftarrow$) This is given by Lemma 2.2.1.

($\Longrightarrow$) We show this by contradiction. If not, there exists a $t_0 \in (a, b)$ such that $W(t_0) = 0$. Consider the system

$$(2.2.3) \qquad c_1 x_1(t_0) + c_2 x_2(t_0) + \cdots + c_n x_n(t_0) = 0$$

as a system for $c_1, c_2, \ldots, c_n$. Since the coefficient determinant $W(t_0) = 0$, Eq. (2.2.3) has a nonzero solution $\tilde{c}_1, \tilde{c}_2, \ldots, \tilde{c}_n$. Let

$$x(t) = \tilde{c}_1 x_1(t) + \tilde{c}_2 x_2(t) + \cdots + \tilde{c}_n x_n(t).$$

By Theorem 2.2.1, $x(t)$ is a solution of Eq. (H), and

$$x(t_0) = \tilde{c}_1 x_1(t_0) + \tilde{c}_2 x_2(t_0) + \cdots + \tilde{c}_n x_n(t_0) = 0.$$

By the uniqueness of solutions of IVPs associated with Eq. (H), we see that $x(t) \equiv 0$ on (a, b). That is,

$$\tilde{c}_1 x_1(t) + \tilde{c}_2 x_2(t) + \cdots + \tilde{c}_n x_n(t) \equiv 0 \quad \text{on } (a, b).$$

This leads to the conclusion that $x_1, \ldots, x_n$ are linearly dependent on (a, b) which contradicts the assumption. $\qquad\square$

As a direct consequence of Lemma 2.2.2 we have

COROLLARY 2.2.1. *Let $x_i(t)$, $i = 1, \ldots, n$, be solutions of Eq. (H) and $W(t)$ their Wronskian. Then either $W(t) \equiv 0$ on (a, b) or $W(t) \neq 0$ on (a, b). Moreover,*

$W(t) \equiv 0$ *on* $(a, b) \Longleftrightarrow x_1, \ldots, x_n$ *are linearly dependent on* (a, b),   *and*
$W(t) \neq 0$ *for all* $t \in (a, b) \Longleftrightarrow x_1, \ldots, x_n$ *are linearly independent on* (a, b).

REMARK 2.2.3. (i) Corollary 2.2.1 implies that the Wronskian $W(t)$ of n linearly independent solutions of Eq. (H) satisfies

$W(t) \equiv 0$ on $(a, b) \quad \Longleftrightarrow \quad W(t_0) = 0$ for some $t_0 \in (a, b)$,   and
$W(t) \neq 0$ on $(a, b) \quad \Longleftrightarrow \quad W(t_0) \neq 0$ for some $t_0 \in (a, b)$.

Therefore, the linear dependence or independence of the solutions can be determined by the value of $W(t)$ at one point $t_0 \in (a, b)$.

(ii) Equation (H) has at least n linearly independent solutions. For example, let $x_i(t)$, $i = 1, \ldots, n$, be solutions of Eq. (H) satisfying the ICs

$$x_1(t_0) = \begin{bmatrix} 1 \\ 0 \\ 0 \\ \vdots \\ 0 \end{bmatrix}, \; x_2(t_0) = \begin{bmatrix} 0 \\ 1 \\ 0 \\ \vdots \\ 0 \end{bmatrix}, \; \ldots, \; x_n(t_0) = \begin{bmatrix} 0 \\ 0 \\ 0 \\ \vdots \\ 1 \end{bmatrix},$$

respectively. Then

$$W(t_0) = \det \begin{bmatrix} 1 & & & \\ & 1 & & \\ & & \ddots & \\ & & & 1 \end{bmatrix} = 1 \neq 0,$$

from which it follows that $W(t) \neq 0$ for any $t \in (a, b)$. By Corollary 2.2.1, $x_i(t)$, $i = 1, \ldots, n$, are linearly independent on (a, b).

(iii) Let $x_1(t), \ldots, x_n(t)$ be linearly independent solutions of Eq. (H), and $x(t)$ any solution of Eq. (H). Then there exist $c_1, \ldots, c_n$ such that

(2.2.4) $$x(t) \equiv c_1 x_1(t) + \cdots + c_n x_n(t) \quad \text{on } (a, b).$$

In fact, for $t_0 \in (a, b)$, $\{x_1(t_0), \ldots, x_n(t_0)\}$ is a basis for $\mathbb{R}^n$. Since $x(t_0) \in \mathbb{R}^n$, there exist $c_1, \ldots, c_n$ such that

$$x(t_0) = c_1 x_1(t_0) + \cdots + c_n x_n(t_0).$$

Let

$$y(t) = c_1 x_1(t) + \cdots + c_n x_n(t).$$

Then $y(t)$ is a solution of Eq. (H) satisfying

$$y(t_0) = c_1 x_1(t_0) + \cdots + c_n x_n(t_0) = x(t_0).$$

By the uniqueness of solutions of IVPs associated with Eq. (H), we see that $y(t) \equiv x(t)$ on (a, b). This shows that (2.2.4) holds.

Now, we have the structure of the solution space for Eq. (H).

THEOREM 2.2.2. *The set of solutions of Eq. (H) is an n-dimensional vector space.*

DEFINITION 2.2.3. (a) An $n \times n$ matrix-valued function $X(t) = \begin{bmatrix} x_1 & \cdots & x_n \end{bmatrix}(t)$ is said to be a matrix solution of Eq. (H) if its columns $x_i(t)$, $i = 1, \ldots, n$, are solutions of Eq. (H).

(b) An $n \times n$ matrix-valued function $X(t) = \begin{bmatrix} x_1 & \cdots & x_n \end{bmatrix}(t)$ is said to be a fundamental matrix solution of Eq. (H) if its columns $x_i(t)$, $i = 1, \ldots, n$, are linearly independent solutions of Eq. (H).

(c) A fundamental matrix solution $X(t)$ of Eq. (H) is said to be the principal matrix solution of Eq. (H) at $t_0 \in (a, b)$ if $X(t_0) = I$, the identity matrix.

Using Corollary 2.2.1 we can easily prove the following properties of matrix solutions of Eq. (H).

LEMMA 2.2.3. *(a) An $n \times n$ matrix-valued function $X(t)$ is a matrix solution of Eq. (H) $\iff$ $X'(t) = A(t)X(t)$ for any $t \in (a, b)$.*

(b) An $n \times n$ matrix-valued function $X(t)$ is a fundamental matrix solution of Eq. (H) $\iff$ $X'(t) = A(t)X(t)$ and $\det X(t) \neq 0$ for any $t \in (a, b)$.

(c) Let $X(t)$ be a fundamental matrix solution of Eq. (H). Then $x(t)$ is a solution of Eq. (H) $\iff$ there exists a $c \in \mathbb{R}^n$ such that $x(t) = X(t)c$ for any $t \in (a, b)$; in fact, $c = X^{-1}(t_0)x(t_0)$.

(d) Let $X(t)$ be a fundamental matrix solution of Eq. (H). Then $Y(t)$ is a fundamental matrix solution of Eq. (H) $\iff$ there exists a nonsingular $C \in \mathbb{R}^{n \times n}$ such that $Y(t) = X(t)C$ for any $t \in (a, b)$; in fact, $C = X^{-1}(t_0)Y(t_0)$.

REMARK 2.2.4. (i) Let $X(t)$ be a fundamental matrix solution of Eq. (H) and $C \in \mathbb{R}^{n \times n}$. Then $Z(t) = CX(t)$ is not a matrix solution of Eq. (H) in general.

(ii) Let $X(t)$ be a principal matrix solution of Eq. (H) at t_0 and $Y(t)$ a matrix solution of Eq. (H). Then $Y(t) = X(t)Y(t_0)$ for any $t \in (a, b)$.

The next result can be used to obtain the determinant of a fundamental matrix solution without finding the fundamental matrix solution itself.

THEOREM 2.2.3 (Liouville). *Let $X(t)$ be a matrix solution of Eq. (H) and $W(t) = \det X(t)$. Then*

$$(2.2.5) \qquad W'(t) = [\operatorname{tr} A(t)]W(t), \ t \in (a, b),$$

where $\operatorname{tr} A(t) = \sum_{i=1}^{n} a_{ii}(t)$. *Consequently,*

$$(2.2.6) \qquad W(t) = W(t_0)e^{\int_{t_0}^{t} \operatorname{tr} A(s)\, ds} \qquad \text{for } t_0, t \in (a, b).$$

PROOF. Let $t \in (a, b)$. Since

$$X'(t) = \lim_{\tau \to t} \frac{X(\tau) - X(t)}{\tau - t},$$

for any $\tau \in (a, b)$ we have

$$X(\tau) = X(t) + X'(t)(\tau - t) + o(\tau - t),$$

where $o(\tau - t)/(\tau - t) \to 0$ as $\tau \to t$. Note that $X(t)$ satisfies Eq. (H). Then

$$(2.2.7) \qquad \begin{aligned} X(\tau) &= X(t) + A(t)X(t)(\tau - t) + o(\tau - t) \\ &= [I + A(t)(\tau - t)]X(t) + o(\tau - t). \end{aligned}$$

By the definition of determinant we have

$$\det[I + A(t)(\tau - t)]$$

$$= \begin{vmatrix} 1 + a_{11}(t)(\tau - t) & a_{12}(t)(\tau - t) & \cdots & a_{1n}(t)(\tau - t) \\ a_{21}(t)(\tau - t) & 1 + a_{22}(t)(\tau - t) & \cdots & a_{2n}(t)(\tau - t) \\ \cdots & \cdots & \cdots & \cdots \\ a_{n1}(t)(\tau - t) & a_{n2}(t)(\tau - t) & \cdots & 1 + a_{nn}(t)(\tau - t) \end{vmatrix}$$

$$= 1 + \sum_{i=1}^{n} a_{ii}(t)(\tau - t) + o(\tau - t) = 1 + \operatorname{tr} A(t)(\tau - t) + o(\tau - t).$$

This together with (2.2.7) shows that

$$W(\tau) = (1 + \operatorname{tr} A(t)(\tau - t))W(t) + o(\tau - t),$$

from which it follows that

$$W'(t) = \lim_{\tau \to t} \frac{W(\tau) - W(t)}{\tau - t} = [\operatorname{tr} A(t)]W(t).$$

This confirms (2.2.5). By solving (2.2.5) as a first-order differential equation for $W(t)$ and applying the IC, we obtain (2.2.6). □

We comment that there are several different proofs of Theorem 2.2.3, and ours is from Grimshaw [18]. For other proofs, we refer the reader to Walter [44, pp 166] and Liu [37].

It is implied by (2.2.6) that $W(t) \neq 0$ for any $t \in (a, b)$ if and only if $W(t_0) \neq 0$ for some $t_0 \in (a, b)$. Thus, Theorem 2.2.3 is a further development of the result in Corollary 2.2.1.

2.3. Nonhomogeneous Linear Equations

Based on the results for homogeneous linear equation

(H) $$x' = A(t)x$$

obtained in Sect. 2.2, we study the nonhomogeneous linear equation

(NH) $$x' = A(t)x + f(t),$$

where $A \in C((a, b), \mathbb{R}^{n \times n})$ and $f \in C((a, b), \mathbb{R}^n)$ with $-\infty \leq a < b \leq \infty$.

It is easy to verify by direct substitutions that if $x_1(t)$ and $x_2(t)$ are solutions of Eq. (NH), then $(x_1 + x_2)(t)$ is not a solution of Eq. (NH) unless $f(t) \equiv 0$. This means that the solutions of Eq. (NH) do not form a vector space. However, the following properties are satisfied by the solutions of Eq. (NH) and Eq. (H).

LEMMA 2.3.1. (a) $x_1(t)$ and $x_2(t)$ are solutions of Eq. (NH) $\implies$ $(x_1 - x_2)(t)$ is a solution of Eq. (H).

(b) $x_1(t)$ is a solution of Eq. (NH) and $x_2(t)$ is a solution of Eq. (H) $\implies$ $(x_1 + x_2)(t)$ is a solution of Eq. (NH).

(c) $X(t)$ is a fundamental matrix solution of Eq. (H) and $x_1(t)$ is a solution of Eq. (NH) $\implies$ $x(t) = X(t)c + x_1(t)$ is a solution of Eq. (NH) for any $c \in \mathbb{R}^n$.

(d) $X(t)$ is a fundamental matrix solution of Eq. (H) and $x_1(t)$ is a solution of Eq. (NH) $\implies$ for any solution $x(t)$ of Eq. (NH), there exists a $c \in \mathbb{R}^n$ such that $x(t) = X(t)c + x_1(t)$.

PROOF. Parts (a)–(c) can be easily verified by substitutions. We now prove Part (d). Since $x(t)$ and $x_1(t)$ are solutions of Eq. (NH), $(x - x_1)(t)$ is a solution of Eq. (H) by Part (a). From Lemma 2.2.3, Part (c) we see that $x(t) - x_1(t) = X(t)c$ for some $c \in \mathbb{R}^n$. This completes the proof of Part (d). $\qquad\qquad\square$

Lemma 2.3.1 reveals the structure of the solutions of Eq. (NH). From it we see that in order to find the general solution of Eq. (NH), we need to find a fundamental matrix solution of Eq. (H) and a particular solution of Eq. (NH). However, it does not tell us how a particular solution of Eq. (NH) can be obtained. The next theorem provides us with a formula for the solutions of Eq. (NH) which makes use of a fundamental matrix solution of Eq. (H) only.

THEOREM 2.3.1 (Variation of Parameters Formula). *Let $X(t)$ be a fundamental matrix solution of Eq. (H) and $t_0 \in (a, b)$. Then the general solution of Eq. (NH) is*

$$(2.3.1) \qquad x = X(t)c + \int_{t_0}^{t} X(t)X^{-1}(s)f(s)\,ds.$$

In particular, the solution of the IVP consisting of Eq. (NH) and the IC $x(t_0) = x_0$ is

$$(2.3.2) \qquad x = X(t)X^{-1}(t_0)x_0 + \int_{t_0}^{t} X(t)X^{-1}(s)f(s)\,ds.$$

PROOF. Let $x(t)$ be any solution of Eq. (NH). We look for a function $u(t)$ such that

$$(2.3.3) \qquad x(t) = X(t)u(t).$$

Clearly, such a function $u(t)$ always exists and $u(t) = X^{-1}(t)x(t)$. To find $u(t)$, we differentiate both sides of (2.3.3) to get that

$$\begin{aligned} x'(t) &= X'(t)u(t) + X(t)u'(t) = A(t)X(t)u(t) + X(t)u'(t) \\ &= A(t)x(t) + X(t)u'(t). \end{aligned}$$

On the other hand, since $x(t)$ is a solution of Eq. (NH), we have

$$x'(t) = A(t)x(t) + f(t).$$

The combination of the above two equalities shows that $X(t)u'(t) = f(t)$ and hence $u'(t) = X^{-1}(t)f(t)$. Integrating both sides from t_0 to t, we have

$$u(t) = c + \int_{t_0}^{t} X^{-1}(s)f(s)\,ds$$

for $c \in \mathbb{R}^n$. Thus, (2.3.1) follows from (2.3.3).

Applying the IC $x(t_0) = x_0$ in (2.3.1), we find that $c = X^{-1}(t_0)x_0$. Thus, (2.3.2) holds. □

Note that for the scalar case, i.e., when $n = 1$, the principal matrix solution of the homogeneous linear equation $x' = a(t)x$ at t_0 is $x = e^{\int_{t_0}^{t} a(s)\,ds}$. From Theorem 2.3.1 we see that the general solution of the scalar linear equation $x' = a(t)x + f(t)$, where $a \in C((a,b), \mathbb{R})$, is given by

$$x = ce^{\int_{t_0}^{t} a(s)\,ds} + \int_{t_0}^{t} e^{\int_{s}^{t} a(\tau)\,d\tau} f(s)\,ds.$$

This is consistent with the formula that we learned earlier from an elementary differential equations course.

Now, we apply Theorem 2.3.1 to derive the variation of parameters formula for second-order and higher-order scalar linear equations.

We first consider the second-order linear equation

(nh-2) $$x'' + a_1(t)x' + a_2(t)x = f(t),$$

where $a_1, a_2, f \in C((a,b), \mathbb{R})$. Clearly, the corresponding homogeneous linear equation is

(h-2) $$x'' + a_1(t)x' + a_2(t)x = 0.$$

COROLLARY 2.3.1. *Let $\phi_1(t)$ and $\phi_2(t)$ be two linearly independent solutions of Eq. (h-2). Then the general solution of Eq. (nh-2) is*

$$x = c_1\phi_1(t) + c_2\phi_2(t) + \int_{t_0}^{t} \frac{f(s)}{W(s)}\left[-\phi_1(t)\phi_2(s) + \phi_2(t)\phi_1(s)\right]ds,$$

where $W(s) = \det\begin{bmatrix} \phi_1 & \phi_2 \\ \phi_1' & \phi_2' \end{bmatrix}(s)$.

PROOF. Let $x_1 = x$ and $x_2 = x'$. Then Eq. (nh-2) can be written as the vector-valued first-order linear equation

(NH-2) $$\begin{bmatrix} x_1 \\ x_2 \end{bmatrix}' = \begin{bmatrix} 0 & 1 \\ -a_2(t) & -a_1(t) \end{bmatrix}\begin{bmatrix} x_1 \\ x_2 \end{bmatrix} + \begin{bmatrix} 0 \\ f(t) \end{bmatrix}.$$

Clearly, the corresponding homogeneous linear equation to Eq. (NH-2) is

(H-2) $$\begin{bmatrix} x_1 \\ x_2 \end{bmatrix}' = \begin{bmatrix} 0 & 1 \\ -a_2(t) & -a_1(t) \end{bmatrix}\begin{bmatrix} x_1 \\ x_2 \end{bmatrix}.$$

Note that $\begin{bmatrix} \phi_1 \\ \phi_1' \end{bmatrix}(t)$ and $\begin{bmatrix} \phi_2 \\ \phi_2' \end{bmatrix}(t)$ are linearly independent solutions of Eq. (H-2) with the Wronskian $W(t)$. Thus, $X(t) = \begin{bmatrix} \phi_1 & \phi_2 \\ \phi_1' & \phi_2' \end{bmatrix}(t)$ is a fundamental matrix solution of Eq. (H-2) and $X^{-1}(t) = \dfrac{1}{W(t)} \begin{bmatrix} \phi_2' & -\phi_2 \\ -\phi_1' & \phi_1 \end{bmatrix}(t)$. By Theorem 2.3.1, the general solution of Eq. (NH-2) is

$$\begin{bmatrix} x_1 \\ x_2 \end{bmatrix} = X(t)\begin{bmatrix} c_1 \\ c_1 \end{bmatrix} + \int_{t_0}^{t} X(t)X^{-1}(s)\begin{bmatrix} 0 \\ f(s) \end{bmatrix} ds$$

$$= \begin{bmatrix} c_1\phi_1(t) + c_2\phi_2(t) \\ c_1\phi_1'(t) + c_2\phi_2'(t) \end{bmatrix} + \int_{t_0}^{t} \frac{f(s)}{W(s)} \begin{bmatrix} -\phi_1(t)\phi_2(s) + \phi_2(t)\phi_1(s) \\ -\phi_1'(t)\phi_2(s) + \phi_2'(t)\phi_1(s) \end{bmatrix} ds.$$

Since the first component x_1 of this solution is the general solution of Eq. (nh-2), we have reached the conclusion. □

EXAMPLE 2.3.1. Consider the second-order linear equation $x'' + x = f(t)$, where $f \in C(\mathbb{R}, \mathbb{R})$. It is easy to see that $\phi_1(t) = \cos t$ and $\phi_2(t) = \sin t$ are linearly independent solutions of the homogeneous linear equation $x'' + x = 0$ with the Wronskian $W(t) \equiv 1$. By Corollary 2.3.1, the general solution of the nonhomogeneous linear equation is

$$x = c_1 \cos t + c_2 \sin t + \int_{t_0}^{t} f(s)[-\cos t \sin s + \sin t \cos s]\, ds$$

$$= c_1 \cos t + c_2 \sin t + \int_{t_0}^{t} f(s) \sin(t - s)\, ds.$$

The variation of parameters formula for second-order linear equations given in Corollary 2.3.1 can be extended to higher-order linear equations with a similar proof. Here, we only state the result. The proof is left as an exercise, see Exercise 2.10.

Consider the nth-order linear equation

(nh-n) $\qquad x^{(n)} + a_1(t)x^{(n-1)} + \cdots + a_{n-1}(t)x' + a_n(t)x = f(t),$

where $a_1, \ldots, a_2, f \in C((a, b), \mathbb{R})$. Clearly, the corresponding homogeneous linear equation is

(h-n) $\qquad x^{(n)} + a_1(t)x^{(n-1)} + \cdots + a_{n-1}(t)x' + a_n(t)x = 0.$

COROLLARY 2.3.2. *Let* $\phi_1(t), \ldots, \phi_n(t)$ *be linearly independent solutions of Eq.* (h-n) *and*

$$W[\phi_1, \ldots, \phi_n](t) = \det \begin{bmatrix} \phi_1 & \phi_2 & \cdots & \phi_n \\ \phi_1' & \phi_2' & \cdots & \phi_n' \\ \cdots & \cdots & \cdots & \cdots \\ \phi_1^{(n-1)} & \phi_2^{(n-1)} & \cdots & \phi_n^{(n-1)} \end{bmatrix}(t).$$

Then the general solution of Eq. (nh-n) is

$$x(t) = \sum_{i=1}^{n} c_i \phi_i(t) + \sum_{k=1}^{n} \phi_k(t) \int_{t_0}^{t} f(s) \frac{W_k[\phi_1, \ldots, \phi_n](s)}{W[\phi_1, \ldots, \phi_n](s)} \, ds,$$

where for $k = 1, \ldots, n$, $W_k[\phi_1, \ldots, \phi_n]$ *is the determinant obtained from*

$W[\phi_1, \ldots, \phi_n]$ *where the* k*-th column is replaced by* $\begin{bmatrix} 0 \\ \vdots \\ 0 \\ 1 \end{bmatrix}$.

From Theorem 2.3.1 we see that to solve the nonhomogeneous linear equation (NH), we must solve the corresponding homogeneous linear equation (H). However, the general homogeneous linear equation (H) has not been completely solved yet. In the next two sections, we will discuss two special cases of Eq. (H): Eq. (H) with constant coefficients and periodic coefficients, respectively.

2.4. Homogeneous Linear Equations with Constant Coefficients

In this section, we study the homogeneous linear equation with constant coefficients

(H-c) $x' = Ax,$

where $A \in \mathbb{R}^{n \times n}$. By Theorem 2.1.1, all solutions of Eq. (H-c) exist on the whole real number line $(-\infty, \infty)$. First we introduce certain properties of Eq. (H-c) which are not satisfied by the general homogeneous linear equation (H).

LEMMA 2.4.1. *(a) If* $x(t)$ *is a solution of Eq. (H-c), then* $x^{(k)}(t)$ *exists and is a solution of Eq. (H-c) for each* $k \in \mathbb{N}$.

(b) If $x(t)$ *is a solution of Eq. (H-c), then* $x(t + \alpha)$ *is a solution of Eq. (H-c) for each* $\alpha \in \mathbb{R}$.

PROOF. (a) Since $x(t)$ is a solution of Eq. (H-c), $x'(t)$ exists on $\mathbb{R}$. This shows that the right-hand side of Eq. (H-c) is differentiable on $\mathbb{R}$, and so is the left-hand side. By differentiating both sides of Eq. (H-c) we have that $(x'(t))' = Ax'(t)$. This means that $x'(t)$ is also a solution of Eq. (H-c). The general conclusion is obtained by induction.

(b) This can be verified by a direct substitution.

□

It follows from Theorem 2.2.2 that among the solutions $\{x^{(k)}(t) : k \in \mathbb{N}_0\}$ and among the solutions $\{x(t + \alpha) : k \in \mathbb{N}_0\}$, respectively, there are at most n linearly independent solutions.

Motivated by the fact that the solutions of first-order scalar homogeneous linear equations are given by exponential functions, to derive a formula for a fundamental matrix solution of Eq. (H-c), we need to employ the concept of matrix exponential.

DEFINITION 2.4.1. Let $A \in \mathbb{C}^{n \times n}$. Define the exponential of A by

$$e^A = \sum_{k=0}^{\infty} \frac{A^k}{k!} = I + \frac{A}{1!} + \frac{A^2}{2!} + \cdots + \frac{A^k}{k!} + \cdots,$$

where I is the $n \times n$ identity matrix.

Since $\left| A^k / k! \right| \leq |A|^k / k!$ and the scalar series $\sum_{k=0}^{\infty} |A|^k / k!$ is convergent, the matrix series $\sum_{k=0}^{\infty} A^k / k!$ is norm-convergent and hence is convergent. This means that e^A is well defined.

Now, we present some properties of matrix exponentials.

LEMMA 2.4.2. (a) Let $\mathbf{0}$ be the $n \times n$ zero matrix. Then $e^{\mathbf{0}} = I$.
   (b) Let $A, B \in \mathbb{C}^{n \times n}$. Then $e^{A+B} = e^A e^B$ if A and B commute, i.e., if $AB = BA$. However, $e^{A+B} \neq e^A e^B$ in general.
   (c) For any $A \in \mathbb{C}^{n \times n}$, e^A is nonsingular and $(e^A)^{-1} = e^{-A}$.
   (d) Let $T \in \mathbb{C}^{n \times n}$ be nonsingular. Then $e^{T^{-1}AT} = T^{-1}e^A T$.

PROOF. (a) This follows directly from the definition.
   (b) With the assumption that $AB = BA$, the matrices A and B satisfy the same rules as numbers in matrix multiplications. Thus from the definition of matrix exponential, the proof for $e^{A+B} = e^A e^B$ is the same as that for the scalar exponentials.
   (c) Since A and $-A$ commute, by Part (b) we see that $e^A e^{-A} = I$ which implies that $(e^A)^{-1}$ exists and $(e^A)^{-1} = e^{-A}$.
   (d) From the observation that $(T^{-1}AT)^k = T^{-1}A^K T$ for any $k \in \mathbb{N}_0$ we have

$$e^{T^{-1}AT} = \sum_{k=0}^{\infty} \frac{(T^{-1}AT)^k}{k!} = \sum_{k=0}^{\infty} \frac{T^{-1}A^k T}{k!} = T^{-1} \left(\sum_{k=0}^{\infty} \frac{A^k}{k!} \right) T = T^{-1}e^A T.$$

$\square$

Then we present a result on the solutions of Eq. (H-c).

THEOREM 2.4.1. $X(t) = e^{A(t-t_0)}$ is the principal matrix solution of Eq. (H-c) at t_0.

PROOF. We first show that the $X(t)$ given in the theorem is a matrix solution of Eq. (H-c) on $\mathbb{R}$. In fact, by definition,

$$X(t) = \sum_{k=0}^{\infty} \frac{A^k (t - t_0)^k}{k!}, \quad t \in \mathbb{R}.$$

For any $r > 0$ and $t \in [t_0 - r, t_0 + r]$,

$$\left| \frac{A^k(t - t_0)^k}{k!} \right| \le \frac{(|A|r)^k}{k!}.$$

Since the scalar series $\sum_{k=0}^{\infty} |Ar|^k / k!$ is convergent, the matrix-valued function series $\sum_{k=0}^{\infty} A^k(t - t_0)^k / k!$ is uniformly convergent on $[t_0 - r, t_0 + r]$. Consequently, for any $t \in (t_0 - r, t_0 + r)$,

$$
\begin{aligned}
X'(t) &= \left(\sum_{k=0}^{\infty} \frac{A^k(t - t_0)^k}{k!} \right)' = \sum_{k=0}^{\infty} \frac{A^k \left((t - t_0)^k \right)'}{k!} = \sum_{k=1}^{\infty} \frac{A^k(t - t_0)^{k-1}}{(k-1)!} \\
&= A \sum_{k=1}^{\infty} \frac{A^{k-1}(t - t_0)^{k-1}}{(k-1)!} = A \sum_{k=0}^{\infty} \frac{A^k(t - t_0)^k}{k!} = AX(t).
\end{aligned}
$$

Thus, $X(t)$ is a matrix solution of Eq. (H-c) on $(t_0 - r, t_0 + r)$. Since $r > 0$ is arbitrary, $X(t)$ is a matrix solution of Eq. (H-c) on $\mathbb{R}$. Clearly, $X(t_0) = I$, and hence $X(t)$ is the principal matrix solution of Eq. (H-c) at t_0. $\square$

REMARK 2.4.1. The result in Theorem 2.4.1 cannot be simply extended to Eq. (H) with variable matrix $A(t)$. However, it has been shown that if $A \in C((a, b), \mathbb{R}^{n \times n})$ such that

$$A(t) \left(\int_{t_0}^{t} A(s) \, ds \right) = \left(\int_{t_0}^{t} A(s) \, ds \right) A(t) \quad \text{for any } t \in (a, b),$$

where $t_0 \in (a, b)$, then $e^{\int_{t_0}^{t} A(s) \, ds}$ is the principal matrix solution of Eq. (H) at t_0.

The corollary below follows from the combination of Theorems 2.3.1 and 2.4.1.

COROLLARY 2.4.1. Let $A \in \mathbb{R}^{n \times n}$, $f \in C((a, b), \mathbb{R}^n)$, and $t_0 \in (a, b)$ with $-\infty \le a < b \le \infty$. Then the solution of the IVP

$$x' = Ax + f(t), \quad x(t_0) = x_0$$

is

$$x = e^{A(t - t_0)} x_0 + \int_{t_0}^{t} e^{A(t - s)} f(s) \, ds.$$

To find the principal matrix solution of Eq. (H-c) given in Theorem 2.4.1, we need to know how to compute the matrix exponential e^{At}. We first look at some special cases:

1. $A = \begin{bmatrix} a_1 & & \\ & \ddots & \\ & & a_n \end{bmatrix}$ is a diagonal matrix, where $a_i \in \mathbb{R}$, $i=1, 2, \ldots, n$.

In this case, $A^k = \begin{bmatrix} a_1^k & & \\ & \ddots & \\ & & a_n^k \end{bmatrix}$, $k = 1, 2, \ldots$. Thus,

$$e^{At} = \sum_{k=0}^{\infty} \frac{A^k t^k}{k!} = \sum_{k=0}^{\infty} \begin{bmatrix} a_1^k t^k / k! & & \\ & \ddots & \\ & & a_n^k t^k / k! \end{bmatrix} = \begin{bmatrix} e^{a_1 t} & & \\ & \ddots & \\ & & e^{a_n t} \end{bmatrix}.$$

Note that this result can be extended to the case where $A = \begin{bmatrix} A_1 & & \\ & \ddots & \\ & & A_l \end{bmatrix}$ with A_i, $i = 1, \ldots, l$, square matrix blocks. In fact, similar to the above, we have

$$e^{At} = \begin{bmatrix} e^{A_1 t} & & \\ & \ddots & \\ & & e^{A_l t} \end{bmatrix}.$$

2. $A = \begin{bmatrix} a & 1 & & \\ & \ddots & \ddots & \\ & & \ddots & 1 \\ & & & a \end{bmatrix}$ is a bidiagonal matrix, where $a \in \mathbb{R}$. In this

case, $A = aI + B$ with $B = \begin{bmatrix} & 1 & & \\ & & \ddots & \\ & & & 1 \\ & & & \end{bmatrix}$. Clearly, $e^{aIt} = e^{at}I$,

and with some calculations we see that $B^n = \mathbf{0}$ and hence

$$e^{Bt} = I + Bt + \frac{B^2 t^2}{2!} + \cdots + \frac{B^{n-1} t^{n-1}}{(n-1)!} = \begin{bmatrix} 1 & t & \frac{t^2}{2!} & \cdots & \frac{t^{n-1}}{(n-1)!} \\ & \ddots & \ddots & \ddots & \vdots \\ & & \ddots & \ddots & \frac{t^2}{2!} \\ & & & \ddots & t \\ & & & & 1 \end{bmatrix}.$$

Since aIt and B commute, by Lemma 2.4.2, Part (b),

$$e^{At} \;=\; e^{aIt}e^{Bt} = e^{at}\left(I + Bt + \frac{B^2 t^2}{2!} + \cdots + \frac{B^{n-1}t^{n-1}}{(n-1)!}\right)$$

$$= \; e^{at}\begin{bmatrix} 1 & t & \frac{t^2}{2!} & \cdots & \frac{t^{n-1}}{(n-1)!} \\ & \ddots & \ddots & \ddots & \vdots \\ & & \ddots & \ddots & \frac{t^2}{2!} \\ & & & \ddots & t \\ & & & & 1 \end{bmatrix}.$$

For a general matrix $A \in \mathbb{R}^{n\times n}$, from linear algebra, see [27] for the detail, we know that there is a nonsingular matrix $T \in \mathbb{C}^{n\times n}$ such that $T^{-1}AT = J$, where $J = \begin{bmatrix} J_0 \\ & J_1 \\ & & \ddots \\ & & & J_l \end{bmatrix}$ is the Jordan canonical form of A with

$$J_0 = \begin{bmatrix} \lambda_1 \\ & \ddots \\ & & \lambda_q \end{bmatrix}_{q\times q} \quad \text{and} \quad J_i = \begin{bmatrix} \lambda_{q+i} & 1 \\ & \ddots & \ddots \\ & & \ddots & 1 \\ & & & \lambda_{q+i} \end{bmatrix}_{r_i\times r_i} \quad i = 1,\ldots,l,$$

satisfying $q + r_1 + \cdots + r_l = n$. We also know that all the λ's in J_i, $i = 0,\ldots,l$, are eigenvalues of A, i.e., they are roots of the equation $\det(\lambda I - A) = 0$.

From the above, once the Jordan canonical form J and the transformation matrix T are found, we can compute e^{At} by applying Lemma 2.4.2, Part (d) and the results in the above special cases 1 and 2. In fact, since $A = TJT^{-1}$, from Lemma 2.4.2, Part (d),

$$(2.4.1) \qquad e^{At} = e^{(TJT^{-1})t} = Te^{Jt}T^{-1} = T\begin{bmatrix} e^{J_0 t} \\ & e^{J_1 t} \\ & & \ddots \\ & & & e^{J_l t} \end{bmatrix} T^{-1},$$

where

$$e^{J_0 t} = \begin{bmatrix} e^{\lambda_1 t} \\ & \ddots \\ & & e^{\lambda_q t} \end{bmatrix}.$$

and

$$e^{J_i t} = e^{\lambda_{q+i} t} \begin{bmatrix} 1 & t & \frac{t^2}{2!} & \cdots & \frac{t^{r_i-1}}{(r_i-1)!} \\ & \ddots & \ddots & \ddots & \vdots \\ & & \ddots & \ddots & \frac{t^2}{2!} \\ & & & \ddots & t \\ & & & & 1 \end{bmatrix}, \quad i = 1, \ldots, l.$$

As a special case, when A is diagonalizable, i.e., $J = J_0 = \begin{bmatrix} \lambda_1 & \\ & \ddots & \\ & & \lambda_n \end{bmatrix}$,

we let $T = [v_1 \ldots v_n]$, where v_i, $i = 1, \ldots, n$, are linearly independent eigenvectors associated with λ_i, $i = 1, \ldots, n$, respectively. Then T is the transformation matrix, i.e., $T^{-1}AT = J_0$. By Lemma 2.4.2, Part (d),

$$T^{-1} e^{At} T = e^{T^{-1}ATt} = e^{J_0 t} = \begin{bmatrix} e^{\lambda_1 t} & \\ & \ddots & \\ & & e^{\lambda_n t} \end{bmatrix}.$$

This shows that

$$(2.4.2) \qquad e^{At} = T \begin{bmatrix} e^{\lambda_1 t} & \\ & \ddots & \\ & & e^{\lambda_n t} \end{bmatrix} T^{-1}$$

is the principal matrix solution of Eq. (H-c) at 0.

We note that A is diagonalizable if its eigenvalues λ_i, $i = 1, \ldots, n$, are distinct. In this case, the above method can be implemented by the following steps:

(i) Find all eigenvalues λ_i, $i = 1, \ldots, n$, of A and verify that they are distinct.

(ii) For $i = 1, \ldots, n$, solve the equation $(\lambda I_i - A)v_i = 0$ to find an eigenvector v_i associated with λ_i, and write $T = [v_1 \ldots v_n]$.

(iii) Use (2.4.2) to compute e^{At}.

EXAMPLE 2.4.1. Find e^{At} for $A = \begin{bmatrix} 1 & -1 \\ -6 & 2 \end{bmatrix}$.

Since

$$\det(\lambda I - A) = \det \begin{bmatrix} \lambda - 1 & 1 \\ 6 & \lambda - 2 \end{bmatrix} = (\lambda + 1)(\lambda - 4),$$

we see that $\lambda_1 = -1$ and $\lambda_2 = 4$ are the eigenvalues of matrix A. Note that $\lambda_1 \neq \lambda_2$ and hence matrix A is diagonizable. By solving the equation $(\lambda I - A)v = 0$ for $\lambda = \lambda_1, \lambda_2$, respectively, we obtain eigenvectors

$$v_1 = \begin{bmatrix} 1 \\ 2 \end{bmatrix} \quad \text{and} \quad v_2 = \begin{bmatrix} 1 \\ -3 \end{bmatrix}.$$

associated with λ_1 and λ_2, respectively. Let $T = \begin{bmatrix} 1 & 1 \\ 2 & -3 \end{bmatrix}$. Then $T^{-1} = \frac{1}{5}\begin{bmatrix} 3 & 1 \\ 2 & -1 \end{bmatrix}$. Thus,

$$e^{At} = \frac{1}{5}\begin{bmatrix} 1 & 1 \\ 2 & -3 \end{bmatrix}\begin{bmatrix} e^{-t} & 0 \\ 0 & e^{4t} \end{bmatrix}\begin{bmatrix} 3 & 1 \\ 2 & -1 \end{bmatrix} = \frac{1}{5}\begin{bmatrix} 3e^{-t} + 2e^{4t} & e^{-t} - e^{4t} \\ 6e^{-t} - 6e^{4t} & 2e^{-t} + 3e^{4t} \end{bmatrix}.$$

REMARK 2.4.2. When A is not diagonalizable, i.e., $J \neq J_0$, although the transformation matrix T exists, it is hard to compute. In this case, the Putzer algorithm, which employs the so-called generalized eigenvectors of matrices, can be used to compute the principal matrices of Eq. (H-c), see [18, Section 2.5].

Here, to avoid the involvement of generalized eigenvectors, we introduce an alternative version of the Putzer algorithm for the computation of e^{At} for a general $A \in \mathbb{R}^{n \times n}$. We recall that the characteristic polynomial of matrix A is defined to be the n-th order polynomial $p(\lambda) = \det(\lambda I - A)$; and for any eigenvalue λ_i of A, the algebraic multiplicity of λ_i is the multiplicity of λ_i as a root of the characteristic polynomial. The following theorem is a modification of Theorem 16.2 in [4], where the proof was not given.

THEOREM 2.4.2. Let λ_i, $i = 1, \ldots, h$, be the eigenvalues of a matrix $A \in \mathbb{R}^{n \times n}$ whose algebraic multiplicities are m_i, $i = 1, \ldots, h$, respectively. Then

$$e^{At} = \sum_{k=0}^{n-1} \alpha_k(t) A^k,$$

where $\alpha_k(t)$, $k = 0, \ldots, n-1$, are uniquely determined by the following system of n linear equations:
(2.4.3)

$$e^{\lambda_i t} = r(\lambda_i, t), \ te^{\lambda_i t} = \frac{\partial r}{\partial \lambda}(\lambda_i, t), \ \ldots, \ t^{m_i-1}e^{\lambda_i t} = \frac{\partial^{m_i-1} r}{\partial \lambda^{m_i-1}}(\lambda_i, t), \ i=1, \ldots, h,$$

with

$$r(\lambda, t) = \sum_{k=0}^{n-1} \alpha_k(t) \lambda^k.$$

To prove Theorem 2.4.2, we need the Cayley–Hamilton theorem below from linear algebra.

LEMMA 2.4.3 (Cayley–Hamilton Theorem). Let the characteristic polynomial of matrix $A \in \mathbb{R}^{n \times n}$ be

$$p(\lambda) = \det(\lambda I - A) = \lambda^n - \sum_{i=1}^{n-1} c_i \lambda^i,$$

where $c_i \in \mathbb{R}$, $i = 1, \ldots, n-1$. Then $p(A) = \mathbf{0}$, i.e.,

$$A^n - \sum_{i=1}^{n-1} c_i A^i = \mathbf{0}.$$

PROOF OF THEOREM 2.4.2. By Lemma 2.4.3 we see that for any $A \in \mathbb{R}^{n \times n}$, there are $c_i \in \mathbb{R}$, $i = 1, \ldots, n-1$, such that $A^n = \sum_{i=1}^{n-1} c_i A^i$. It follows that for any $j \in \mathbb{N}_0$, A^j can be expressed as a linear combination of $I, A, \ldots, A^{n-1}$. Let $A^j = \sum_{k=0}^{n-1} d_{kj} A^k$ for $d_{kj} \in \mathbb{R}$. Then

$$e^{At} = \sum_{j=0}^{\infty} \frac{A^j t^j}{j!} = \sum_{j=0}^{\infty} \frac{t^j}{j!} \left(\sum_{k=0}^{n-1} d_{kj} A^k \right).$$

Note that the double sum above is norm-convergent for any $t \in \mathbb{R}$. We can interchange the order of summation to obtain that

$$e^{At} = \sum_{k=0}^{n-1} \left(\sum_{j=0}^{\infty} \frac{d_{kj} t^j}{j!} \right) A^k := \sum_{k=0}^{n-1} \alpha_k(t) A^k,$$

where

$$\alpha_k(t) = \sum_{j=0}^{\infty} \frac{d_{kj} t^j}{j!}, \quad k = 0, \ldots, n-1.$$

Let J be the Jordan canonical form of matrix A. Then J and A have exactly the same eigenvalues. Using the same argument as in (2.4.1) we have

(2.4.4)
$$e^{Jt} = \sum_{k=0}^{n-1} \alpha_k(t) J^k.$$

Let λ_i be an eigenvalue of A (and hence eigenvalue of J) with algebraic multiplicity m_i. It follows from (2.4.4) that λ_i is a root of the equation $e^{\lambda t} = r(\lambda, t)$ with algebraic multiplicity m_i. Therefore, (2.4.3) holds. Since $m_1 + \cdots + m_h = n$, there are n linear equations in system (2.4.3) for the n unknowns $\alpha_0, \ldots, \alpha_{n-1}$. It can be verified that the coefficient determinant of the system for $\alpha_0, \ldots, \alpha_{n-1}$ is not zero. Hence $\alpha_0, \ldots, \alpha_{n-1}$ are uniquely determined. □

To practice the method in Theorem 2.4.2, we give three examples. The first one is a revisit of Example 2.4.1.

EXAMPLE 2.4.2. Find e^{At} for $A = \begin{bmatrix} 1 & -1 \\ -6 & 2 \end{bmatrix}$ using the method in Theorem 2.4.2.

Here $n = 2$. By Theorem 2.4.2,

(2.4.5)
$$e^{At} = \alpha_0(t) I + \alpha_1(t) A.$$

From Example 2.4.1, $\lambda_1 = -1$ and $\lambda_2 = 4$ are the eigenvalues of matrix A which are algebraically simple. Let

$$r(\lambda, t) = \alpha_0(t) + \alpha_1(t) \lambda.$$

Then $\alpha_0(t)$ and $\alpha_1(t)$ satisfy $e^{\lambda_i t} = r(\lambda_i, t)$ for $i = 1, 2$. Solving the system

$$e^{-t} = \alpha_0(t) - \alpha_1(t), \quad e^{4t} = \alpha_0(t) + 4\alpha_1(t),$$

we find that

$$\alpha_0(t) = \frac{1}{5}(4e^{-t} + e^{4t}) \quad \text{and} \quad \alpha_1(t) = \frac{1}{5}(-e^{-t} + e^{4t}).$$

Thus,

$$
\begin{aligned}
e^{At} &= \frac{1}{5}(4e^{-t} + e^{4t})\begin{bmatrix} 1 & 0 \\ 0 & 1 \end{bmatrix} + \frac{1}{5}(-e^{-t} + e^{4t})\begin{bmatrix} 1 & -1 \\ -6 & 2 \end{bmatrix} \\
&= \frac{1}{5}\begin{bmatrix} 3e^{-t} + 2e^{4t} & e^{-t} - e^{4t} \\ 6e^{-t} - 6e^{4t} & 2e^{-t} + 3e^{4t} \end{bmatrix}.
\end{aligned}
$$

The second example involves a multiple real eigenvalue of matrix A.

EXAMPLE 2.4.3. Find e^{At} for $A = \begin{bmatrix} 3 & 2 & 0 \\ 0 & 3 & -1 \\ 0 & 0 & 1 \end{bmatrix}$.

Here $n = 3$. By Theorem 2.4.2,

$$(2.4.6) \qquad e^{At} = \alpha_0(t)I + \alpha_1(t)A + \alpha_2(t)A^2.$$

Clearly, $\lambda_1 = 3$ and $\lambda_2 = 1$ are the eigenvalues of matrix A with algebraic multiplicities $m_1 = 2$ and $m_2 = 1$. Let

$$r(\lambda, t) = \alpha_0(t) + \alpha_1(t)\lambda + \alpha_2(t)\lambda^2.$$

Then

$$\frac{\partial r}{\partial \lambda}(\lambda, t) = \alpha_1(t) + 2\alpha_2(t)\lambda.$$

Hence $\alpha_i(t)$, $i = 0, 1, 2$, satisfy the system

$$(2.4.7) \qquad \begin{cases} e^{3t} &= \alpha_0(t) + 3\alpha_1(t) + 9\alpha_2(t) \\ te^{3t} &= \alpha_1(t) + 6\alpha_2(t) \\ e^{t} &= \alpha_0(t) + \alpha_1(t) + \alpha_2(t). \end{cases}$$

By solving system (2.4.7) we find that

$$\begin{cases} \alpha_0(t) &= \frac{9}{4}e^t - \frac{5}{4}e^{3t} + \frac{3}{2}te^{3t} \\ \alpha_1(t) &= -\frac{3}{2}e^t + \frac{3}{2}e^{3t} - 2te^{3t} \\ \alpha_2(t) &= \frac{1}{4}e^t - \frac{1}{4}e^{3t} + \frac{1}{2}te^{3t}. \end{cases}$$

Note that

$$A^2 = \begin{bmatrix} 3 & 2 & 0 \\ 0 & 3 & -1 \\ 0 & 0 & 1 \end{bmatrix}\begin{bmatrix} 3 & 2 & 0 \\ 0 & 3 & -1 \\ 0 & 0 & 1 \end{bmatrix} = \begin{bmatrix} 9 & 12 & -2 \\ 0 & 9 & -4 \\ 0 & 0 & 1 \end{bmatrix}.$$

Thus

$$
e^{At} = \left(\frac{9}{4}e^t - \frac{5}{4}e^{3t} + \frac{3}{2}te^{3t}\right)\begin{bmatrix} 1 & 0 & 0 \\ 0 & 1 & 0 \\ 0 & 0 & 1 \end{bmatrix}
$$

$$
+ \left(-\frac{3}{2}e^t + \frac{3}{2}e^{3t} - 2te^{3t}\right)\begin{bmatrix} 3 & 2 & 0 \\ 0 & 3 & -1 \\ 0 & 0 & 1 \end{bmatrix}
$$

$$
+ \left(\frac{1}{4}e^t - \frac{1}{4}e^{3t} + \frac{1}{2}te^{3t}\right)\begin{bmatrix} 9 & 12 & -2 \\ 0 & 9 & -4 \\ 0 & 0 & 1 \end{bmatrix}
$$

$$
= \begin{bmatrix} e^{3t} & te^{3t} & -\frac{1}{2}e^t + \frac{1}{2}e^{3t} - te^{3t} \\ 0 & e^{3t} & \frac{1}{2}e^t - \frac{1}{2}e^t \\ 0 & 0 & e^t \end{bmatrix}.
$$

The last example involves multiple non-real eigenvalues of matrix A.

EXAMPLE 2.4.4. Find e^{At} for $A = \begin{bmatrix} 0 & -1 & 0 & 0 \\ 1 & 0 & 0 & 0 \\ 0 & 0 & 0 & -1 \\ 2 & 0 & 1 & 0 \end{bmatrix}$.

Here $n = 4$. By Theorem 2.4.2,

$$(2.4.8) \qquad e^{At} = \alpha_0(t)I + \alpha_1(t)A + \alpha_2(t)A^2 + \alpha_3(t)A^3.$$

It is easy to find that $\lambda_1 = \mathbf{i}$ and $\lambda_2 = -\mathbf{i}$ are the eigenvalues of matrix A with algebraic multiplicities $m_1 = m_2 = 2$. Let

$$r(\lambda, t) = \alpha_0(t) + \alpha_1(t)\lambda + \alpha_2(t)\lambda^2 + \alpha_3(t)\lambda^3.$$

Then

$$\frac{\partial r}{\partial \lambda}(\lambda, t) = \alpha_1(t) + 2\alpha_2(t)\lambda + 3\alpha_3(t)\lambda^2.$$

Hence α_i, $i = 0, 1, 2, 3$, satisfy the system

$$(2.4.9) \qquad \begin{cases} e^{\mathbf{i}t} &= \alpha_0(t) + \alpha_1(t)\mathbf{i} - \alpha_2(t) - \alpha_3(t)\mathbf{i} \\ te^{\mathbf{i}t} &= \alpha_1(t) + 2\alpha_2(t)\mathbf{i} - 3\alpha_3(t) \\ e^{-\mathbf{i}t} &= \alpha_0(t) - \alpha_1(t)\mathbf{i} - \alpha_2(t) + \alpha_3(t)\mathbf{i} \\ te^{-\mathbf{i}t} &= \alpha_1(t) - 2\alpha_2(t)\mathbf{i} - 3\alpha_3(t). \end{cases}$$

Solving system (2.4.9) and using the relation

$$\cos t = \frac{1}{2}(e^{\mathbf{i}t} + e^{-\mathbf{i}t}) \quad \text{and} \quad \sin t = \frac{1}{2\mathbf{i}}(e^{\mathbf{i}t} - e^{-\mathbf{i}t}),$$

we find that

$$\alpha_0(t) = \frac{1}{2}t\sin t + \cos t, \quad \alpha_1(t) = \frac{3}{2}\sin t - \frac{1}{2}t\cos t,$$

$$\alpha_2(t) = \frac{1}{2}t\sin t, \qquad \alpha_3(t) = \frac{1}{2}\sin t - \frac{1}{2}t\cos t.$$

By matrix multiplications we see that

$$A^2 = \begin{bmatrix} -1 & 0 & 0 & 0 \\ 0 & -1 & 0 & 0 \\ -2 & 0 & -1 & 0 \\ 0 & -2 & 0 & -1 \end{bmatrix} \quad \text{and} \quad A^3 = \begin{bmatrix} 0 & 1 & 0 & 0 \\ -1 & 0 & 0 & 0 \\ 0 & 2 & 0 & 1 \\ -4 & 0 & -1 & 0 \end{bmatrix}.$$

From (2.4.8) and by a simple computation we obtain that

$$e^{At} = \begin{bmatrix} \cos t & -\sin t & 0 & 0 \\ \sin t & \cos t & 0 & 0 \\ -t\sin t & \sin t - t\cos t & \cos t & -\sin t \\ \sin t + t\cos t & -t\sin t & \sin t & \cos t \end{bmatrix}.$$

We omit the details.

Finally, we use the form of e^{At} to study the behavior of solutions of Eq. (H-c), especially as $t \to \infty$. Note that for any solution $x(t)$ of Eq. (H-c), there exists a $c \in \mathbb{R}^n$ such that $x(t) = e^{At}c$ for $t \in \mathbb{R}$. Thus from Theorem 2.4.2 we can derive the form of solutions of Eq. (H-c) as follows.

LEMMA 2.4.4. *Let λ_i be a simple eigenvalue of A for $i = 1, \ldots, q$ and λ_{q+i} an eigenvalue of A with algebraic multiplicity r_i for $i = 1, \ldots, l$ such that $q + r_1 + \cdots + r_l = n$. Then all solutions of Eq. (H-c) are of the form*

$$x(t) = \sum_{i=1}^{q} c_i e^{\lambda_i t} + \sum_{i=1}^{l} p_i(t) e^{\lambda_{q+i} t},$$

where

$c_i \in \mathbb{C}^n$ *is a constant vector, $i = 1, \ldots, q$;   and*

$p_i(t)$ *is a complex-valued vector polynomial of t with degree $\leq r_i - 1$, $i = 1, \ldots, l$.*

Recall that the eigenvalues of A may not be all real. Then Lemma 2.4.4 leads to the following result.

THEOREM 2.4.3. *Let $\lambda_1, \ldots, \lambda_n$ be the eigenvalues of the matrix A. Then*

(a) *all solutions of Eq. (H-c) satisfy $\lim_{t\to\infty} x(t) = 0 \iff \operatorname{Re}\lambda_i < 0$, $i = 1, \ldots, n$;*

(b) *all solutions of Eq. (H-c) are bounded on $[0, \infty) \iff \operatorname{Re}\lambda_i \leq 0$, $i = 1, \ldots, n$, and $\operatorname{Re}\lambda_i = 0$ occurs only when the λ_i's are in the diagonal Jordan block of the matrix A.*

2.5. Homogeneous Linear Equations with Periodic Coefficients

In this section, we study the homogeneous linear equation with periodic coefficients

(H-p) $$x' = A(t)x,$$

where $A \in C(\mathbb{R}, \mathbb{R}^{n \times n})$ is a non-constant matrix function satisfying $A(t+\omega) = A(t)$ for some $\omega > 0$ and all $t \in \mathbb{R}$. The smallest such $\omega > 0$ is called the period of the equation. An ω-periodic solution of the equation is similarly defined. By Theorem 2.1.1, all solutions of Eq. (H-p) exist on the whole real number line $(-\infty, \infty)$. However, we should note that the solutions of Eq. (H-p) are not necessarily periodic. First we introduce certain properties of Eq. (H-p) which are not satisfied by the general homogeneous linear equation (H).

LEMMA 2.5.1. *If $x(t)$ is a solution of Eq.* (H-p), *then $x(t + k\omega)$ is also a solution of Eq.* (H-p) *for each $k \in \mathbb{Z} := \{0, \pm 1, . \pm 2, \dots\}$.*

PROOF. Since $x(t)$ is a solution of Eq. (H-p), we have $x'(t) = A(t)x(t)$ for $t \in \mathbb{R}$. Then for $t \in \mathbb{R}$ and $k \in \mathbb{Z}$,

$$[x(t + k\omega)]' = x'(t + k\omega) = A(t + k\omega)x(t + k\omega) = A(t)x(t + k\omega).$$

Thus, $x(t + k\omega)$ is a solution of Eq. (H-p). $\qquad\square$

It follows from Theorem 2.2.2 that among the solutions $\{x(t + k\omega) : k \in \mathbb{Z}\}$, there are at most n linearly independent solutions.

Unlike the constant coefficient case, there is no general way to find a fundamental matrix solution for the linear equation with periodic coefficients (H-p). However, the structure of the fundamental matrices of Eq. (H-c) is revealed by the Floquet theory as shown below.

THEOREM 2.5.1. *Let $X(t)$ be a fundamental matrix solution of Eq.* (H-p). *Then $X(t + \omega)$ is also a fundamental matrix solution of Eq.* (H-p), *and there exists a nonsingular matrix $V \in \mathbb{R}^{n \times n}$, called the transition matrix for $X(t)$, such that*

$$(2.5.1) \qquad\qquad X(t + \omega) = X(t)V \quad \text{for all } t \in \mathbb{R}.$$

Moreover,

$$(2.5.2) \quad X(t + m\omega) = X(t)V^m \quad \text{for all } m = 0, \pm 1, \pm 2, \dots \text{ and } t \in \mathbb{R};$$

where $V^{-m} = (V^{-1})^m$. In fact, $V = X^{-1}(0)X(\omega)$.

PROOF. By Lemma 2.5.1, for any $m = 0, \pm 1, \pm 2, \dots$, $X(t + m\omega)$ is a matrix solution of Eq. (H-p). Since $\det X(t) \neq 0$ for all $t \in \mathbb{R}$, $\det X(t + m\omega) \neq 0$ for all $t \in \mathbb{R}$. This means that $X(t + m\omega)$ is a fundamental matrix solution of Eq. (H-p). From Lemma 2.2.3, Part (d), there exists a nonsingular matrix $V \in \mathbb{R}^{n \times n}$ such that (2.5.1) holds. If we let $t = 0$ in (2.5.1), we see that $V = X^{-1}(0)X(\omega)$. By induction, we have

$$X(t + m\omega) = X(t)V^m \quad \text{for all } m = 0, 1, 2, \dots \text{ and } t \in \mathbb{R}.$$

For $m = -k$ with $k \in \mathbb{N}$, since $X(t) = X(t - k\omega)V^k$ for $t \in \mathbb{R}$, we have

$$X(t - k\omega) = X(t)(V^k)^{-1} = X(t)(V^{-1})^k = X(t)V^{-k},$$

i.e., $X(t + m\omega) = X(t)V^m$ for $t \in \mathbb{R}$. Therefore, (2.5.2) holds. $\qquad\square$

REMARK 2.5.1. The significance of Theorem 2.5.1 lies in the fact that although a fundamental matrix solution of Eq. (H-p) may not be periodic, its values on the whole domain $\mathbb{R}$ is determined by its values on $[0, \omega]$.

We note that the transition matrix V depends on the choice of a fundamental matrix solution $X(t)$. However, as consequences of Theorem 2.5.1, the following corollaries show that certain features of the transition matrices are independent of the choice of the fundamental matrix solution.

COROLLARY 2.5.1. *For any fundamental matrix solution $X(t)$ of Eq. (H-p), the transition matrix V satisfies* $\det V = e^{\int_0^\omega \operatorname{tr} A(s)\,ds}$.

PROOF. By Theorem 2.2.3 we have $\det X(\omega) = \det X(0) e^{\int_0^\omega \operatorname{tr} A(s)\,ds}$. Thus,

$$\det V = \det X^{-1}(0) \det X(\omega) = \det X^{-1}(0) \det X(0) e^{\int_0^\omega \operatorname{tr} A(s)\,ds} = e^{\int_0^\omega \operatorname{tr} A(s)\,ds}.$$

$\square$

COROLLARY 2.5.2. *Let $X(t)$ and $Y(t)$ be fundamental matrix solutions of Eq. (H-p) with transition matrices U and V, respectively. Then U and V are similar matrices and hence have the same set of eigenvalues.*

PROOF. Since $X(t)$ and $Y(t)$ are fundamental matrices of Eq. (H-p), by Lemma 2.2.3, Part (d), there exists a nonsingular $C \in \mathbb{R}^{n \times n}$ such that $X(t) = Y(t)C$ for all $t \in \mathbb{R}$. This implies that $X(t+\omega) = Y(t+\omega)C$ for all $t \in \mathbb{R}$. Note that

$$X(t+\omega) = X(t)U \text{ and } Y(t+\omega) = Y(t)V \text{ for all } t \in \mathbb{R}.$$

Therefore,

$$X(t)U = X(t+\omega) = Y(t+\omega)C = Y(t)VC = X(t)C^{-1}VC.$$

Since $X(t)$ is nonsingular on $\mathbb{R}$, we have that $U = C^{-1}VC$, i.e., $U \sim V$. $\square$

COROLLARY 2.5.3. *Let $X(t)$ be a fundamental matrix solution of Eq. (H-p). Then there exists a nonsingular $C \in \mathbb{C}^{n \times n}$ such that $\tilde{X}(t) := X(t)C$ satisfies $\tilde{X}(t+\omega) = \tilde{X}(t)J$, where $J \in \mathbb{C}^{n \times n}$ is of the Jordan canonical form.*

PROOF. Let V be the transition matrix for $X(t)$, i.e., $X(t+\omega) = X(t)V$ for all $t \in \mathbb{R}$. Then there exists a $C \in \mathbb{C}^{n \times n}$ such that $C^{-1}VC = J$ with $J \in \mathbb{C}^{n \times n}$ of the Jordan canonical form. Thus for this C,

$$\tilde{X}(t+\omega) = X(t+\omega)C = X(t)VC = \tilde{X}(t)C^{-1}VC = \tilde{X}(t)J.$$

$\square$

We observe that the matrix solution $\tilde{X}(t)$ of Eq. (H-p) defined in Corollary 2.5.3 may not be real-valued due to the fact that the constant matrix C is complex-valued in general.

By Corollary 2.5.2 we see that the eigenvalues of the transition matrices are independent of the choice of the fundamental matrix solution. Thus, the definition below makes sense.

DEFINITION 2.5.1. Let $\mu_1, \ldots, \mu_n$ be the eigenvalues of the transition matrix V of any fundamental matrix solution of Eq. (H-p). Then $\mu_1, \ldots, \mu_n$ are called the characteristic multipliers of Eq. (H-p).

The following is a direct consequence of Corollary 2.5.1.

COROLLARY 2.5.4. *The characteristic multipliers* $\mu_1, \ldots, \mu_n$ *of Eq.* (H-p) *satisfy* $\mu_1 \cdots \mu_n = e^{\int_0^\omega \operatorname{tr} A(s)\, ds}$.

Using the characteristic multipliers we can characterize the behavior of solutions of Eq. (H-p).

THEOREM 2.5.2. *Let* $\mu_1, \ldots, \mu_n$ *be the characteristic multipliers of Eq.* (H-p). *Then*

    (a) all solutions of Eq. (H-p) *satisfy* $\lim_{t \to \infty} x(t) = 0 \iff |\mu_i| < 1$, $i = 1, \ldots, n$.

    (b) all solutions of Eq. (H-p) *are bounded on* $[0, \infty) \iff |\mu_i| \leq 1$, $i = 1, \ldots, n$, *and* $|\mu_i| = 1$ *occurs only when the* μ_i's *are in the diagonal Jordan block of the transition matrix* V.

PROOF. Let $X(t)$ be a fundamental matrix solution of Eq. (H-p) and $\tilde{X}(t) := (x_1, \ldots, x_n)(t)$ be defined as in Corollary 2.5.3 whose transition

matrix is $J = \begin{bmatrix} J_0 & & & \\ & J_1 & & \\ & & \ddots & \\ & & & J_l \end{bmatrix}$ with

$$J_0 = \begin{bmatrix} \mu_1 & & \\ & \ddots & \\ & & \mu_q \end{bmatrix}_{q \times q} \quad \text{and } J_i = \begin{bmatrix} \mu_{q+i} & 1 & & \\ & \ddots & \ddots & \\ & & \ddots & 1 \\ & & & \mu_{q+i} \end{bmatrix}_{r_i \times r_i}, \quad i = 1, \ldots, l,$$

satisfying $q + r_1 + \cdots + r_l = n$. We note from Corollary 2.5.3 that $\tilde{X}(t)$ may not be real-valued. However, this does not affect our discussion since

$$X(t) \to \mathbf{0} \Leftrightarrow \tilde{X}(t) \to \mathbf{0} \quad \text{and} \quad X(t) \text{ is bounded} \Leftrightarrow \tilde{X}(t) \text{ is bounded},$$

no matter if $\tilde{X}(t)$ is real or nonreal. By the uniqueness of solutions of IVPs associated with Eq. (H-p) we know that $x_i(t) \neq 0$ for $t \in \mathbb{R}$ and $i = 1, \ldots, n$. Let $t \in [0, \omega]$ be fixed. Then by Theorem 2.5.1, (2.5.2) holds, i.e.,

(2.5.3) $\qquad (x_1, \ldots, x_n)(t + m\omega) = (x_1, \ldots, x_n)(t) J^m, \quad m \in \mathbb{Z}.$

    (i) Let μ_i be in J_0 for some $i \in \{1, \ldots, q\}$. By (2.5.3), for this i we have

$$x_i(t + m\omega) = x_i(t)\mu_i^m \to 0 \text{ as } m \to \infty \quad \Leftrightarrow \quad |\mu_i| < 1,$$
$$\text{the set } \{x_i(t + m\omega)\}_{m=0}^\infty \text{ is bounded} \quad \Leftrightarrow \quad |\mu_i| \leq 1.$$

    (ii) Let $\mu = \mu_{q+i}$ be in J_i for some $i \in \{1, \ldots, l\}$. Without loss of generality we denote by $x_1, \ldots, x_j$ the columns of $X(t)$ corresponding to this μ. Then by (2.5.3) we have

$$(x_1, \ldots, x_j)(t + \omega) = (x_1, \ldots, x_j)(t) \begin{bmatrix} \mu & 1 & & \\ & \ddots & \ddots & \\ & & \ddots & 1 \\ & & & \mu \end{bmatrix},$$

i.e.,

$$\begin{cases} x_1(t + \omega) = \mu x_1(t), \\ x_2(t + \omega) = \mu x_2(t) + x_1(t), \\ \quad \cdots\cdots \\ x_j(t + \omega) = \mu x_j(t) + x_{j-1}(t). \end{cases}$$

By induction we obtain that for $m \geq j - 1$,

$$\begin{cases} x_1(t + m\omega) = \mu^m x_1(t), \\ x_2(t + m\omega) = \mu^m x_2(t) + \binom{m}{1} \mu^{m-1} x_1(t), \\ \quad \cdots\cdots \\ x_j(t + m\omega) = \mu^m x_j(t) + \binom{m}{1} \mu^{m-1} x_{j-1}(t) + \cdots \\ \qquad\qquad + \binom{m}{j-1} \mu^{m-j+1} x_1(t), \end{cases}$$

where $\binom{m}{k} = \dfrac{m!}{k!(m-k)!}$. Therefore, for $i = 1, \ldots, j$,

$$x_i(t + m\omega) \to 0 \text{ as } m \to \infty \Leftrightarrow |\mu_i| < 1;$$

also,

the set $\{x_1(t + m\omega)\}_{m=0}^{\infty}$ is bounded $\Leftrightarrow |\mu_i| \leq 1$,

and for $i = 2, \ldots, j$,

the set $\{x_i(t + m\omega))\}_{m=0}^{\infty}$ is bounded $\Leftrightarrow |\mu_i| < 1$.

It is easy to see that for $i = 1, \ldots, n$, the behavior of $x_i(t)$ as $t \to \infty$ is the same as that of $x_i(t + m\omega)$ as $m \to \infty$ for fixed $t \in [0, \omega]$. Combining the results in Parts (i) and (ii), we get that

$$x(t) \to 0 \text{ as } t \to \infty \quad \Leftrightarrow \quad |\mu_i| < 1, \ i = 1, \ldots, n;$$
$$x(t) \text{ is bounded on } [0, \infty) \quad \Leftrightarrow \quad |\mu_i| \leq 1, \ i = 1, \ldots, n, \text{ and}$$
$$|\mu_i| = 1 \text{ occurs only for } \mu_i \in J_0.$$

$\square$

THEOREM 2.5.3 (Floquet Theorem). *Let $X(t)$ be a fundamental matrix solution of Eq. (H-p). Then there exists an $R \in \mathbb{C}^{n \times n}$ and a nonsingular ω-periodic $P \in C^1(\mathbb{R}, \mathbb{C}^{n \times n})$ such that $X(t) = P(t)e^{Rt}$.*

This theorem shows that the structure of the fundamental matrices of Eq. (H-p) is similar to that of the fundamental matrices of Eq. (H-c) with constant coefficients, but is subjected to a periodic fluctuation. To prove this theorem, we need the lemma below for the matrix logarithm. The reader may refer to Coddington and Levinson [11, p. 65–66] for the proof.

LEMMA 2.5.2. *For any nonsingular $A \in \mathbb{R}^{n \times n}$, there exists a $B \in \mathbb{C}^{n \times n}$, denoted by $B = \log A$, such that $A = e^B$. Furthermore, $\lambda_i(B) = \log \lambda_i(A) [mod\ 2\pi\mathbf{i}]$, $i = 1, \ldots, n$.*

PROOF OF THEOREM 2.5.3. Let V be the transition matrix of $X(t)$, i.e., $V \in \mathbb{R}^{n \times n}$ is nonsingular and satisfies (2.5.1). Let $R = (\log V)/\omega$. By Lemma 2.5.2, we have $V = e^{R\omega}$. Then it follows from (2.5.1) that $X(t+\omega) = X(t)e^{R\omega}$. Let $P(t) = X(t)e^{-Rt}$. Clearly, $P \in C^1(\mathbb{R}, \mathbb{C}^{n \times n})$ and $X(t) = P(t)e^{Rt}$. Note that for $t \in \mathbb{R}$

$$P(t + \omega) = X(t + \omega)e^{-R(t+\omega)} = X(t)e^{R\omega}e^{-R(t+\omega)} = X(t)e^{-Rt} = P(t).$$

This shows that $P(t)$ is ω-periodic.   □

Clearly, the eigenvalues of the matrix $R = (\log V)/\omega$ are independent of the choice of the fundamental matrix solution. Thus, the definition below makes sense.

DEFINITION 2.5.2. Let V be the transition matrix of a fundamental matrix solution of Eq. (H-p) and $R = (\log V)/\omega$. Then the eigenvalues $\rho_1, \ldots, \rho_n$ of R are called the characteristic exponents of Eq. (H-p).

Recall that $\mu_1, \ldots, \mu_n$ are eigenvalues of V. From Lemma 2.5.2 we have that for $i = 1, \ldots, n$,

$$\rho_i = (\log \mu_i)/\omega = (\ln |\mu_i| + \mathbf{i} \arg \mu_i)/\omega \ [\text{mod } 2\pi\mathbf{i}]$$

and hence $\operatorname{Re} \rho_i = (\ln |\mu_i|)/\omega$. We observe that

$$|\mu_i| < 1 \Leftrightarrow \operatorname{Re} \rho_i < 0 \quad \text{and} \quad |\mu_i| \leq 1 \Leftrightarrow \operatorname{Re} \rho_i \leq 0.$$

Then Theorem 2.5.3 leads to the following result.

COROLLARY 2.5.5. *Let $\rho_1, \ldots, \rho_n$ be the characteristic exponents of Eq. (H-p). Then*

  (a) *all solutions of Eq. (H-p) satisfy $\lim_{t \to \infty} x(t) = 0 \iff \operatorname{Re} \rho_i < 0$, $i = 1, \ldots, n$.*

  (b) *all solutions of Eq. (H-p) are bounded on $[0, \infty) \iff \operatorname{Re} \rho_i \leq 0$, $i = 1, \ldots, n$, and $\operatorname{Re} \rho_i = 0$ occurs only when the ρ_i's are in the diagonal Jordan block of the matrix R.*

We comment that the characteristic exponents play the same role for Eq. (H-p) as that of eigenvalues of the matrix A for Eq. (H-c) in determining the behavior of solutions as $t \to \infty$.

We also observe that in Theorem 2.5.3, the matrices R and $P(t)$ may not be real-valued even though the transition matrix V is real. However, if we allow $P(t)$ to be 2ω-periodic, it can be guaranteed that such R and $P(t)$ are real-valued. The following result was established in [**5**, p. 97].

THEOREM 2.5.4. *Let $X(t)$ be a fundamental matrix solution of Eq. (H-p). Then there exists an $R \in \mathbb{R}^{n \times n}$ and a nonsingular 2ω-periodic $P \in C^1(\mathbb{R}, \mathbb{R}^{n \times n})$ such that $X(t) = P(t)e^{Rt}$.*

PROOF. This proof is based on a result from linear algebra: for any nonsingular $A \in \mathbb{R}^{n \times n}$, there exists a $B \in \mathbb{R}^{n \times n}$ such that $A^2 = e^B$. By Theorem 2.5.1, $X(t + 2\omega) = X(t)V^2$. Note that V is nonsingular. Hence there exists an $R \in \mathbb{R}^{n \times n}$ such that $V^2 = e^{2R\omega}$. This implies that $X(t+2\omega) = X(t)e^{2R\omega}$. Let $P(t) = X(t)e^{-Rt}$. Then $X(t) = P(t)e^{Rt}$ and

$$P(t + 2\omega) = X(t + 2\omega)e^{-R(t+2\omega)} = X(t)e^{2R\omega}e^{-R(t+2\omega)} = X(t)e^{-Rt} = P(t).$$

This shows that $P(t)$ is 2ω-periodic.   □

Exercises

2.1. Let

(a) $x_1 = \begin{bmatrix} 1 \\ t \end{bmatrix}$, $x_2 = \begin{bmatrix} t \\ 2t^2 \end{bmatrix}$;   (b) $x_1 = \begin{bmatrix} \cos t \\ \sin t \end{bmatrix}$, $x_2 = \begin{bmatrix} \sin t \\ \cos t \end{bmatrix}$.

For each of the cases (a) and (b), is it possible for both $x_1(t)$ and $x_2(t)$ to be solutions of Eq. (H)? Justify your answer.

2.2. Let $x_1(t)$ and $x_2(t)$ be solutions of Eq. (H) with $n = 2$. Assume there exist $\alpha, \beta \in \mathbb{R}$ with $\alpha^2 + \beta^2 \neq 0$ such that $\alpha x_1(t_0) + \beta x_2(t_0) = 0$ for some $t_0 \in \mathbb{R}$. Show that $x_1(t)$ and $x_2(t)$ are linearly dependent on $\mathbb{R}$.

2.3. Let $\begin{bmatrix} x_1 \\ y_1 \end{bmatrix}$ and $\begin{bmatrix} x_2 \\ y_2 \end{bmatrix}$ be solutions of the system $\begin{cases} x' = a(t)x + b(t)y \\ y' = c(t)x - a(t)y, \end{cases}$

where $a(t), b(t)$, and $c(t)$ are continuous on $\mathbb{R}$, such that $\begin{bmatrix} x_1 \\ y_1 \end{bmatrix}(0) = \begin{bmatrix} 0 \\ 1 \end{bmatrix}$ and $\begin{bmatrix} x_2 \\ y_2 \end{bmatrix}(0) = \begin{bmatrix} 1 \\ -1 \end{bmatrix}$.

   (a) Find the Wronskian as a function of t.
   (b) Determine if the two solutions are linearly independent on $\mathbb{R}$.

2.4. Let $x_1(t)$ and $x_2(t)$ be solutions of the equation $x'' + p(t)x' + q(t)x = 0$ with $p, q \in C(\mathbb{R}, \mathbb{R})$. Assume both of the two solutions satisfy the same condition: either $x(0) = 0$ or $x'(0) = 0$. Show that they are linearly dependent.

2.5. Consider the equation $y'' - ty' + (1 + t)y = 0$.

   (a) Write this equation as a system of first-order equations.
   (b) Let $y_1(t)$ and $y_2(t)$ be solutions of the equation satisfying $y_1(0) = 1, y_1'(0) = 0$; $y_2(0) = 1, y_2'(0) = 3$. Find the Wronskian $W(t)$ of y_1 and y_2 (i.e., the Wronskian of the corresponding solutions of the system obtained in Part (a)).

2.6. Let $X(t)$ be the principal matrix solution of Eq. (H-c) at $t_0 = 0$. Show that

$$X(t+s) = X(t)X(s) \quad \text{for any } t, s \in \mathbb{R}$$

and

$$X^{-1}(t) = X(-t) \quad \text{for any } t \in \mathbb{R}.$$

2.7. Assume $A \in C(\mathbb{R}, \mathbb{R}^{n \times n})$. Let $\phi(t) = (\phi_1, \cdots, \phi_n)^T(t)$ be a solution of the equation $x' = A(t)x$ and $\psi(t) = (\psi_1, \cdots, \psi_n)^T(t)$ be a solution of the equation $x' = -A^T(t)x$. Show that the inner product $\langle \phi(t), \psi(t) \rangle := \sum_{i=1}^n \phi_i(t)\psi_i(t)$ is constant on $\mathbb{R}$.

2.8. Let $x_1(t), \ldots, x_{n+1}(t)$ be $n+1$ linearly independent solutions of equation (NH). Show that for any solution $x(t)$ of equation (NH) we have

$$(2.6.1) \qquad x(t) = \alpha_1 x_1(t) + \cdots + \alpha_{n+1} x_{n+1}(t)$$

where $\alpha_1, \ldots, \alpha_{n+1}$ are constants satisfying

$$(2.6.2) \qquad \alpha_1 + \cdots + \alpha_{n+1} = 1.$$

Furthermore, for any constants $\alpha_1, \ldots, \alpha_{n+1}$ satisfying (2.6.2), the function $x(t)$ given by (2.6.1) is a solution of equation (NH).

For those with knowledge of affine geometry, this means that the solutions of Eq. (NH) form an n-dimensional affine subspace of an $(n+1)$-dimensional linear space.

2.9. Find the general solutions of the following second-order equations:

    (a) $x'' + x = \tan t$.

    (b) $x'' - 4x' + 3x = \dfrac{1}{1 + e^{-t}}$.

2.10. Prove Corollary 2.3.2.

2.11. Assume $A, B \in C(\mathbb{R}, \mathbb{R}^{n \times n})$. Let $X(t)$ be a fundamental matrix solution of the equation $x' = A(t)x$ such that both $|X(t)|$ and $|X^{-1}(t)|$ are bounded on $[0, \infty)$. Let $y(t)$ be a solution of the equation $y' = B(t)y$ with $\int_0^\infty |B(s) - A(s)|\, ds < \infty$. Show that $y(t)$ is bounded on $[0, \infty)$.

2.12. Find e^{At} for each of the following matrices A:

    (a) $\begin{bmatrix} -1 & 1 \\ 0 & -2 \end{bmatrix}$,   (b) $\begin{bmatrix} 1 & -1 \\ 4 & 3 \end{bmatrix}$,   (c) $\begin{bmatrix} 1 & -1 \\ 1 & 3 \end{bmatrix}$,   (d) $\begin{bmatrix} 0 & 2 & 0 \\ 0 & 0 & 1 \\ 0 & -1 & 0 \end{bmatrix}$,

    (e) $\begin{bmatrix} 1 & 0 & -1 \\ 0 & 2 & 1 \\ 0 & 0 & 2 \end{bmatrix}$,   (f) $\begin{bmatrix} 0 & 1 & 0 & 0 \\ 1 & 0 & 0 & 0 \\ 0 & 0 & 1 & -1 \\ 0 & 0 & 1 & 1 \end{bmatrix}$,   (g) $\begin{bmatrix} 1 & -1 & 0 & 0 \\ 1 & 1 & 0 & 0 \\ 0 & 0 & 1 & -1 \\ 2 & 0 & 1 & 1 \end{bmatrix}$.

2.13. Let $A \in \mathbb{R}^{2 \times 2}$ and assume $\lambda = 2 - \mathbf{i}$ is an eigenvalue of A with eigenvector $\begin{bmatrix} 1 \\ -\mathbf{i} \end{bmatrix}$. Find e^{At}.

2.14. Let $A = \begin{bmatrix} 0 & -\beta \\ \beta & 0 \end{bmatrix}$ with $\beta \neq 0$. Show that

$$e^{At} = (\cos \beta t)I + \frac{\sin \beta t}{\beta} A,$$

where I is the 2×2 identity matrix.

2.15. Find e^{At} for $A = \begin{bmatrix} 1 & 2 \\ 4 & 3 \end{bmatrix}$, and then solve the IVP $x' = Ax$, $x(0) = \begin{bmatrix} 1 \\ 1 \end{bmatrix}$.

2.16. (a) Find the principal matrix solution of the equation $x' = \begin{bmatrix} 1 & 1 \\ 0 & 1 \end{bmatrix} x$ at $t_0 = 0$.

 (b) Solve the IVP $x' = \begin{bmatrix} 1 & 1 \\ 0 & 1 \end{bmatrix} x + \begin{bmatrix} e^{-t} \\ 0 \end{bmatrix}$, $x(0) = \begin{bmatrix} -1 \\ 1 \end{bmatrix}$.

2.17. Assume $x(t)$ is a nontrivial periodic solution of Eq. (H-p). Suppose that $x(t)$ is $n\omega$-periodic and $x(t), x(t + \omega), \ldots, x(t + (n - 1)\omega)$ are linearly independent. Show that every solution of (H-p) is $n\omega$-periodic.

2.18. (a) Show that Eq. (H-p) has a nontrivial solution $x(t)$ satisfying the property $x(t + \omega) = kx(t)$ for some constant $k \iff k \in \mathbb{R}$ is a characteristic multiplier of the equation.

 (b) Show that Eq. (H-p) has a nontrivial ω-periodic solution $\iff 1$ is a characteristic multiplier of the equation.

2.19. (a) Let $x(t)$ be a nontrivial solution of Eq. (H-p) satisfying $x(t_0 + \omega) = x(t_0)$ for some $t_0 \in \mathbb{R}$. Show that $x(t)$ is ω-periodic.

 (b) Let $x(t)$ be a nontrivial solution of Eq. (H-p) satisfying $x(t_0 + k\omega) = x(t_0)$ for some $t_0 \in \mathbb{R}$ and $k \in \mathbb{Z}$ with $k \neq 0$. What can you say about the solution $x(t)$? Justify your answer.

2.20. (a) Let $X(t)$ be a fundamental matrix solution of Eq. (H-p). Show that all solutions of Eq. (H-p) are ω-periodic if and only if

$$X(t_0 + \omega) = X(t_0) \quad \text{for some } t_0 \in \mathbb{R}.$$

 (b) If $A \in C(\mathbb{R}, \mathbb{R}^{n \times n})$ is ω-periodic and odd (i.e., $A(-t) = -A(t)$ for all t), show that all solutions of Eq. (H-p) are ω-periodic.

2.21. Consider the nonhomogeneous linear equation (NH), where

$$A(t + \omega) \equiv A(t) \quad \text{and} \quad f(t + \omega) \equiv f(t) \quad \text{on } \mathbb{R}.$$

Show that

 (a) a solution $x(t)$ of Eq. (NH) is ω-periodic $\iff x(0) = x(\omega)$;
 (b) for $f(t) \not\equiv 0$, Eq. (NH) has a unique ω-periodic solution $\iff 1$ is not a characteristic multiplier of Eq. (H) $\iff$ Eq. (H) has no nontrivial ω-periodic solution.

2.22. Consider Eq. (H-p). Show that
   (a) if $\int_0^\omega tr\, A(s)\, ds > 0$, then $\limsup_{t\to\infty} |x(t)| = \infty$ for some solution $x(t)$;
   (b) if $\int_0^\omega tr\, A(s)\, ds < 0$, then $\lim_{t\to\infty} |x(t)| = 0$ for some nontrivial solution $x(t)$.

2.23. Let $A(t) = \begin{bmatrix} -1 & a \\ b & \sin(\pi t) \end{bmatrix}$ for some $a, b \in \mathbb{R}$. Let μ_1, μ_2 be the characteristic multipliers of Eq. (H-p).
   (a) Show that $\mu_1 \mu_2 = e^{-2}$.
   (b) Assume $\mu_1 = -1$. What can you say about the boundedness of solutions of Eq. (H-p) on $[0, \infty)$?

2.24. Consider the equation $x'' + a(t)x' + b(t)x = 0$, where $a, b \in C(\mathbb{R}, \mathbb{R})$ are ω-periodic with some $\omega > 0$. Let μ_1, μ_2 be the characteristic multipliers of the equation (i.e., the characteristic multipliers of the corresponding system of first-order equations).
   (a) Show that $\mu_1 \mu_2 = e^{-\int_0^\omega a(s)\, ds}$.
   (b) Discuss the asymptotic behavior of the solutions of the equation based on the sign of $\int_0^\omega a(s)\, ds$.

2.25. Consider Hill's equation $x'' + p(t)x = 0$, where $p \in C(\mathbb{R}, \mathbb{R})$ is ω-periodic. Show that if this equation has a nontrivial periodic solution of period $n\omega$ for $n > 2$, but has no nontrivial periodic solutions of period ω or 2ω, then all solutions are periodic of period $n\omega$.

CHAPTER 3

Lyapunov Stability Theory

3.1. Basic Concepts

Different from the continuous dependence of solutions of IVPs on the initial conditions discussed in Sect. 1.5, stability characterizes the behavior of solutions of differential equations as $t \to \infty$. Before introducing the concept of stability, we look at the following example.

EXAMPLE 3.1.1. Consider the well known logistic equation

$$(3.1.1) \qquad x' = ax - bx^2,$$

where $a, b > 0$ are constants. Obviously, $x_1 \equiv a/b$ and $x_2 \equiv 0$ are constant solutions of Eq. (3.1.1). Now, we solve Eq. (3.1.1) for the solution satisfying the IC $x(0) = x_0 \neq 0$ or a/b. For this purpose, we rewrite Eq. (3.1.1) to the following:

$$\frac{dx}{x(a - bx)} = dt.$$

By employing the partial fractions method and then doing integration we have

$$\ln \left| \frac{x}{a - bx} \right| = at + c \ \text{ or } \ \frac{x}{a - bx} = c_1 e^{at}.$$

From the IC $x(0) = x_0$ we find that $c_1 = x_0/(a - bx_0)$. With some calculations we get that

$$(3.1.2) \qquad x = \frac{a}{b + \left(\frac{a}{x_0} - b \right) e^{-at}}.$$

We denote by $g(t)$ the denominator $b + (a/x_0 - b) e^{-at}$ of (3.1.2). Then we have the following observations:

1. When $x_0 > 0$, $g(t) > b(1 - e^{-at}) > 0$ for all $t \geq 0$. Hence the solution $x(t)$ exists on the whole half-line $[0, \infty)$ and $x(t) \to a/b$ as $t \to \infty$.

2. When $x_0 < 0$, $g(0) = a/x_0 < 0$ and $g(t) \to b > 0$ as $t \to \infty$. This implies that there exists a $t^* \in (0, \infty)$ such that $g(t^*) = 0$ and hence $\lim_{t \to t^*-} x(t) = -\infty$. By Theorem 1.4.1, $x(t)$ exists on $[0, t^*)$ but cannot be extended further to the right.

© Springer International Publishing Switzerland 2014
Q. Kong, *A Short Course in Ordinary Differential Equations*, Universitext,
DOI 10.1007/978-3-319-11239-8_3

From the above discussion we notice that when the initial value x_0 is sufficiently close to a/b, the corresponding solution of the IVP stays close to and eventually approaches the constant solution $x_1 \equiv a/b$ as $t \to \infty$. However, no matter how close the initial value x_0 is to 0, the corresponding solution of the IVP either blows up at a finite time or is getting away from the constant solution $x_2 \equiv 0$ as $t \to \infty$. See Fig. 3.1.

As we will see later, the solution $x_1 \equiv a/b$ is asymptotically stable, and the solution $x_2 \equiv 0$ is unstable.

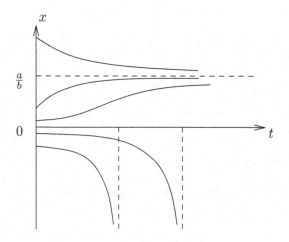

FIGURE 3.1. Solutions of Eq. (3.1.1)

Now, we discuss the stability of solutions of general equations. Consider the equation

$$\text{(E)} \qquad\qquad\qquad x' = f(t, x),$$

where $f \in C([0, \infty) \times \mathbb{R}^n, \mathbb{R}^n)$. Here, we do not require uniqueness for solutions of IVPs associated with Eq. (E). Let $x = \phi(t)$ be a solution of Eq. (E) which exists on the whole interval $[t_*, \infty)$ for some $t_* \geq 0$. The definitions below for stabilities of solutions are in the sense of Lyapunov stability.

DEFINITION 3.1.1. Denote by $x(t; t_0, x_0)$ any solution of Eq. (E) satisfying $x(t_0; t_0, x_0) = x_0$, where $t_0 \geq t_*$ and $x_0 \in \mathbb{R}^n$. Then

(a) $x = \phi(t)$ is said to be stable if for any $\epsilon > 0$ and any $t_0 \geq t_*$, there exists a $\delta = \delta(\epsilon, t_0) > 0$ such that $|x_0 - \phi(t_0)| < \delta$ implies that $|x(t; t_0, x_0) - \phi(t)| < \epsilon$ for all $t \geq t_0$.

(b) $x = \phi(t)$ is said to be uniformly stable if it is stable with $\delta = \delta(\epsilon)$ independent of the choice of t_0; i.e., for any $\epsilon > 0$, there exists a $\delta = \delta(\epsilon) > 0$ such that for any $t_0 \geq t_*$, $|x_0 - \phi(t_0)| < \delta$ implies that $|x(t; t_0, x_0) - \phi(t)| < \epsilon$ for all $t \geq t_0$.

(c) $x = \phi(t)$ is said to be asymptotically stable if it is stable and for
any $t_0 \geq t_*$, there exists a $\delta_1 = \delta_1(t_0) > 0$ such that $|x_0 - \phi(t_0)| < \delta_1$
implies that $|x(t; t_0, x_0) - \phi(t)| \to 0$ as $t \to \infty$.

(d) $x = \phi(t)$ is said to be unstable if it is not stable.

The geometric interpretation for a stable solution $x = \phi(t)$ is given by
Fig. 3.2.

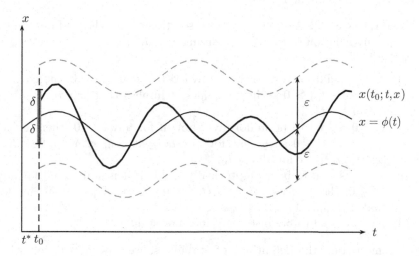

FIGURE 3.2. Stable solution

We comment that some other stabilities such as exponentially asymptotic stability and uniformly asymptotic stability can be defined in a similar
way. However, in this book, we will concentrate on the stabilities defined in
Definition 3.1.1.

We point out that the stability of the zero solution of certain differential
equations plays a special role in the study of stability of any solutions of
differential equations. This is characterized by the following lemma.

LEMMA 3.1.1. *Let $x = \phi(t)$ be a solution of Eq. (E) which exists on the
whole interval $[t_*, \infty)$ for some $t_* \geq 0$. Then the stability problem of the
solution $x = \phi(t)$ of Eq. (E) can be transformed to the stability problem of
the zero solution of some other equation.*

PROOF. Let $x(t)$ be any solution of Eq. (E) and $y(t) = x(t) - \phi(t)$. Then
$$\begin{aligned} y'(t) &= x'(t) - \phi'(t) = f(t, x(t)) - f(t, \phi(t)) \\ &= f(t, y(t) + \phi(t)) - f(t, \phi(t)), \end{aligned}$$
i.e.,

(3.1.3) $y'(t) = g(t, y(t)) := f(t, y(t) + \phi(t)) - f(t, \phi(t)).$

We note that $y = 0$ is a solution of Eq. (3.1.3). Moreover, the closeness of
$x(t)$ to the solution $\phi(t)$ of Eq. (E) is the same as the closeness of $y(t)$ to the

solution $y = 0$ of Eq. (3.1.3). Therefore, the stability of the solution $\phi(t)$ of Eq. (E) is the same as the stability of the solution $y = 0$ of Eq. (3.1.3).   □

By Lemma 3.1.1, without loss of generality, we may assume that $f(t, 0) \equiv 0$ for $t \in [0, \infty)$ in Eq. (E) and study the stability of the zero solution. To highlight this case, we state the definition of stabilities of the zero solution for the convenience of the reader.

DEFINITION 3.1.2. Assume $x = 0$ is a solution of Eq. (E) and denote by $x(t; t_0, x_0)$ any solution of Eq. (E) satisfying $x(t_0; t_0, x_0) = x_0$, where $t_0 \geq 0$ and $x_0 \in \mathbb{R}^n$. Then

(a) $x = 0$ is said to be stable if for any $\epsilon > 0$ and any $t_0 \geq 0$, there exists a $\delta = \delta(\epsilon, t_0) > 0$ such that $|x_0| < \delta$ implies that $|x(t; t_0, x_0)| < \epsilon$ for all $t \geq t_0$.

(b) $x = 0$ is said to be uniformly stable if for any $\epsilon > 0$, there exists a $\delta = \delta(\epsilon) > 0$ such that for any $t_0 \geq 0$, $|x_0| < \delta$ implies that $|x(t; t_0, x_0)| < \epsilon$ for all $t \geq t_0$.

(c) $x = 0$ is said to be asymptotically stable if it is stable and for any $t_0 \geq 0$, there exists a $\delta_1 = \delta_1(t_0) > 0$ such that $|x_0| < \delta_1$ implies that $|x(t; t_0, x_0)| \to 0$ as $t \to \infty$.

(d) $x = 0$ is said to be unstable if it is not stable.

To demonstrate the definitions of stabilities, we look at the following examples:

EXAMPLE 3.1.2. Consider the scalar equation

$$x' = kx,$$

where $k \in \mathbb{R}$. Clearly, $x = 0$ is a solution. With any IC $x(t_0) = x_0$, the solution of the equation is $x(t) = x_0 e^{k(t-t_0)}$. It is easy to see that $x = 0$ is uniformly stable and asymptotically stable when $k < 0$, uniformly stable but not asymptotically stable when $k = 0$, and unstable when $k > 0$.

EXAMPLE 3.1.3. Consider the scalar equation

$$x' = x^2.$$

Clearly, $x = 0$ is a solution. By solving the equation we see that with any IC $x(t_0) = x_0 \neq 0$, the solution of the equation is

$$x(t) = \frac{x_0}{1 - x_0(t - t_0)}.$$

It is easy to see that when $x_0 > 0$, the solution exists on the interval $[t_0, T)$ with $T = 1/x_0 + t_0$ and $\lim_{t \to T^-} x(t) = \infty$; and when $x_0 < 0$, the solution exists for $t \geq t_0$ and $\lim_{t \to \infty} x(t) = 0$. By definition, $x = 0$ is unstable.

The example below is chosen from Coppel [12, p. 52].

EXAMPLE 3.1.4. Consider the scalar equation

$$x' = a(t)x,$$

where $a \in C([0,\infty), \mathbb{R})$. Clearly, $x = 0$ is a solution. With any IC $x(t_0) = x_0$, the solution of the equation is $x(t) = x_0 e^{\int_{t_0}^{t} a(s)\,ds}$ for $t \in \mathbb{R}$. By definition it is easy to see the following:

(i) $x = 0$ is stable $\Longleftrightarrow$ for any $t_0 \geq 0$, there exists an $M(t_0) \geq 0$ such that $\int_{t_0}^{t} a(s)\,ds \leq M(t_0)$ for any $t \geq t_0$ $\Longleftrightarrow$ there exists an $M \geq 0$ such that $\int_{0}^{t} a(s)\,ds \leq M$ for any $t \geq 0$.

(ii) $x = 0$ is uniformly stable $\Longleftrightarrow$ there exists an $M \geq 0$ such that $\int_{t_0}^{t} a(s)\,ds \leq M$ for any $0 \leq t_0 \leq t < \infty$.

(iii) $x = 0$ is asymptotically stable $\Longleftrightarrow$ $\lim_{t\to\infty} \int_{0}^{t} a(s)\,ds = -\infty$.

(iv) $x = 0$ is unstable $\Longleftrightarrow$ $\limsup_{t\to\infty} \int_{0}^{t} a(s)\,ds = \infty$.

As a special case, let

$$a(t) = \sin(\ln t) + \cos(\ln t) - \alpha,$$

where $1 < \alpha < \sqrt{2}$. Then

$$\int_{0}^{t} a(s)\,ds = t\sin(\ln t) - \alpha t \to -\infty \quad \text{as } t \to \infty.$$

This shows that $x = 0$ is asymptotically stable and hence is stable. However, we claim that there exist two sequences $\{t_{0n}\}_{n=0}^{\infty}$ and $\{t_n\}_{n=0}^{\infty}$ in $[0,\infty)$ such that $t_{0n} < t_n$ for $n \in \mathbb{N}$, $t_{0n} \to \infty$ and $\int_{t_{0n}}^{t_n} a(s)\,ds \to \infty$ as $n \to \infty$. Therefore, $x = 0$ is not uniformly stable. This can be intuitively seen from Fig. 3.3:

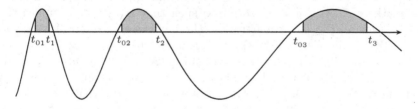

FIGURE 3.3. Graph of function $a(t)$

To prove this result mathematically, we first note that $\sin(\pi/4) + \cos(\pi/4) = \sqrt{2} > \alpha$. This implies that there exist a $\beta \in (\alpha, \sqrt{2})$ and an interval $[\theta_1, \theta_2]$ satisfying $\theta_1 < \pi/4 < \theta_2$ and $\sin\theta + \cos\theta \geq \beta$ for $\theta \in [\theta_1, \theta_2]$. For $n \in \mathbb{N}$, define

$$t_{0n} = e^{2n\pi + \theta_1} \quad \text{and} \quad t_n = e^{2n\pi + \theta_2}.$$

Then

$$\int_{t_{0n}}^{t_n} a(s)\, ds = \int_{e^{2n\pi+\theta_1}}^{e^{2n\pi+\theta_2}} (\sin(\ln s) + \cos(\ln s) - \alpha)\, ds$$

$$= \int_{\theta_1}^{\theta_2} (\sin\theta + \cos\theta - \alpha)e^{2n\pi+\theta}\, d\theta \geq e^{2n\pi} \int_{\theta_1}^{\theta_2} (\beta - \alpha)e^{\theta}\, d\theta$$

$$= e^{2n\pi}(\beta - \alpha)(e^{\theta_2} - e^{\theta_1}) \to \infty$$

as $n \to \infty$. Therefore, $x = 0$ is not uniformly stable.

This example shows that stability does not imply uniform stability, and even asymptotic stability does not imply uniform stability. However, as shown below, for autonomous differential equations, i.e., for Eq. (E) with the right-hand side function f explicitly independent of t, stability means uniform stability.

Consider the autonomous equation

(A) $x' = f(x),$

where $f \in C(\mathbb{R}^n, \mathbb{R}^n)$ such that $f(0) = 0$. By a simple verification we see that Eq. (A) satisfies the following property:

LEMMA 3.1.2. *If $x(t)$ is a solution of Eq. (A) on an interval (a, b), then for any $\alpha \in \mathbb{R}$, $x(t+\alpha)$ is a solution of Eq. (A) on the interval $(a+\alpha, b+\alpha)$.*

Using Lemma 3.1.2 we derive the equivalent relation between the stability and uniform stability of the zero solution of Eq. (A).

LEMMA 3.1.3. *The zero solution of Eq. (A) is stable if and only if it is uniformly stable.*

PROOF. We only need to prove that if $x = 0$ is stable, then it is uniformly stable. Assume $x = 0$ is stable. Then by definition, for any $\epsilon > 0$ and any $t_0 \geq 0$, there exists a $\delta = \delta(\epsilon, t_0) > 0$ such that $|x(t_0)| < \delta$ implies that $|x(t)| < \epsilon$ for all $t \geq t_0$. In particular, there exists a $\delta_0 = \delta(\epsilon, 0) > 0$ such that $|x(0)| < \delta_0$ implies that $|x(t)| < \epsilon$ for all $t \geq 0$.

For any $t_0 \geq 0$, let $y(t)$ be a solution of Eq. (A) satisfying $|y(t_0)| < \delta_0$ and let $z(t) = y(t + t_0)$. Then by Lemma 3.1.2, $z(t)$ is also a solution of Eq. (A) satisfying $|z(0)| = |y(t_0)| < \delta_0$. From the above we have that $|z(t)| < \epsilon$ for all $t \geq 0$. Thus $|y(t + t_0)| < \epsilon$ for all $t \geq 0$ or $|y(t)| < \epsilon$ for all $t \geq t_0$. This means that $x = 0$ is uniformly stable.   □

For autonomous equation (A), $a \in \mathbb{R}^n$ is said to be an equilibrium if $f(a) = 0$. In this case, $x(t) = a, t \in \mathbb{R}$, is a constant solution of Eq. (A). We comment that the stability of an equilibrium a of Eq. (A) can always be transformed to the stability of the equilibrium 0 for another autonomous equation. In fact, if we let $y = x - a$, then Eq. (A) becomes the equation $y' = g(y) := f(y + a)$, where 0 is an equilibrium with the same stability property as the equilibrium a for Eq. (A).

We conclude this section with the result that for a nonautonomous equation with the right-hand side function periodic in t, stability also means uniform stability as presented below.

LEMMA 3.1.4. *Assume that $f \in C([0,\infty)\times\mathbb{R}^n, \mathbb{R}^n)$ such that $f(t+\omega, x) = f(t,x)$ for some $\omega > 0$ and all $(t,x) \in [0,\infty) \times \mathbb{R}^n$. Then the zero solution of Eq. (E) is stable if and only if it is uniformly stable.*

The idea for the proof of Lemma 3.1.4 is essentially the same as that of Lemma 3.1.3 for autonomous equations. However, a slightly more detailed argument is involved. We omit it here. An interested reader is referred to [**48**, Theorem 7.3 in p. 30] for the proof.

3.2. Stability of Linear Equations

Based on the definitions of stabilities in Sect. 3.1, we study the stabilities of the homogeneous linear equation

(H) $$x' = A(t)x$$

and the nonhomogeneous linear equation

(NH) $$x' = A(t)x + f(t),$$

where $A \in C([0,\infty), \mathbb{R}^{n\times n})$ and $f \in C([0,\infty), \mathbb{R}^n)$. By Theorem 2.1.1, all solutions of Eqs. (H) and (NH) exist on the whole interval $[0,\infty)$.

LEMMA 3.2.1. *All solutions of Eqs. (H) and (NH) have the same stability property as the zero solution of Eq. (H).*

PROOF. Since Eq. (H) can be regarded as a special case of Eq. (NH) with $f \equiv 0$, without loss of generality we only prove the conclusion for Eq. (NH).

Let $\phi(t)$ be a given solution of Eq. (NH) and $x(t)$ any solution of Eq. (NH). Let $y(t) = x(t) - \phi(t)$. Then $y(t)$ is a solution of Eq. (H). Moreover, the closeness of $x(t)$ to $\phi(t)$ is the same as the closeness of $y(t)$ to the zero solution of Eq. (H). Therefore, the stability of the solution $\phi(t)$ of Eq. (NH) is the same as the stability of the zero solution of Eq. (H). □

REMARK 3.2.1. From Lemma 3.2.1 we see that the stability of solutions of Eq. (NH) is independent of the right-hand side function f, and hence we only need to consider the stability of solutions of Eq. (H). Since all solutions of Eq. (H) have the same stability property, we may speak of the stability of Eq. (H) instead of the stability of a specific solution.

However, for a nonlinear differential equation, we must speak of the stability of a given solution, in particular, the stability of the zero solution.

Since all solutions of Eq. (H) are determined by a fundamental matrix solution, the stability of Eq. (H) can be determined by a fundamental matrix solution, as shown in the following theorem.

THEOREM 3.2.1. *Let $X(t)$ be a fundamental matrix solution of Eq. (H) and $Y(t,t_0)$ the principal matrix solution of Eq. (H) at t_0. Then*

(a) *Equation* (H) *is stable* $\iff$ *there exists a* $k > 0$ *such that* $|X(t)| \leq k$
for all $t \geq 0$ $\iff$ *all solutions of Eq.* (H) *are bounded on* $[0, \infty)$;

(b) *Equation* (H) *is uniformly stable* $\iff$ *there exists a* $k > 0$ *such that*
$|X(t)X^{-1}(t_0)| \leq k$ *for all* $t \geq t_0 \geq 0$ $\iff$ $Y(t, t_0)$ *is uniformly*
bounded for all $t \geq t_0 \geq 0$;

(c) *Equation* (H) *is asymptotically stable* $\iff$ $X(t) \to \mathbf{0}$ *as* $t \to \infty$ $\iff$
$\lim_{t \to \infty} x(t) = 0$ *for all solutions* $x(t)$ *of Eq.* (H).

PROOF. (a) We note that the second equivalence in the statement is
obviously true.

($\Longleftarrow$) Assume that there exists a $k > 0$ such that $|X(t)| \leq k$
for all $t \geq 0$. Let $t_0 \geq 0$ and $x_0 \in \mathbb{R}^n$. Then $x(t; t_0, x_0) =$
$X(t)X^{-1}(t_0)x_0$. For any $\epsilon > 0$ and any $t_0 \geq 0$, let $\delta = \epsilon/(k|X^{-1}(t_0)|)$.
Then when $|x_0| < \delta$, we have that for all $t \geq t_0$

$$
\begin{aligned}
|x(t; t_0, x_0)| &= |X(t)X^{-1}(t_0)x_0| \leq |X(t)||X^{-1}(t_0)||x_0| \\
&\leq k|X^{-1}(t_0)||x_0| < k|X^{-1}(t_0)|\delta < \epsilon.
\end{aligned}
$$

Thus Eq. (H) is stable.

($\Longrightarrow$) Assume Eq. (H) is stable. Then for any $t_0 \geq 0$, there
exists a $\delta = \delta(t_0) > 0$ such that $|x_0| < \delta$ implies that $|x(t; t_0, x_0)| <$
1 for all $t \geq t_0$. Hence $x(t; t_0, x_0)$ is bounded on $[0, \infty)$.

Let $y(t)$ be any solution of Eq. (H) with $y(t_0) = y_0 \neq 0$, and let
$z(t) = (\delta/(2|y_0|))y(t)$. Then $z(t)$ is also a solution of Eq. (H), and
$|z(t_0)| = \delta/2 < \delta$. By the above, $|z(t)| < 1$ for all $t \geq t_0$. It follows
that $|y(t)| < 2|y_0|/\delta$ and hence is bounded on $[t_0, \infty)$. Since $y(t)$
exists and is bounded on $[0, t_0]$, $y(t)$ is bounded on $[0, \infty)$.

(b) We note that the second equivalence in the statement is obviously
true.

($\Longleftarrow$) Assume that there exists a $k > 0$ such that $|X(t)X^{-1}(t_0)|$
$\leq k$ for all $t \geq t_0 \geq 0$. Let $t_0 \geq 0$ and $x_0 \in \mathbb{R}^n$. Then $x(t; t_0, x_0) =$
$X(t)X^{-1}(t_0)x_0$. For any $\epsilon > 0$, let $\delta = \epsilon/k$. Then when $|x_0| < \delta$,
we have that for all $t \geq t_0$

$$
|x(t; t_0, x_0)| = |X(t)X^{-1}(t_0)x_0| \leq |X(t)X^{-1}(t_0)||x_0| \leq k|x_0| < k\delta < \epsilon.
$$

Thus Eq. (H) is uniformly stable.

($\Longrightarrow$) Assume Eq. (H) is uniformly stable. Then there exists a
$\delta > 0$ such that for any $t_0 \geq 0$, $|x_0| < \delta$ implies that $|x(t; t_0, x_0)| < 1$
for all $t \geq t_0 \geq 0$.

For any $t_0 \geq 0$, let $x_i(t)$ be the solution of Eq. (H) satisfying
$x_i(t_0) = \alpha_i := (\delta/2)e_i$, $i = 1, \ldots, n$, where e_i is ith basic vector in
$\mathbb{R}^n$, i.e., the vector whose ith component is 1 and all others are 0.
Then $x_i(t) = X(t)X^{-1}(t_0)\alpha_i$ and $|x_i(t_0)| = \delta/2 < \delta$. By the above,
for $i = 1, \ldots, n$,

$$
|X(t)X^{-1}(t_0)\alpha_i| = |x_i(t)| < 1 \quad \text{for all } t \geq t_0 \geq 0.
$$

It follows that

$$|X(t)X^{-1}(t_0)| = \frac{2}{\delta}|X(t)X^{-1}(t_0)[\alpha_1 \cdots \alpha_n]| = \frac{2}{\delta}\sum_{i=1}^{n}|X(t)X^{-1}(t_0)\alpha_i| < \frac{2n}{\delta}$$

for all $t \geq t_0 \geq 0$.

(c) We note that the second equivalence in the statement is obviously true. Let $X(t) = [x_1, \cdots, x_n](t)$.

($\Longleftarrow$) Assume that $X(t) \to \mathbf{0}$ as $t \to \infty$. Then it is easy to see that there exists a $k > 0$ such that $|X(t)| \leq k$ for all $t \geq 0$. By Part (a), Eq. (H) is stable. Since every solution $x(t)$ is a linear combination of $x_1(t), \ldots, x_n(t)$, we have that $\lim_{t\to\infty} x(t) = 0$. This shows that Eq. (H) is asymptotically stable.

($\Longrightarrow$) Assume Eq. (H) is asymptotically stable. Then for any $t_0 \geq 0$, there exists a $\delta_1 = \delta_1(t_0)$ such that $|x_0| < \delta_1$ implies that $|x(t; t_0, x_0)| \to 0$ as $t \to \infty$.

For $i = 1, \ldots, n$, let $y_i(t) := (\delta_1/(2|x_i(t_0)|))x_i(t)$. Then $y_i(t)$ is a solution of Eq. (H) and $|y_i(t_0)| = \delta_1/2 < \delta_1$. By the above, $\lim_{t\to\infty} y_i(t) = 0$. It follows that $\lim_{t\to\infty} x_i(t) = 0$ for $i = 1, \ldots, n$, and hence $\lim_{t\to\infty} X(t) = \mathbf{0}$.

$\square$

Now, we apply Theorem 3.2.1 to obtain specific criteria for the stability of the linear equation with constant coefficients

(H-c) $$x' = Ax,$$

where $A \in \mathbb{R}^{n\times n}$. Since Eq. (H-c) is autonomous, it is stable if and only if it is uniformly stable. Thus, the theorem below is a direct consequence of Theorem 3.2.1, Lemma 2.4.4, and Theorem 2.4.2.

THEOREM 3.2.2. Let λ_i, $i = 1, \ldots, n$, be the eigenvalues of matrix A. Then

(a) Equation (H-c) is uniformly stable $\Longleftrightarrow$ $\mathrm{Re}\,\lambda_i \leq 0$, $i = 1, \ldots, n$, and $\mathrm{Re}\,\lambda_i = 0$ occurs only when the λ_i's are in the diagonal Jordan block of the matrix A;

(b) Equation (H-c) is asymptotically stable $\Longleftrightarrow$ $\mathrm{Re}\,\lambda_i < 0$, $i = 1, \ldots, n$ $\Longleftrightarrow$ there exist $k > 0$ and $0 < \alpha < \min\{-\mathrm{Re}\,\lambda_i, i = 1, \ldots, n\}$ such that $|e^{At}| \leq ke^{-\alpha t}$, $t \geq 0$;

(c) Equation (H-c) is unstable $\Longleftrightarrow$ there exists an $i \in \{1, \ldots, n\}$ such that either $\mathrm{Re}\,\lambda_i > 0$, or $\mathrm{Re}\,\lambda_i = 0$ which occurs when λ_i is not in the diagonal Jordan block of A.

We then apply Theorem 3.2.1 to obtain specific criteria for the stability of the linear equation with periodic coefficients

(H-p) $$x' = A(t)x,$$

where $A \in C(\mathbb{R}, \mathbb{R}^{n\times n})$ such that $A(t + \omega) = A(t)$ for some $\omega > 0$ and all $t \in \mathbb{R}$.

The following lemma on the relation between the stability and uniform stability of Eq. (H-p) is a special case of Lemma 3.1.4, but it can be proved directly using the Floquet theory. This is left as an exercise, see Exercise 3.3.

LEMMA 3.2.2. *Equation* (H-p) *is stable if and only if it is uniformly stable.*

The theorem below is then a direct consequence of Theorems 3.2.1 and 2.5.1.

THEOREM 3.2.3. *Let μ_i, $i = 1, \ldots, n$, be the characteristic multipliers of Eq.* (H-p). *Then*

(a) *Equation* (H-p) *is uniformly stable $\Longleftrightarrow |\mu_i| \leq 1$, $i = 1, \ldots, n$, and $|\mu_i| = 1$ occurs only when the μ_i's are in the diagonal Jordan block of the transition matrix V;*

(b) *Equation* (H-p) *is asymptotically stable $\Longleftrightarrow |\mu_i| < 1$, $i = 1, \ldots, n$;*

(c) *Equation* (H-p) *is unstable $\Longleftrightarrow$ there exists an $i \in \{1, \ldots, n\}$ such that either $|\mu_i| > 1$, or $|\mu_i| = 1$ which occurs when μ_i is not in the diagonal Jordan block of the transition matrix V.*

By combining Theorem 3.2.1 and Corollary 2.5.5, we also obtain an equivalent version of Theorem 3.2.3 using the characteristic exponents of Eq. (H-p).

THEOREM 3.2.4. *Let ρ_i, $i = 1, \ldots, n$, be the exponents of Eq.* (H-p), *i.e., the eigenvalues of the matrix $R = (\log V)/\omega$ for a transition matrix V of Eq.* (H-p). *Then*

(a) *Equation* (H-p) *is uniformly stable $\Longleftrightarrow \operatorname{Re} \rho_i \leq 0$, $i = 1, \ldots, n$, and $\operatorname{Re} \rho_i = 0$ occurs only when the ρ_i's are in the diagonal Jordan block of the matrix R;*

(b) *Equation* (H-p) *is asymptotically stable $\Longleftrightarrow \operatorname{Re} \rho_i < 0$, $i = 1, \ldots, n$;*

(c) *Equation* (H-p) *is unstable $\Longleftrightarrow$ there exists an $i \in \{1, \ldots, n\}$ such that either $\operatorname{Re} \rho_i > 0$, or $\operatorname{Re} \rho_i = 0$ which occurs when ρ_i is not in the diagonal Jordan block of R.*

3.3. Stability of Linear Equations by Lozinskii Measures

From Sect. 3.2 we see that for the homogeneous linear equation with constant coefficients (H-c), the stability is completely determined by the eigenvalues λ_i, $i = 1, \ldots, n$, of matrix A. In particular, Eq. (H-c) is asymptotically stable if $\operatorname{Re} \lambda_i < 0$ for all $i = 1, \ldots, n$. Naturally, we would ask a question for the general homogeneous linear equation (H): Is Eq. (H) asymptotically stable if $\operatorname{Re} \lambda_i(A(t)) \leq \delta < 0$ for all $t \geq 0$ and $i = 1, \ldots, n$? Unfortunately, the answer is no. The following example is taken from [**43**, p. 184].

EXAMPLE 3.3.1. Consider Eq. (H) with

$$A(t) = \begin{bmatrix} -1 + a\cos^2 t & 1 - a\sin t \cos t \\ -1 - a\sin t \cos t & -1 + a\sin^2 t \end{bmatrix},$$

where $a \in \mathbb{R}$. It is easy to verify that the principal matrix solution of Eq. (H) at $t_0 = 0$ is

$$X(t) = \begin{bmatrix} e^{(a-1)t}\cos t & e^{-t}\sin t \\ -e^{(a-1)t}\sin t & e^{-t}\cos t \end{bmatrix}.$$

When $a > 1$, $X(t)$ is clearly unbounded on $[0, \infty)$. By Theorem 3.2.1, Part (a), Eq. (H) is unstable. On the other hand, by a simple computation we find that for each $t \geq 0$

$$\det[A(t) - \lambda I] = \lambda^2 + (2-a)\lambda + (2-a).$$

It follows that the eigenvalues $\lambda_i(t)$, $i = 1, 2$, of $A(t)$ are independent of t. Moreover, when $a < 2$,

$$\lambda_i(t) = \frac{-(2-a) \pm \sqrt{a^2 - 4}}{2} = -\frac{2-a}{2} \pm \frac{\sqrt{4-a^2}}{2}\mathbf{i}, \quad i = 1, 2.$$

Therefore, when $1 < a < 2$, $\operatorname{Re}\lambda_i(t) = -(1 - a/2) < 0$, $i = 1, 2$, but Eq. (H) is unstable.

Theoretically, the stability of the general equation (H) can be studied by investigating the behavior of a fundamental matrix solution as $t \to \infty$. However, the fundamental matrices of Eq. (H) are generally unknown. Thus, we would prefer to obtain sufficient conditions for stability using the coefficient matrix $A(t)$ only. In this section, we will derive stability criteria for Eq. (H) using the so-called Lozinskii measures introduced first by Lozinskii [36] and Dahlquist [13]. The results in this section are mainly excerpted from Coppel [12] and Vidyasagar [43]. In order to restrict the length of discussions, we will introduce the basic concepts and results on Lozinskii measures but omit their technical proofs.

We start with a discussion on matrix norms. In Sect. 2.1, we have defined a matrix norm, i.e., for any $A \in \mathbb{R}^{n \times n}$, we denote $|A| = \sum_{i,j=1}^{n} |a_{ij}|$. Now, we introduce some other norms for matrices in $\mathbb{R}^{n \times n}$ which are induced by the norms of vectors in $\mathbb{R}^n$. Recall from Sect. 1.1 that for $i = 1, 2, \infty$, the i-norm of a vector $x \in \mathbb{R}^n$ is defined by

$$|x|_1 := \sum_{i=1}^{n} |x_i|, \quad |x|_2 := \left(\sum_{i=1}^{n} |x_i|^2\right)^{1/2}, \quad \text{and} \quad |x|_\infty := \max_{1 \leq i \leq n} |x_i|.$$

DEFINITION 3.3.1. Let $A \in \mathbb{R}^{n \times n}$. Then for $i = 1, 2, \infty$, the i-norm of A induced by the i-norm on $\mathbb{R}^n$ is

$$|A|_i := \max_{x \in \mathbb{R}^n, |x|_i = 1} |Ax|_i.$$

For $A \in \mathbb{R}^{n \times n}$, we denote by $\lambda_i(A)$, $i = 1, \ldots, n$, the eigenvalues of A satisfying

$$\operatorname{Re}\lambda_1(A) \geq \operatorname{Re}\lambda_2(A) \geq \cdots \geq \operatorname{Re}\lambda_n(A).$$

Then by the standard method for computing operator norms we find that

$$|A|_1 = \max_j \sum_{i=1}^{n} |a_{ij}|, \quad |A|_2 = \sqrt{\lambda_1(A^T A)}, \quad \text{and} \quad |A|_\infty = \max_i \sum_{j=1}^{n} |a_{ij}|.$$

We note that the induced i-norm of a matrix A is different for different i and all the induced norms of A are different from the norm $|A|$ given in Sect. 2.1. However, all the induced norms satisfy the same properties for the norm $|A|$: For $i = 1, 2, \infty$,

(a) $|A|_i \geq 0$ for any $A \in \mathbb{R}^{n \times n}$, and $|A|_i = 0 \iff A = \mathbf{0}$;
(b) $|kA|_i = |k||A|_i$ for any $A \in \mathbb{R}^{n \times n}$ and $k \in \mathbb{R}$;
(c) $|A + B|_i \leq |A|_i + |B|_i$ and $|AB|_i \leq |A|_i|B|_i$ for any $A, B \in \mathbb{R}^{n \times n}$;
(d) $|Ax|_i \leq |A|_i|x|_i$ for any $A \in \mathbb{R}^{n \times n}$ and $x \in \mathbb{R}^n$.

Moreover, we also have that for $i = 1, 2, \infty$,

(e) $|I|_i = 1$ with I the $n \times n$ the identity matrix in $\mathbb{R}^{n \times n}$.

We point out that Property (e) is not satisfied by the norm $|A|$ given in Sect. 2.1 due to the fact that it is not an induced norm.

Based on the induced norms, we can define the Lozinskii measures or logarithmic norms of matrices in $\mathbb{R}^{n \times n}$.

DEFINITION 3.3.2. Let $A \in \mathbb{R}^{n \times n}$. For $i = 1, 2, \infty$, the Lozinskii measure $\mu_i(A)$ of A is defined by

$$\mu_i(A) = \lim_{h \to 0+} \frac{|I + hA|_i - 1}{h}.$$

REMARK 3.3.1. (i) For $i = 1, 2, \infty$, the Lozinskii measure $\mu_i(A)$ exists for any matrix $A \in \mathbb{R}^{n \times n}$. To prove this, we let $g(h) := (|I+hA|_i-1)/h$. We first show that $g(h)$ is increasing for $h > 0$, i.e.,

$$\frac{|I + \theta h A|_i - 1}{\theta h} \leq \frac{|I + hA|_i - 1}{h} \quad \text{for } h > 0 \text{ and } 0 < \theta \leq 1.$$

This is true due to the fact that

$$|I + \theta h A|_i = |\theta(I + hA) + (1 - \theta)I|_i \leq \theta|I + hA|_i + (1 - \theta).$$

Then we show that $g(h)$ is bounded for $h > 0$. In fact,

$$
\begin{aligned}
|g(h)| &= \frac{1}{h}\big||I + hA|_i - 1\big| = \frac{1}{h}\big||I + hA|_i - |I|_i\big| \\
&\leq \frac{1}{h}\big|(I + hA) - I\big|_i = |A|_i.
\end{aligned}
$$

(ii) A Lozinskii measure is not a matrix norm. In fact, it is not always nonnegative. In particular, for $i = 1, 2, \infty$,

$$\mu_i(\pm I) = \lim_{h \to 0+} \frac{|1 \pm h|_i - 1}{h} = \pm 1.$$

(iii) For $i = 1, 2, \infty$, $\mu_i(-A) \neq -\mu_i(A)$ in general.

For any matrix $A \in \mathbb{R}^{n \times n}$, we denote

$$\nu_i(A) = -\mu_i(-A) \quad \text{for } i = 1, 2, \infty.$$

Then by some calculations we obtain the explicit forms of $\mu_i(A)$ and $\nu_i(A)$ for $i = 1, 2, \infty$:

$$\mu_1(A) = \max_j \left\{ a_{jj} + \sum_{i,i\neq j} |a_{ij}| \right\}, \quad \nu_1(A) = \min_j \left\{ a_{jj} - \sum_{i,i\neq j} |a_{ij}| \right\};$$

$$\mu_2(A) = \lambda_1 \left(\tfrac{1}{2}(A + A^T) \right), \quad \nu_2(A) = \lambda_n \left(\tfrac{1}{2}(A + A^T) \right);$$

$$\mu_\infty(A) = \max_i \left\{ a_{ii} + \sum_{j,j\neq i} |a_{ij}| \right\}, \quad \nu_\infty(A) = \min_i \left\{ a_{ii} - \sum_{j,j\neq i} |a_{ij}| \right\}.$$

EXAMPLE 3.3.2. Let $A = \begin{bmatrix} 2 & 3 \\ -1 & -2 \end{bmatrix}$. Then

(i) $\mu_1(A) = \max\{2 + 1, -2 + 3\} = 3$ and
$\nu_1(A) = \min\{2 - 1, -2 - 3\} = -5$,

(ii) $\mu_2(A) = \lambda_1(\tfrac{1}{2}(A + A^T)) = \lambda_1 \left(\begin{bmatrix} 2 & 1 \\ 1 & -2 \end{bmatrix} \right) = \sqrt{5}$ and
$\nu_2(A) = \lambda_2(\tfrac{1}{2}(A + A^T)) = \lambda_2 \left(\begin{bmatrix} 2 & 1 \\ 1 & -2 \end{bmatrix} \right) = -\sqrt{5}$,

(iii) $\mu_\infty(A) = \max\{2 + 3, -2 + 1\} = 5$ and
$\nu_\infty(A) = \min\{2 - 3, -2 - 1\} = -3$.

In the sequel, without specification, we denote by $\mu(A)$ and $\nu(A)$ any pair of $\mu_i(A)$ and $\nu_i(A)$ with $|A| = |A|_i$ the associated induced norm of A, $i = 1, 2, \infty$. Then for any $A, B \in \mathbb{R}^{n \times n}$ we have

(a) $\mu(\alpha A) = \alpha \mu(A)$ and $\nu(\alpha A) = \alpha \nu(A)$ for $\alpha \geq 0$;
(b) $\mu(A + B) \leq \mu(A) + \mu(B)$ and $\nu(A + B) \geq \nu(A) + \nu(B)$;
(c) $-|A| \leq \nu(A) \leq \operatorname{Re} \lambda_n(A) \leq \operatorname{Re} \lambda_1(A) \leq \mu(A) \leq |A|$.

From Properties (b) and (c) we derive that

(d) $\mu(A) - \mu(B) \leq |A - B|$ and $\nu(A) - \nu(B) \geq -|A - B|$.

Now, we are going to use the Lozinskii measures to study the stability of Eq. (H). Let $x \in C^1([0, \infty), \mathbb{R}^n)$ and $|x(t)|$ an i-norm of $x(t)$ for any $i = 1, 2, \infty$ and fixed $t \in [0, \infty)$. We note that $|x(t)|$ may not be differentiable everywhere. So we will make use of the right-derivative of $|x(t)|$

$$|x|'_+(t) := \lim_{h \to 0+} \frac{|x(t + h)| - |x(t)|}{h}.$$

We have the following lemmas.

LEMMA 3.3.1. Let $x \in C^1([0, \infty), \mathbb{R}^n)$. Then $|x|'_+(t)$ exists for any $t \in [0, \infty)$ and satisfies

(3.3.1) $$|x|'_+(t) = \lim_{h \to 0+} \frac{|x(t) + hx'(t)| - |x(t)|}{h}.$$

PROOF. With a similar argument to that in Remark 3.3.1, Part (i), we can show that for any $x, y \in \mathbb{R}^n$, the limit

$$\lim_{h \to 0+} \frac{|x + hy| - |x|}{h}$$

exists. We leave it as an exercise, see Exercise 3.4. For $t \in [0, \infty)$, by replacing x and y by $x(t)$ and $x'(t)$, respectively, we see that the right-hand side of (3.3.1) exists.

Let $t \in [0, \infty)$. We observe that

$$\left| [|x(t + h)| - |x(t)|] - [|x(t) + hx'(t)| - |x(t)|] \right|$$
$$= \left| |x(t + h)| - |x(t) + hx'(t)| \right|$$
$$\leq |x(t + h) - x(t) - hx'(t)| = o(h) \quad \text{as } h \to 0 + .$$

By dividing both sides by h and taking limits as $h \to 0+$ we see that (3.3.1) holds. $\qquad\square$

LEMMA 3.3.2. *Assume $x(t)$ is a solution of Eq. (H). Then*

$$|x|'_+(t) \leq \mu(A(t))|x(t)|, \quad t \in [0, \infty).$$

PROOF. Clearly, $x \in C^1([0, \infty), \mathbb{R}^n)$. By Lemma 3.3.1, $x(t)$ satisfies (3.3.1). Since $x(t)$ is a solution of Eq. (H),

$$|x(t) + hx'(t)| = |x(t) + hA(t)x(t)| \leq |I + hA(t)||x(t)|.$$

By (3.3.1),

$$|x|'_+(t) \leq \lim_{h \to 0+} \frac{|I + hA(t)| - 1}{h}|x(t)| = \mu(A(t))|x(t)|.$$

$\qquad\square$

Based on Lemma 3.3.2, we obtain lower and upper bounds for $|x(t)|$ on $[0, \infty)$ for any solution $x(t)$ of Eq. (H).

LEMMA 3.3.3. *Let $x(t)$ be the solution of Eq. (H) satisfying the IC $x(t_0) = x_0$ for $t_0 \geq 0$ and $x_0 \in \mathbb{R}^n$. Then for $t \geq t_0$*

$$(3.3.2) \quad |x_0| \exp\left(\int_{t_0}^t \nu(A(s))\, ds \right) \leq |x(t)| \leq |x_0| \exp\left(\int_{t_0}^t \mu(A(s))\, ds \right).$$

PROOF. From Lemma 3.3.2 we see that $|x|'_+(t) - \mu(A(t))|x(t)| \leq 0$ for $t \geq t_0$. Multiplying both sides by $\exp\left(-\int_{t_0}^t \mu(A(s))\, ds \right)$ we have

$$\left[|x(t)| \exp\left(-\int_{t_0}^t \mu(A(s))\, ds \right) \right]'_+ \leq 0,$$

and hence $|x(t)| \exp\left(-\int_{t_0}^{t} \mu(A(s))\, ds\right)$ is decreasing for $t \geq t_0$. Therefore,

$$|x(t)| \exp\left(-\int_{t_0}^{t} \mu(A(s))\, ds\right) \leq |x(t_0)| = |x_0|, \quad t \geq t_0.$$

This shows that the right inequality of (3.3.2) holds.

To prove the left inequality, we make the transformation $s = -t$ and $y(s) = x(t)$. Let $s_0 = -t_0$. Then $y(s)$ is the solution of the IVP

$$(3.3.3) \qquad \frac{dy}{ds} = -A(-s)y(s), \quad y(s_0) = x_0.$$

We note that $t \geq t_0 \iff s \leq s_0$. For any $s \leq s_0$, by applying the right inequality of (3.3.2) to Eq. (3.3.3) on the interval $[s, s_0]$ we find that

$$|y(s_0)| \leq |y(s)| \exp\left(\int_{s}^{s_0} \mu(-A(-\sigma))\, d\sigma\right),$$

i.e.,

$$|y(s)| \geq |y(s_0)| \exp\left(\int_{s_0}^{s} \mu(-A(-\sigma))\, d\sigma\right) = |x_0| \exp\left(\int_{s_0}^{s} \mu(-A(-\sigma))\, d\sigma\right).$$

By replacing $y(s)$ by $x(t)$ in the above inequality and making the change of variable $\sigma = -\tau$ in the integration we obtain that

$$|x(t)| \geq |x_0| \exp\left(\int_{-s_0}^{-s} -\mu(-A(\tau))\, d\tau\right) = |x_0| \exp\left(\int_{t_0}^{t} \nu(A(\tau))\, d\tau\right).$$

This shows the left inequality of (3.3.2) holds. $\qquad\square$

We are ready to present the results on the stability of Eq. (H) using Lozinskii measures.

THEOREM 3.3.1. *(a)* $x = 0$ *is stable if there exists an* $M \geq 0$ *such that* $\int_{0}^{t} \mu(A(s))\, ds \leq M$ *for any* $t \geq 0$.

(b) $x = 0$ *is uniformly stable if there exists an* $M \geq 0$ *such that* $\int_{t_0}^{t} \mu(A(s))\, ds \leq M$ *for any* $0 \leq t_0 \leq t < \infty$.

(c) $x = 0$ *is asymptotically stable if* $\int_{0}^{\infty} \mu(A(s))\, ds = -\infty$.

(d) $x = 0$ *is unstable if* $\limsup_{t \to \infty} \int_{0}^{t} \nu(A(s))\, ds = \infty$.

PROOF. This follows immediately from Lemma 3.3.3 and Definition 3.1.2 for stability. $\qquad\square$

Now we demonstrate the applications of Theorem 3.3.1 by examples.

EXAMPLE 3.3.3. Consider Eq. (H) with $A = \begin{bmatrix} -1 & 2 + 4e^{-t} \\ -e^{-3t} & -2.5 \end{bmatrix}$. It is easy to see that for all large t, $\mu_1(A(t)) = -0.5 + 4e^{-t} \to -0.5$ as $t \to \infty$. Hence $\int_{0}^{\infty} \mu_1(A(t)) = -\infty$. By Theorem 3.3.1, Part (c), Eq. (H) is asymptotically stable.

EXAMPLE 3.3.4. Consider Eq. (H) with $A = \begin{bmatrix} \cos t & -r(t) \\ r(t) & \cos t \end{bmatrix}$ with $r \in C([0, \infty), \mathbb{R})$. It is easy to see that

$$\mu_2(A(t)) = \lambda_1(\frac{1}{2}(A + A^T)) = \lambda_1\left(\begin{bmatrix} \cos t & 0 \\ 0 & \cos t \end{bmatrix}\right) = \cos t.$$

Hence for any $0 \leq t_0 \leq t$

$$\int_{t_0}^{t} \mu_2(A(s)) \, ds = \sin t - \sin t_0 \leq 2.$$

By Theorem 3.3.1, Part (b), Eq. (H) is uniformly stable.

EXAMPLE 3.3.5. Consider Eq. (H) with $A(t) = \begin{bmatrix} -t & t/2 & 1 \\ t^2/2 & -t^2 + a(t) & -t^2/2 \\ 1/2 & -1/2 & -2 \end{bmatrix}$,

where $a(t) = \sin(\ln t) + \cos(\ln t) - \alpha$ with $1 < \alpha < \sqrt{2}$. Clearly, for all large t

$$\mu_\infty(A(t)) = -t^2 + a(t) + t^2 = a(t).$$

From Example 3.1.4 we see that $\int_0^\infty a(t) \, dt = -\infty$, but $\int_{t_0}^t a(s) \, ds$ is not uniformly bounded for $0 \leq t_0 \leq t < \infty$. Therefore, by Theorem 3.3.1, Eq. (H) is asymptotically stable. However, Theorem 3.3.1 cannot be used to determine uniform stability.

EXAMPLE 3.3.6. Consider Eq. (H) with $A = \begin{bmatrix} 1 & \cos t \\ -t \sin t & t \end{bmatrix}$. It is easy to see that for $t \geq 1$,

$$\nu_\infty(A(t)) = \min\{1 - |\cos t|, t(1 - |\sin t|)\} > 1 - \max\{|\cos t|, |\sin t|\}$$

and

$$\lim_{t \to \infty} \int_0^t (1 - \max\{|\cos s|, |\sin s|\}) \, ds = \infty.$$

Hence $\lim_{t \to \infty} \int_0^t \nu_\infty(A(t)) = \infty$. By Theorem 3.3.1, Part (d), Eq. (H) is unstable.

REMARK 3.3.2. From Theorem 3.3.1 we have the following observations:
  (i) By applying the i-Lozinskii measure for $i = 1$, we see that Eq. (H) is asymptotically stable [unstable] if for $j = 1, \ldots, n$, $a_{jj}(t)$ is negative [positive] and dominates all other entries in the same column.
  (ii) By applying the i-Lozinskii measure for $i = \infty$, we see that Eq. (H) is asymptotically stable [unstable] if for $i = 1, \ldots, n$, $a_{ii}(t)$ is negative [positive] and dominates all other entries in the same row.
  (iii) By applying the i-Lozinskii measure for $i = 2$, we see that
    1. Equation (H) is stable if $\lambda_1(\frac{1}{2}(A + A^T)(t)) \leq 0$ for all $t \geq 0$,
    2. Equation (H) is asymptotically stable if $\lambda_1(\frac{1}{2}(A + A^T)(t)) \leq -\delta < 0$ for all $t \geq 0$,
    3. Equation (H) is unstable if $\lambda_n(\frac{1}{2}(A + A^T)(t)) \geq \delta > 0$ for all $t \geq 0$.

In particular, when $A(t)$ is symmetric, since $\frac{1}{2}(A + A^T) = A$, these results generalize those in Theorem 3.2.2 for Eq. (H-c) with constant coefficients. However, as shown in Example 3.3.1, we cannot expect Theorem 3.3.1 to generalize Theorem 3.2.2 for the case when A is non-symmetric. This is due to the fact that $\operatorname{Re} \lambda_1(A) \leq \lambda_1(\frac{1}{2}(A + A^T))$ in general. Therefore, the conditions for stability given in Theorem 3.3.1 are sufficient but not necessary.

3.4. Perturbations of Linear Equations

In this section we discuss the stability of the zero solution of the nonlinear differential equation

(E) $$x' = f(t, x),$$

where $f \in C([0, \infty) \times \mathbb{R}^n, \mathbb{R}^n)$ such that $f(t, 0) \equiv 0$. The main idea is to utilize the results on the stability of the linear equations to investigate the stability of the zero solution of Eq. (E). For this purpose, we need to find a linear equation which is "close" to Eq. (E) when x is small. One way of doing so is to use the linearization method. Note that when $f \in C^1([0, \infty) \times \mathbb{R}^n, \mathbb{R}^n)$, the function f can be linearized in the following way:

$$f(t, x) = f(t, 0) + \frac{\partial f}{\partial x}(t, 0)\, x + r(t, x) = A(t)x + r(t, x),$$

where $A(t) := \partial f / \partial x(t, 0)$ is the Jacobian matrix of f with respect to x at $x = 0$ and $r \in C([0, \infty) \times \mathbb{R}^n, \mathbb{R}^n)$ is the error term of the linearization at $x = 0$. Thus, $r(t, 0) \equiv 0$ and $|r(t, x)| = o(|x|)$ as $x \to 0$, i.e., $|r(t, x)|/|x| \to 0$ as $x \to 0$, for any fixed $t \in [0, \infty)$. Here we call the linear equation

(H) $$x' = A(t)x$$

the linearization of Eq. (E) and call Eq. (E) a perturbation of Eq. (H). We study the stability of Eq. (E) in two cases.

Case I. Assume the Jacobian matrix $\partial f / \partial x(t, 0)$ of f at $x = 0$ is a constant matrix. In this case, equation (E) is of the form

(3.4.1) $$x' = Ax + r(t, x),$$

where $A \in \mathbb{R}^{n \times n}$ and $r \in C([0, \infty) \times \mathbb{R}^n, \mathbb{R}^n)$ satisfying $r(t, 0) \equiv 0$ for all $t \in [0, \infty)$.

THEOREM 3.4.1. *Let* λ_i, $i = 1, \ldots, n$, *be the eigenvalues of the matrix* A. *Assume* $|r(t, x)| = o(|x|)$ *as* $x \to 0$ *uniformly for* $t \in [0, \infty)$, *i.e.,* $\lim_{x \to 0} |r(t, x)|/|x| = 0$ *uniformly for* $t \in [0, \infty)$.

    (a) *If* $\operatorname{Re} \lambda_i < 0$ *for all* $i = 1, \ldots, n$, *then the zero solution of Eq. (3.4.1) is uniformly stable and asymptotically stable.*

    (b) *If there exists an* $i \in \{1, \ldots, n\}$ *such that* $\operatorname{Re} \lambda_i > 0$, *then the zero solution of Eq. (3.4.1) is unstable.*

PROOF. (a) For any $(t_0, x_0) \in [0, \infty) \times \mathbb{R}^n$, let $x(t)$ be a solution of the IVP consisting of Eq. (3.4.1) and the IC $x(t_0) = x_0$. By Corollary 2.4.1, $x(t)$ satisfies the integral equation

$$(3.4.2) \qquad x(t) = e^{A(t-t_0)}x_0 + \int_{t_0}^t e^{A(t-s)}r(s, x(s))\, ds.$$

Since $\operatorname{Re} \lambda_i < 0$ for all $i = 1, \ldots, n$, by Theorem 3.2.2, Part (b), we have that the equation $x' = Ax$ is asymptotically stable and there exist a $k > 1$ and a $\alpha > 0$ such that

$$|e^{A(t-t_0)}| \le ke^{-\alpha(t-t_0)}, \quad t \ge t_0.$$

Thus from (3.4.2) we see that for $t \ge t_0$

$$(3.4.3) \qquad |x(t)| \le ke^{-\alpha(t-t_0)}|x_0| + \int_{t_0}^t ke^{-\alpha(t-s)}|r(s, x(s))|\, ds.$$

We also observe that the condition $\lim_{x \to 0} |r(t, x)|/|x| = 0$ uniformly for $t \in [0, \infty)$ shows that for sufficiently small $\epsilon > 0$, $|x| < \epsilon$ implies that $|r(t, x)| \le (\alpha/2k)|x|$ for all $t \in [0, \infty)$. Then from (3.4.3) we have that for $t \ge t_0$, as long as $|x(s)| < \epsilon$ for $t_0 \le s \le t$,

$$|x(t)| \le ke^{-\alpha(t-t_0)}|x_0| + \int_{t_0}^t \frac{\alpha}{2}e^{-\alpha(t-s)}|x(s)|\, ds.$$

By multiplying both sides by $e^{\alpha(t-t_0)}$ we get that

$$e^{\alpha(t-t_0)}|x(t)| \le k|x_0| + \int_{t_0}^t \frac{\alpha}{2}e^{\alpha(s-t_0)}|x(s)|\, ds.$$

Then applying the Gronwall inequality to $u(t) := e^{\alpha(t-t_0)}|x(t)|$ we obtain that for $t \ge t_0$, as long as $|x(s)| < \epsilon$ for $t_0 \le s \le t$,

$$e^{\alpha(t-t_0)}|x(t)| \le k|x_0|e^{\frac{\alpha}{2}(t-t_0)},$$

and hence

$$|x(t)| \le k|x_0|e^{-\frac{\alpha}{2}(t-t_0)}.$$

This implies that

(i) $|x(t)| \le k|x_0| < \epsilon$ for all $t \ge t_0$ if $|x_0| < \delta := \epsilon/k$ and hence $x = 0$ is uniformly stable.

(ii) $\lim_{t \to \infty} x(t) = 0$ and hence $x = 0$ is asymptotically stable.

(b) The proof involves a decomposition of the Jordan canonical form of the matrix A and much more technical arguments. We omit it here. The interested reader is referred to [18, p. 95] or [11, p. 317] for the detail.

$\square$

REMARK 3.4.1. For the case when $\operatorname{Re} \lambda_i \le 0$, $i = 1, \ldots, n$, the linearization (H-c) of Eq. (3.4.1) is uniformly stable if $\operatorname{Re} \lambda_i = 0$ occurs only when λ_i's are in the diagonal Jordan block of the matrix A, and is unstable if $\operatorname{Re} \lambda_i = 0$ occurs when at least one λ_i is not in the diagonal Jordan block of the matrix A. However, when $\operatorname{Re}\lambda_i \le 0$, $i = 1,\ldots,n$, the zero solution of Eq. (3.4.1) may be stable, asymptotically stable, or unstable, no matter whether $\operatorname{Re}\lambda_i = 0$ occurs for λ_i's in the diagonal Jordan block or not. This means that the linearization method fails to work for this case.

EXAMPLE 3.4.1. Consider the system of equations

$$
(3.4.4) \qquad \begin{cases} x' = -kxe^{-y} - \sin y \\ y' = -ky\cos x + \tan x, \end{cases}
$$

where $k \in \mathbb{R}$. Then the Jacobian matrix of the right-hand side functions is

$$
A = \left.\frac{\partial(f_1, f_2)}{\partial(x, y)}\right|_{(0,0)} = \left.\begin{bmatrix} -ke^{-y} & kxe^{-y} - \cos y \\ ky\sin x + \sec^2 x & -k\cos x \end{bmatrix}\right|_{(0,0)} = \begin{bmatrix} -k & -1 \\ 1 & -k \end{bmatrix}.
$$

Thus system (3.4.4) can be written as

$$
(3.4.5) \qquad \begin{cases} x' = -kx - y + r_1(x, y) \\ y' = x - ky + r_2(x, y), \end{cases}
$$

where

$$
|r_1(x, y)| + |r_2(x, y)| = o(|x| + |y|) \quad \text{as } (x, y) \to (0, 0).
$$

By a simple computation we see that the eigenvalues of the matrix A are $\lambda_{1,2} = -k \pm \mathbf{i}$. Since $\operatorname{Re}\lambda_{1,2} = -k$, by Theorem 3.4.1, the zero solution of Eq. (3.4.5), and hence the zero solution of Eq. (3.4.4), is uniformly stable and asymptotically stable if $k > 0$ and is unstable if $k < 0$. However, Theorem 3.4.1 fails to work for the case when $k = 0$.

Case II. We consider the equation

$$
(3.4.6) \qquad x' = A(t)x + r(t, x),
$$

where $A \in C([0, \infty), \mathbb{R}^{n \times n})$ and $r \in C([0, \infty) \times \mathbb{R}^n, \mathbb{R}^n)$ satisfying $r(t, 0) \equiv 0$ for all $t \in [0, \infty)$. We derive conditions for uniform stability and asymptotic stability of the zero solution of Eq. (3.4.6) based on those of the following linearized equation

$$
\text{(H)} \qquad x' = A(t)x.
$$

THEOREM 3.4.2. *Assume that there exists a function $p \in C([0, \infty), [0, \infty))$ such that $\int_0^\infty p(t)\,dt < \infty$, and $|r(t, x)| \le p(t)|x|$ for sufficiently small $|x|$ and all $t \in [0, \infty)$.*

    (a) *If Eq. (H) is uniformly stable, then the zero solution of Eq. (3.4.6) is uniformly stable.*

    (b) *If Eq. (H) is uniformly stable and asymptotically stable, then the zero solution of Eq. (3.4.6) is uniformly stable and asymptotically stable.*

PROOF. Let $X(t)$ be a fundamental matrix solution of Eq. (H), and for any $(t_0, x_0) \in [0, \infty) \times \mathbb{R}^n$, let $x(t)$ be a solution of the IVP consisting of

Eq. (3.4.6) and the IC $x(t_0) = x_0$. By Theorem 2.3.1, $x(t)$ satisfies the integral equation

$$(3.4.7) \qquad x(t) = X(t)X^{-1}(t_0)x_0 + \int_{t_0}^{t} X(t)X^{-1}(s)r(s,x(s))\,ds.$$

(a) Assume Eq. (H) is uniformly stable. By Theorem 3.2.1, Part (b), there exists a $k > 1$ such that

$$|X(t)X^{-1}(s)| \le k \quad \text{for all } 0 \le s \le t < \infty.$$

From the condition on $r(t,x)$ we see that for sufficiently small $\epsilon > 0$, $|r(t,x)| \le p(t)|x|$ when $|x| < \epsilon$ and $t \in [0,\infty)$. Then from (3.4.7) we have that for $t \ge t_0$, as long as $|x(s)| < \epsilon$ for $t_0 \le s \le t$,

$$|x(t)| \le k|x_0| + \int_{t_0}^{t} kp(s)|x(s)|\,ds;$$

and hence by the Gronwall inequality,

$$|x(t)| \le k|x_0|e^{\int_{t_0}^{t} kp(s)\,ds} \le k|x_0|e^{\int_{0}^{\infty} kp(s)\,ds}.$$

Let $\delta = \epsilon/(ke^{\int_{0}^{\infty} kp(s)\,ds})$. Then for any $t_0 \ge 0$, $|x_0| \le \delta$ implies that $|x(t)| \le \epsilon$ for all $t \ge t_0$. Therefore, the zero solution of Eq. (3.4.6) is uniformly stable.

(b) Assume Eq. (H) is uniformly stable and asymptotically stable. From Part (a), the zero solution of Eq. (3.4.6) is uniformly stable. Thus for sufficiently small $\epsilon \in (0,1)$, there exists a $\delta > 0$ such that $|x_0| < \delta$ implies that $|x(t)| < \epsilon < 1$ for all $t \ge t_0$, and hence

$$|r(t,x(t))| \le p(t)|x(t)| < p(t) \quad \text{for all } t \ge t_0.$$

Let $t_1 \ge t_0$ such that $\int_{t_1}^{\infty} p(t) < \epsilon/(2k)$. Then from (3.4.7) we have that for any $t \ge t_1$

$$\begin{aligned} x(t) &= X(t)X^{-1}(t_0)x_0 + \int_{t_0}^{t_1} X(t)X^{-1}(s)r(s,x(s))\,ds \\ &\quad + \int_{t_1}^{t} X(t)X^{-1}(s)r(s,x(s))\,ds. \end{aligned}$$

Hence

$$\begin{aligned} |x(t)| &\le |X(t)|\left(|X^{-1}(t_0)x_0| + \int_{t_0}^{t_1} |X^{-1}(s)|p(s)\,ds\right) + \int_{t_1}^{t} kp(s)|x(s)|\,ds \\ &< |X(t)|\left(|X^{-1}(t_0)x_0| + \int_{t_0}^{t_1} |X^{-1}(s)|p(s)\,ds\right) + \epsilon/2. \end{aligned}$$

Since Eq. (H) is asymptotically stable, by Theorem 3.2.1, Part (c), $X(t) \to \mathbf{0}$ as $t \to \infty$. Thus there exists a $t_2 \ge t_1$ such that for $t \ge t_2$

$$|X(t)|\left(|X^{-1}(t_0)x_0| + \int_{t_0}^{t_1} |X^{-1}(s)|p(s)\,ds\right) < \epsilon/2.$$

Combining the two inequalities above, we have that $|x(t)| < \epsilon$ for $t \geq t_2$. This shows that $\lim_{t\to\infty} x(t) = 0$ and hence the zero solution of Eq. (3.4.6) is asymptotically stable.

$\square$

EXAMPLE 3.4.2. We discuss the stability of the zero solution of the second-order differential equation

$$(3.4.8) \qquad x'' + \frac{3e^t}{e^t + 1}x' + 2x - \frac{1}{(t+1)^{3/2}}x^2 = 0.$$

Let $x_1 = x$ and $x_2 = x'$. Then Eq. (3.4.8) becomes the system of first-order equations

$$\begin{cases} x_1' = x_2 \\ x_2' = -2x_1 - \dfrac{3e^t}{e^t + 1}x_2 + \dfrac{1}{(t+1)^{3/2}}x_1^2 \end{cases}$$

which can be written as

$$(3.4.9) \qquad \begin{cases} x_1' = x_2 \\ x_2' = -2x_1 - 3x_2 + \dfrac{3}{e^t + 1}x_2 + \dfrac{1}{(t+1)^{3/2}}x_1^2. \end{cases}$$

We view system (3.4.9) as a perturbation of the linear system

$$(3.4.10) \qquad \begin{cases} x_1' = x_2 \\ x_2' = -2x_1 - 3x_2 \end{cases}$$

with the perturbation terms

$$r_1(x_1, x_2) = 0 \quad \text{and} \quad r_2(x_1, x_2) = \frac{3}{e^t + 1}x_2 + \frac{1}{(t+1)^{3/2}}x_1^2.$$

Then when $|x_1| \leq 1$,

$$\begin{aligned} |r_1(x_1,x_2)| + |r_2(x_1,x_2)| &\leq \frac{1}{(t+1)^{3/2}}|x_1| + \frac{3}{e^t + 1}|x_2| \\ &\leq \left(\frac{1}{(t+1)^{3/2}} + \frac{3}{e^t + 1}\right)(|x_1| + |x_2|). \end{aligned}$$

It is easy to see that

$$\int_0^\infty \left(\frac{1}{(t+1)^{3/2}} + \frac{3}{e^t + 1}\right) dt < \infty.$$

Thus the conditions of Theorem 3.4.2 are satisfied with

$$p(t) = \frac{1}{(t+1)^{3/2}} + \frac{3}{e^t + 1}.$$

Note that the coefficient matrix of system (3.4.10) is $A = \begin{bmatrix} 0 & 1 \\ -2 & -3 \end{bmatrix}$ whose eigenvalues are $\lambda_1 = -1$ and $\lambda_2 = -2$. Thus, system (3.4.10) is uniformly stable and asymptotically stable. By Theorem 3.4.2, the zero solution of system (3.4.9), and hence the zero solution of Eq. (3.4.8), is uniformly stable and asymptotically stable.

REMARK 3.4.2. (i) In Theorem 3.4.2, Part (a), the stability of the zero solution of Eq. (3.4.6) is not guaranteed by the stability of Eq. (H) alone; instead, it requires that Eq. (H) be uniformly stable.

(ii) In Theorem 3.4.2, Part (b), the asymptotic stability of the zero solution of Eq. (3.4.6) is not guaranteed by the asymptotic stability of Eq. (H) alone; instead, it requires that Eq. (H) be uniformly stable and asymptotically stable.

(iii) The instability of Eq. (H) does not guarantee the instability of the zero solution of Eq. (3.4.6) in general.

(iv) We observe that in the conditions of Theorem 3.4.2, we do not require that $|r(t,x)| = o(|x|)$ as $x \to 0$. In fact, the perturbation term $r(t,x)$ may be linear, as shown below.

We study the stability of the linear equation

$$(3.4.11) \qquad x' = [A(t) + B(t)]x$$

using the results for Eq. (H).

THEOREM 3.4.3. *Assume that $B \in C([0, \infty), \mathbb{R}^{n \times n})$ such that $\int_0^\infty |B(t)|\, dt < \infty$.*

(a) If Eq. (H) is uniformly stable, then Eq. (3.4.11) is uniformly stable.

(b) If Eq. (H) is uniformly stable and asymptotically stable, then Eq. (3.4.11) is uniformly stable and asymptotically stable.

PROOF. Let $r(t,x) = B(t)x$. Then $|r(t,x)| \leq |B(t)||x|$ for all $x \in \mathbb{R}^n$ and $t \in [0, \infty)$. Thus, the conditions of Theorem 3.4.2 are satisfied with $p(t) = |B(t)|$. Note that all solutions of Eq. (3.4.11) have the same stability since the equation is linear. Then the conclusions follow from Theorem 3.4.2. □

A variation of Theorem 3.4.3 is for linear perturbations of autonomous linear equation (H-c). Consider the equation

$$(3.4.12) \qquad x' = (A + B(t))x.$$

THEOREM 3.4.4. *assume that $A \in \mathbb{R}^{n \times n}$ and $B \in C([0, \infty) \times \mathbb{R}^n, \mathbb{R}^n)$ such that $\lim_{t \to \infty} |B(t)| = 0$. If $\operatorname{Re} \lambda_i(A) < 0$, $i = 1, \ldots, n$, then Eq. (3.4.12) is uniformly stable and asymptotically stable.*

The proof of Theorem 3.4.4 is similar to those of Theorems 3.4.1 and 3.4.2 using the variation of parameters formula and the Gronwall inequality. We omit it here. The interested reader is referred to [**18**, p. 88] for the details.

Note that the condition for the matrix B in Theorem 3.4.3 cannot replace nor be replaced by that in Theorem 3.4.4.

3.5. Lyapunov Function Method for Autonomous Equations

The linearization method for stability discussed in the previous section does not work for all nonlinear problems. To see this, let's look at the problem of the mathematical pendulum. As shown in Fig. 3.4, a point with mass m is attached to a ceiling with a string of length l. Denote by ϕ the angle

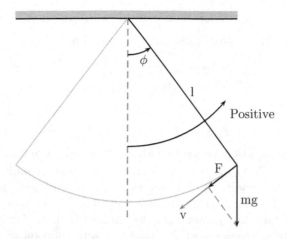

FIGURE 3.4. Mathematical pendulum

between the string and the downward vertical line, v the line velocity of the mass point, and F the component of the gravity force mg in the tangent direction of the motion circle. Considering the counter-clockwise direction as the positive direction, we have

$$F = -mg \sin \phi.$$

Assume there is no damping in this motion. Note that $v = l\,d\phi/dt$. By Newton's second law for motion we also have

$$F = m\frac{dv}{dt} = ml\frac{d^2\phi}{dt^2}.$$

Combining the two equalities above, we obtain that

$$\frac{d^2\phi}{dt^2} + \frac{g}{l}\sin\phi = 0.$$

Let $x = \phi$, $x'' = d^2\phi/dt^2$, and $k = g/l$. Then the equation becomes the equation for the free mathematical pendulum

$$(3.5.1) \qquad\qquad x'' + k\sin x = 0.$$

We observe that $x = 0$ is a solution of Eq. (3.5.1) representing the standstill position of point mass.

Now, we discuss the stability of the zero solution of Eq. (3.5.1) by linearization. To do so, let $x_1 = x$ and $x_2 = x'$. Then Eq. (3.5.1) is changed to the system of equations

$$(3.5.2) \qquad\qquad \begin{cases} x_1' = x_2 \\ x_2' = -k\sin x_1. \end{cases}$$

It is easy to see that the Jacobi matrix of the right-hand side functions is $A = \begin{bmatrix} 0 & 1 \\ -k & 0 \end{bmatrix}$. Thus, system (3.5.2) can be written as

$$(3.5.3) \qquad \begin{cases} x_1' = x_2 + r_1(x_1, x_2) \\ x_2' = -kx_1 + r_2(x_1, x_2), \end{cases}$$

where

$$r_1(x_1, x_2) = 0 \quad \text{and} \quad r_2(x_1, x_2) = k(x_1 - \sin x_1) = o(|x_1|) \quad \text{as } x_1 \to 0.$$

By a simple computation we see that the eigenvalues of matrix A are $\lambda_{1,2} = \pm\sqrt{k}\mathbf{i}$ and hence $\operatorname{Re} \lambda_{1,2} = 0$. In this case, neither the conditions of Theorem 3.4.1 nor the conditions of Theorem 3.4.2 are satisfied. Therefore, the stability of the zero solution $(0,0)$ of system (3.5.3), and hence the zero solution $x = 0$ of Eq. (3.5.1), cannot be determined by linearization.

However, from physics we know that the zero solution of Eq. (3.5.1) is uniformly stable. Can we prove it in a mathematical way? A different idea has been developed from an exploration of the mechanical energy function.

Let $x(t)$ be a solution of Eq. (3.5.1). Multiplying both sides of Eq. (3.5.1) by $x'(t)$ we have that for $t \in [0, \infty)$,

$$x'(t)x''(t) + k(\sin x(t))x'(t) = 0.$$

As a result of integration with respect to t, we obtain that

$$(3.5.4) \qquad \frac{1}{2}x'^2(t) + k(1 - \cos x(t)) = c$$

for some constant c. Note that the left-hand side of Eq. (3.5.4) represents the mechanical energy of the point mass. Thus, Eq. (3.5.4) reveals the fact that the mechanical energy of the point mass remains constant during the whole process of the vibration. Denote

$$(3.5.5) \qquad V(x, x') = \frac{1}{2}x'^2 + k(1 - \cos x).$$

Then $V(x, x')$ satisfies

$$V(0,0) = 0 \quad \text{and} \quad V(x, x') > 0 \text{ for } (x, x') \neq (0,0) \text{ but near } (0,0).$$

Thus, in a neighborhood of $(0,0)$, $V(x, x')$ as a function of two variables is of the shape as in Fig. 3.5.

For any nontrivial solution $x(t)$ of Eq. (3.5.1), we denote

$$\dot{V}(x, x') = \frac{dV}{dt}(x(t), x'(t)).$$

Then along the solution,

$$\begin{aligned} \dot{V}(x, x') &= V_x(x(t), x'(t))x'(t) + V_{x'}(x(t), x'(t))x''(t) \\ &= k(\sin x(t))x'(t) + x'(t)(-k\sin x(t)) = 0. \end{aligned}$$

Therefore, $V(x(t), x'(t)) = V(x(0), x'(0))$ for all $t \in [0, \infty)$. This shows that the set $\{(x(t), x'(t)) : t \in [0, \infty)\}$ is on the level curve $V(x, x') = c$ for some $c > 0$. Clearly, all such level curves of the function $V(x, x')$ are closed curves.

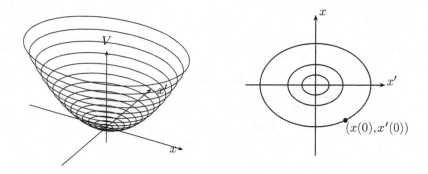

FIGURE 3.5. Function V defined by (3.5.4)

From this it is not difficult to show that the definition for stability is satisfied and hence the zero solution of Eq. (3.5.1) is stable. Here, the function V plays a critical role in the study of the stability. We call V a Lyapunov function and call the method using Lyapunov functions to study the stability problems the Lyapunov function method or the Lyapunov direct method.

Now, we present the Lyapunov function method for the general nonlinear autonomous equation

(A)
$$x' = f(x),$$

where $f \in C(\mathbb{R}^n, \mathbb{R}^n)$ such that $f(0) = 0$. Here we do not assume uniqueness of solutions of IVPs associated with Eq. (A). To discuss the stability of the zero solution of Eq. (A), we first define several types of Lyapunov functions.

DEFINITION 3.5.1. Let $D = \{x \in \mathbb{R}^n : |x| \leq l\}$ for some $l > 0$ and $V \in C(D, \mathbb{R})$ such that $V(0) = 0$. Then V is said to be

(a) positive semi-definite if $V(x) \geq 0$ for $x \in D$, and negative semi-definite if $V(x) \leq 0$ for $x \in D$;

(b) positive definite if $V(x) > 0$ for $x \in D$ such that $x \neq 0$, and negative definite if $V(x) < 0$ for $x \in D$ such that $x \neq 0$;

(c) indefinite if $V(x)$ changes sign in any neighborhood of $x = 0$.

EXAMPLE 3.5.1. On $D = \mathbb{R}^2$,
$V(x, y) = (x + y)^2$ is positive semi-definite but not positive definite;
$V(x, y) = 3x^2 + 2y^6$ and $V(x, y) = (x - y)^2 + y^4$ are positive definite;
$V(x, y) = x^2 - 5y^4$ and $V(x, y) = xy$ are indefinite.

We also see that the quadratic function $V(x, y) = ax^2 + bxy + cy^2$ is

(i) positive definite if $4ac - b^2 > 0$ and $a > 0$, and negative definite if $4ac - b^2 > 0$ and $a < 0$;

(ii) positive semi-definite if $4ac - b^2 = 0$ and $a > 0$, and negative semi-definite if $4ac - b^2 = 0$ and $a < 0$;

(iii) indefinite if $4ac - b^2 < 0$.

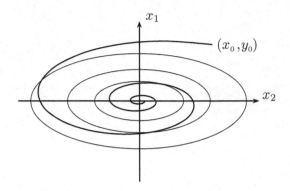

FIGURE 3.6. Solution behavior for case (b) of Theorem 3.5.1

Although the function $V(x, y) = 1 - \cos x + y^2$ is not positive definite on $\mathbb{R}^2$, it is positive definite on $D = \{(x, y) : |x| + |y| \leq \pi\}$.

DEFINITION 3.5.2. Let $D = \{x \in \mathbb{R}^n : |x| \leq l\}$ for some $l > 0$ and $V \in C^1(D, \mathbb{R})$. Then the derivative of the function $V(x)$ along Eq. (A) is defined to be

$$\dot{V}(x) = \left[\sum_{i=1}^{n} \frac{\partial V}{\partial x_i} f_i \right] (x) = [\nabla V \circ f](x),$$

where ∇V stands for the gradient of V.

By definition, $\dot{V}(x)$ reflects the "directional derivative" of $V(x)$ in the direction of f, except that f is not a unit vector in the direction in general. It is easy to see that along any solution $x(t)$ of Eq. (A), $\dot{V}(x(t)) = [V(x(t))]'$.

EXAMPLE 3.5.2. Let $V(x_1, x_2) = k(1 - \cos x_1) + x_2^2/2$. Then the derivative of $V(x)$ along Eq. (3.5.2) is

$$\dot{V}(x) = k(\sin x_1)x_2 + x_2(-k \sin x_1) = 0.$$

Now we are ready to state the results on stability using Lyapunov functions.

THEOREM 3.5.1. Let $D = \{x \in \mathbb{R}^n : |x| \leq l\}$ for some $l > 0$ and $V \in C^1(D, \mathbb{R})$.

    (a) If $V(x)$ is positive definite and $\dot{V}(x)$ is negative semi-definite, then the zero solution of (A) is uniformly stable.

    (b) If $V(x)$ is positive definite and $\dot{V}(x)$ is negative definite, then the zero solution of (A) is uniformly stable and asymptotically stable.

    (c) If $V(0) = 0$ and in any neighborhood of $x = 0$ in D, there exists an x_0 such that $V(x_0) > 0$ and $\dot{V}(x)$ is positive definite, then the zero solution of (A) is unstable.

Note that in Part (c), the function V may be chosen as positive definite or indefinite. The result in Part (c) was further developed by N. G. Chetaev [7] and [8] into the so-called Chetaev instability theorem. We omit the details.

PROOF. (a) For any $\epsilon > 0$, let $\eta := \eta(\epsilon) = \min_{\epsilon \leq |x| \leq l} V(x)$. Then $\eta > 0$, and $V(x) < \eta$ implies that $|x| < \epsilon$. Since $V(0) = 0$ and V is continuous on D, there exists a $\delta := \delta(\epsilon) > 0$ such that $|x| < \delta$ implies that $V(x) < \eta$.

For any $t_0 \geq 0$, let $x(t)$ be a solution of Eq. (A) with $|x(t_0)| < \delta$. Then $V(x(t_0)) < \eta$. Since $\dot{V}(x)$ is negative semi-definite, we have $[V(x(t))]' \leq 0$ for $t \geq t_0$. This shows that $V(x(t)) \leq V(x(t_0)) < \eta$ for $t \geq t_0$. Thus by the above, $|x(t)| < \epsilon$ for $t \geq t_0$. This shows that the zero solution of Eq. (A) is uniformly stable.

(b) It is seen that under the assumptions, the zero solution of Eq. (A) is uniformly stable by Part (a).

Let δ be given in Part (a), and for any $t_0 \geq 0$, let $x(t)$ be a solution of Eq. (A) with $|x(t_0)| < \delta$.

We first show that $\lim_{t\to\infty} V(x(t)) = 0$. In fact, since $\dot{V}(x)$ is negative semi-definite, we have $[V(x(t))]' \leq 0$ for $t \geq t_0$ and hence $V(x(t))$ is nonnegative and nonincreasing. Thus, $\lim_{t\to\infty} V(x(t)) = c$ for some $c \geq 0$. We claim that $c = 0$. Otherwise, $V(x(t)) \geq c > 0$ for $t \geq t_0$. This implies that there exists an $r > 0$ such that $|x(t)| \geq r$ for $t \geq t_0$. Let $m = \max_{r \leq |x| \leq l} \dot{V}(x)$. Then $m < 0$. Hence

$$V(x(t)) - V(x(t_0)) = \int_{t_0}^{t} \frac{d}{ds} V(x(s))\, ds \leq m(t - t_0) \to -\infty.$$

It follows that $V(x(t)) \to -\infty$ as $t \to \infty$, contradicting the assumption that $V(x)$ is positive definite. We then show that $\lim_{t\to\infty} x(t) = 0$. If not, since the zero solution of Eq. (A) is stable, we see that $x(t)$ is bounded on $[t_0, \infty)$. Thus, there exists a sequence $t_k \to \infty$ such that $\lim_{k\to\infty} x(t_k) = x^* \neq 0$. By the continuity of $V(x)$, we have that $\lim_{k\to\infty} V(x(t_k)) = V(x^*) \neq 0$. This contradicts Part (i) and hence confirms that the zero solution of Eq. (A) is asymptotically stable.

(c) From the conditions we see that for any $\delta > 0$, there exists an $x_0 \in D$ such that $|x_0| < \delta$ and $V(x_0) > 0$. For any $t_0 \geq 0$, let $x(t)$ be a solution of Eq. (A) with $x(t_0) = x_0$. We show that $x(t)$ does not always stay in D. Otherwise, $|x(t)| \leq l$ for all $t \geq t_0$. Since $\dot{V}(x)$ is positive definite, $[V(x(t))]' \geq 0$ for $t \geq t_0$ and hence $V(x(t))$ is nondecreasing on $[t_0, \infty)$. Thus, $V(x(t)) \geq V(x_0) > 0$ for $t \geq t_0$. This implies that there exists an $r > 0$ such that $|x(t)| \geq r$ for $t \geq t_0$. Let $m = \min_{r \leq |x| \leq l} \dot{V}(x)$. Then $m > 0$. Hence

$$V(x(t)) - V(x(t_0)) = \int_{t_0}^{t} \frac{d}{ds} V(x(s))\, ds \geq m(t - t_0) \to \infty.$$

This contradicts the assumption that $x(t) \in D$ for all $t \geq t_0$ and hence confirms that the zero solution of Eq. (A) is unstable.

$\square$

To apply Theorem 3.5.1 to study stability, a key problem is to choose an appropriate Lyapunov function. However, this may not always be easy and in fact is often tricky. Practice can help gain insight on how to construct Lyapunov functions. We explain this by several examples of systems of two equations. We comment that in all the examples below, the linearization method fails to apply.

EXAMPLE 3.5.3. Consider the system

$$\begin{cases} x' = -x + xy^2 \\ y' = -2x^2y - y^3. \end{cases}$$

Let $V = ax^2 + by^2$ with $a, b > 0$ to be determined. Then

$$\begin{aligned} \dot{V} &= 2axx' + 2byy' = 2ax(-x + xy^2) + 2by(-2x^2y - y^3) \\ &= -2ax^2 + 2ax^2y^2 - 4bx^2y^2 - 2by^4. \end{aligned}$$

If we let $a = 2$ and $b = 1$, then $V = 2x^2 + y^2$ is positive definite and $\dot{V} = -4x^2 - 2y^2$ is negative definite. Hence by Theorem 3.5.1, Part (b), $(0,0)$ is uniformly stable and asymptotically stable.

EXAMPLE 3.5.4. Consider the system

$$\begin{cases} x' = 2xy + x^3 \\ y' = -x^2 + y^5. \end{cases}$$

Let $V = ax^2 + by^2$ with $a, b > 0$ to be determined. Then

$$\begin{aligned} \dot{V} &= 2axx' + 2byy' = 2ax(2xy + x^3) + 2b(-x^2 + y^5) \\ &= 4ax^2y + 2ax^4 - 2bx^2y + 2by^6. \end{aligned}$$

If we let $a = 1$ and $b = 2$, then $V = x^2 + 2y^2$ is positive definite and $\dot{V} = 2x^4 + 4y^6$ is positive definite. Hence by Theorem 3.5.1, Part (c), $(0,0)$ is unstable.

EXAMPLE 3.5.5. Consider the system

$$\begin{cases} x' = -x^3 - y \\ y' = x^5 - 2y^3. \end{cases}$$

We first try $V = ax^2 + by^2$ with $a, b > 0$ to be determined. Then

$$\begin{aligned} \dot{V} &= 2axx' + 2byy' = 2ax(-x^3 - y) + 2by(x^5 - 2y^3) \\ &= -6ax^4 - 2axy + 2bx^5y - 4by^4. \end{aligned}$$

We see that no matter how we choose the numbers a and b, the "bad terms" $-2axy$ and $2bx^5y$ cannot be canceled out. Therefore, the first choice of V is not good.

To obtain a better Lyapunov function, we may modify the first one by adjusting the powers of x and y so that the "bad terms" will become like-terms and hence can be canceled with an appropriate choice of a and b. Considering the right-hand side functions of the system, it is not difficult to see that $v = ax^6 + by^2$ works for this system. In this case,

$$
\begin{aligned}
\dot{V} &= 6ax^5 x' + 2byy' = 2ax^5(-x^3 - y) + 2by(x^5 - 2y^3) \\
&= -18ax^8 - 6ax^5 y + 2bx^5 y - 4by^4.
\end{aligned}
$$

If we let $a = 1$ and $b = 3$, then $V = x^6 + 3y^2$ is positive definite and $\dot{V} = -18x^8 - 12y^4$ is negative definite. Hence by Theorem 3.5.1, Part (b), $(0,0)$ is uniformly stable and asymptotically stable.

REMARK 3.5.1. For second-order differential equations, the mechanical energy function is often chosen to be a Lyapunov function. To explain this, we consider the equation

(3.5.6) $$u'' + h(u, u') + g(u) = 0,$$

where $h \in C(\mathbb{R}^2, \mathbb{R})$ such that $h(u, u')u' \geq 0$, and $g \in (\mathbb{R}, \mathbb{R})$ such that $g(u)u > 0$ for u in a neighborhood of 0 with $u \neq 0$. Let $x = u$, $y = u'$. Then Eq. (3.5.6) becomes the first-order system of equations

(3.5.7) $$\begin{cases} x' = y \\ y' = -g(x) - h(x, y). \end{cases}$$

We observe that under the assumptions, the mechanical energy function $y^2/2 + \int_0^x g(s)\, ds$ is positive definite on a neighborhood D of $(0,0)$, so we use it as a Lyapunov function. Let

$$V = \frac{y^2}{2} + \int_0^x g(s)\, ds.$$

Then

$$\dot{V} = y(-g(x) - h(x, y)) + g(x)y = -h(x, y)y$$

is negative semi-definite on D. Therefore, the zero solution $(0,0)$ of system (3.5.7) is uniformly stable, and hence the zero solution $u = 0$ of Eq. (3.5.6) is uniformly stable.

EXAMPLE 3.5.6. Consider the equation

(3.5.8) $$u'' + 2u + 3u^2 = 0.$$

Let $x = u$, $y = u'$. Then Eq. (3.5.8) becomes the first-order system of equations

(3.5.9) $$\begin{cases} x' = y \\ y' = -2x - 3x^2. \end{cases}$$

Let

$$V = \frac{y^2}{2} + \int_0^x (2s + 3s^2)\, ds.$$

Then V is positive definite on $D = \{(x, y) : |x| + |y| \leq 1/2\}$, and
$$\dot{V} = y(-2x - 3x^2) + (2x + 3x^2)y = 0$$
which is negative semi-definite on D. By Theorem 3.5.1, Part (a), the zero solution $(0, 0)$ of system (3.5.9) is uniformly stable, and hence the zero solution $u = 0$ of Eq. (3.5.8) is uniformly stable.

EXAMPLE 3.5.7. Consider the equation
$$(3.5.10) \qquad\qquad u'' + 3u' + \sin u = 0.$$
Let $x = u$, $y = u'$. Then Eq. (3.5.10) becomes the first-order system of equations
$$(3.5.11) \qquad\qquad \begin{cases} x' = y \\ y' = -\sin x - 3y. \end{cases}$$
Let
$$(3.5.12) \qquad\qquad V(x, y) = \frac{y^2}{2} + \int_0^x \sin s \, ds.$$
Then V is positive definite on $D = \{(x, y) : |x| + |y| \leq \pi\}$, and
$$\dot{V}(x, y) = y(-\sin x - 3y) + (\sin x)y = -3y^2$$
which is negative semi-definite on D. By Theorem 3.5.1, Part (a), the zero solution $(x, y) = (0, 0)$ of system (3.5.11) is uniformly stable, and hence the zero solution $u = 0$ of Eq. (3.5.10) is uniformly stable.

From physics, we see that Eq. (3.5.10) reflects the mechanical vibration with a damping force negatively proportional to the velocity. Certainly, we know that the zero solution $u = 0$ is asymptotically stable. However, this cannot be derived from Theorem 3.5.1. Thus we hope to establish a new criterion which can be used to show the asymptotic stability of $u = 0$ in Example 3.5.7 with the same Lyapunov function. For this purpose, we need the following concepts for autonomous equations.

DEFINITION 3.5.3. Let $x = x(t)$ be a solution of Eq. (A) defined on its maximal interval of existence (α, β). Then the set $\{(t, x(t)) : t \in (\alpha, \beta)\}$ is called a solution curve, and the set $\{x(t) : t \in (\alpha, \beta)\}$ is called an orbit or a trajectory.

Clearly, a solution curve of Eq. (A) is a curve in $\mathbb{R}^{n+1}$ and an orbit is a curve in $\mathbb{R}^n$. Moreover, an orbit is the projection of a corresponding solution curve into $\mathbb{R}^n$. For example, $x = \sin t$, $y = \cos t$ is a solution of the system
$$\begin{cases} x' = y \\ y' = -x \end{cases}$$
on $(-\infty, \infty)$ whose solution curve is the helix defined in $\mathbb{R}^3$, as shown in Fig. 3.7, and whose orbit is the unit circle $x^2 + y^2 = 1$ in $\mathbb{R}^2$.

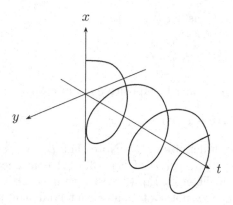

FIGURE 3.7. Helix

The next theorem is a further development of Theorem 3.5.1, Part (a) due to Barbashin, Krasovskii, and LaSalle in [**2, 33, 35**].

THEOREM 3.5.2. *Let* $D = \{x \in \mathbb{R}^n : |x| \leq l\}$ *for some* $l > 0$ *and* $V \in C^1(D, \mathbb{R})$. *Assume that* $V(x)$ *is positive definite and* $\dot{V}(x)$ *is negative semi-definite. Moreover, if the set* $D_0 := \{x \in D : \dot{V}(x) = 0\}$ *does not contain any nontrivial orbit of Eq.* (A), *then the zero solution of Eq.* (A) *is uniformly stable and asymptotically stable.*

An intuitive interpretation of Theorem 3.5.2 is as follows: For any non-trivial solution $x(t)$ with $x(0)$ sufficiently close to 0, we have $[V(x(t))]' \leq 0$ for $t \geq t_0$. Since the orbit corresponding to $x(t)$ is not totally contained in D_0, we expect that $[V(x(t))]' < 0$ for t in a significant part of the interval $[t_0, \infty)$. This will lead to the conclusion that $\lim_{t \to \infty} V(x(t)) = 0$ which means that the zero solution of Eq. (A) is asymptotically stable. However, the mathematical proof of Theorem 3.5.2 uses knowledge of dynamical systems. So we will give the proof at the end of Sect. 4.2.

Now, we rework Example 3.5.7 with the help of Theorem 3.5.2.

EXAMPLE 3.5.7. *Let* $V(x, y)$ be defined by (3.5.12). Then $V(x, y)$ is positive definite and $\dot{V}(x, y)$ is negative semi-definite as shown before. Note that

$$\dot{V}(x, y) = 0 \Longleftrightarrow y = 0.$$

Thus $D_0 = \{(x, 0) : x \in \mathbb{R}\}$. For any solution $(x(t), y(t))$ of Eq. (3.5.11) contained in D_0, we have $y(t) \equiv 0$. Then by Eq. (3.5.11), $x(t) \equiv 0$. This implies that D_0 does not contain any nontrivial orbit of system (3.5.11). Therefore by Theorem 3.5.2, the zero solution $(0, 0)$ of system (3.5.11) is uniformly stable and asymptotically stable, and hence the zero solution $u = 0$ of Eq. (3.5.10) is uniformly stable and asymptotically stable.

In the following two examples we give complete discussions on the stability of the zero solution.

EXAMPLE 3.5.8. Consider the system

(3.5.13)
$$\begin{cases} x' = y - x^3 y^2 \\ y' = -x - x^2 y^5. \end{cases}$$

Let $V(x, y) = x^2 + y^2$. Then V is positive definite on $D = \mathbb{R}^2$, and

$$\dot{V}(x, y) = 2x(y - x^3 y^2) + 2y(-x - x^2 y^5) = -2x^2 y^2 (x^2 + y^4)$$

which is negative semi-definite on D. Note that $D_0 = \{(x, y) : x = 0 \text{ or } y = 0\}$. For any solution $(x(t), y(t))$ of Eq. (3.5.11) contained in D_0, we have $x(t) \equiv 0$ or $y(t) \equiv 0$. Then by Eq. (3.5.13), we have $x(t) \equiv 0$ and $y(t) \equiv 0$. This implies that D_0 does not contain any nontrivial orbit of system (3.5.13). Therefore by Theorem 3.5.2, the zero solution $(0, 0)$ of system (3.5.13) is uniformly stable and asymptotically stable.

EXAMPLE 3.5.9. Consider the system

(3.5.14)
$$\begin{cases} x' = y^3 - x^3 y^4 \\ y' = -xy^2 - x^2 y^7. \end{cases}$$

Let $V(x, y) = x^2 + y^2$. Then V is positive definite on $D = \{(x, y) : |x| + |y| \leq l\}$ for any $l > 0$, and

$$\dot{V}(x, y) = 2x(y^3 - x^3 y^4) + 2y(-xy^2 - x^2 y^7) = -2x^2 y^4 (x^2 + y^4)$$

which is negative semi-definite on D. By Theorem 3.5.1, Part (a), the zero solution $(0, 0)$ of system (3.5.14) is uniformly stable. Note that $D_0 = \{(x, y) \in D : x = 0 \text{ or } y = 0\}$. For any solution $(x(t), y(t))$ of Eq. (3.5.14) contained in D_0, we have $x(t) \equiv 0$ or $y(t) \equiv 0$. If $x(t) \equiv 0$, then by the first equation in (3.5.14) we have $y(t) \equiv 0$; however, if $y(t) \equiv 0$, then from both equations in (3.5.14) we have that $x(t) \equiv c$ for an arbitrary $c \in \mathbb{R}$. This implies that for any $l > 0$, D_0 contains nontrivial equilibria $(c, 0)$ with $0 < |c| \leq l$. Hence Theorem 3.5.2 fails to apply. In fact, from the definition of the asymptotic stability, we see that the zero solution $(0, 0)$ of system (3.5.14) is not asymptotically stable.

3.6. Lyapunov Function Method for Nonautonomous Equations

In this section, we extend the Lyapunov function method introduced in Sect. 3.5 for autonomous equations to the nonautonomous equation

(E) $$x' = f(t, x),$$

where $f \in C([0, \infty) \times \mathbb{R}^n, \mathbb{R}^n)$ such that $f(t, 0) \equiv 0$. This means that $x = 0$ is an equilibrium. In the sequel, we denote by $|x|$ any norm of x for $x \in \mathbb{R}^n$. Similar to Definition 3.5.2 we have

DEFINITION 3.6.1. Let $D = \{(t,x) \in \mathbb{R} \times \mathbb{R}^n : t \in [0,\infty), |x| \le l\}$ for some $l > 0$ and $V \in C^1(D, \mathbb{R})$. Then the derivative of the function $V(t,x)$ along Eq. (E) is defined to be

$$\dot{V}(t,x) = \left[\frac{\partial V}{\partial t} + \sum_{i=1}^{n} \frac{\partial V}{\partial x_i} f_i\right](t,x) = \left[\frac{\partial V}{\partial t} + \nabla_x V \circ f\right](t,x),$$

where $\nabla_x V$ stands for the gradient of V with respect to x.

It is easy to see that along any solution $x(t)$ of Eq. (E), $\dot{V}(t, x(t)) = [V(t, x(t))]'$.

In the sequel, we let $D = \{(t,x) \in \mathbb{R} \times \mathbb{R}^n : t \in [0,\infty), |x| \le l\}$ for some $l > 0$ and $V \in C^1(D, \mathbb{R})$. The following are basic results on the stability of the zero solution of Eq. (E) by the Lyapunov function method.

THEOREM 3.6.1. *Assume that*

(a) $V(t, 0) \equiv 0$ *on* $[0, \infty)$;
(b) $V(t,x) \ge \phi(|x|)$ *for* $(t,x) \in D$, *where* $\phi \in C([0,\infty), [0,\infty))$ *satisfies* $\phi(0) = 0$ *and* $\phi(r) > 0$ *for* $r > 0$;
(c) $\dot{V}(t,x) \le 0$ *for* $(t,x) \in D$.

Then the zero solution of Eq. (E) is stable.

PROOF. Without loss of generality we may assume ϕ is nondecreasing. If not, we define $\phi_*(r) = \min_{r \le p \le l} \phi(p)$. Then $\phi_*(r)$ is nondecreasing and assumption (b) is satisfied with ϕ replaced by ϕ_*.

From assumptions (a) and (b) we have that for any $\epsilon > 0$ and any $t_0 \ge 0$, there exists a $\delta := \delta(\epsilon, t_0) > 0$ such that if $|x_0| < \delta$, then

$$0 \le V(t_0, x_0) < \phi(\epsilon).$$

Let $x(t)$ be a solution of Eq. (E) with $x(t_0) = x_0$. Note that assumption (c) implies that $V(t, x(t))$ is nonincreasing. By assumption (b) we see that for $t \ge t_0$

(3.6.1) $\qquad \phi(|x(t)|) \le V(t, x(t)) \le V(t_0, x_0) < \phi(\epsilon).$

It follows from (3.6.1) and the monotone property of ϕ that $|x(t)| < \epsilon$ for $t \ge t_0$. By definition, the zero solution of Eq. (E) is stable. $\square$

THEOREM 3.6.2. *Assume that*

(a) $\phi_1(|x|) \le V(t,x) \le \phi_2(|x|)$ *for* $(t,x) \in D$, *where for* $i = 1, 2$, $\phi_i \in C([0,\infty), [0,\infty))$ *satisfies* $\phi_i(0) = 0$ *and* $\phi_i(r) > 0$ *for* $r > 0$;
(b) $\dot{V}(t,x) \le 0$ *for* $(t,x) \in D$.

Then the zero solution of Eq. (E) is uniformly stable.

PROOF. As in the proof of Theorem 3.6.1, we may assume ϕ_1 is nondecreasing.

From assumption (a) we have that for any $\epsilon > 0$, there exists a $\delta := \delta(\epsilon) > 0$ such that if $|x_0| < \delta$, then $\phi_2(|x_0|) < \phi_1(\epsilon)$. For any $t_0 \ge 0$, let $x(t)$

be a solution of Eq. (E) with $x(t_0) = x_0$. Note that assumption (b) implies that $V(t, x(t))$ is nonincreasing. By assumption (a) we see that for $t \geq t_0$,

(3.6.2) $\phi_1(|x(t)|) \leq V(t, x(t)) \leq V(t_0, x_0) \leq \phi_2(|x_0|) < \phi_1(\epsilon)$.

Thus, it follows from (3.6.2) and the monotone property of ϕ_1 that $|x(t)| < \epsilon$ for $t \geq t_0$. By definition, the zero solution of Eq. (E) is uniformly stable. □

THEOREM 3.6.3. *Assume that*

   (a) $V(t, 0) \equiv 0$ on $[0, \infty)$;
   (b) $V(t, x) \geq \phi(|x|)$ for $(t, x) \in D$, where $\phi \in C([0, \infty), [0, \infty))$ satisfies $\phi(0) = 0$ and $\phi(r) > 0$ for $r > 0$;
   (c) $\dot{V}(t, x) \leq -\psi(V(t, x))$ for $(t, x) \in D$, where $\psi \in C([0, \infty), [0, \infty))$ satisfies $\psi(0) = 0$ and $\psi(r) > 0$ for $r > 0$.

Then the zero solution of Eq. (E) is asymptotically stable.

PROOF. As in the proof of Theorem 3.6.1, we may assume ψ is nondecreasing.

From assumptions (b) and (c) we have that $\dot{V}(t, x) \leq 0$ for $(t, x) \in D$. Hence all assumptions of Theorem 3.6.1 are satisfied. As a result, Eq. (E) is stable.

Let $(t_0, x_0) \in D$ and $x(t)$ be a solution of Eq. (E) with $x(t_0) = x_0$. Then $V(t, x(t)) \geq 0$ and is nonincreasing for $t \geq t_0$. Thus,

$$\lim_{t \to \infty} V(t, x(t)) = V_0 \geq 0.$$

We claim that $V_0 = 0$. Otherwise, $0 < V_0 \leq V(t, x(t))$ for $t \geq t_0$. Then by assumption (c) and the monotone property of ψ we have for $t \geq t_0$

$$\frac{d}{dt} V(t, x(t)) = \dot{V}(t, x(t)) \leq -\psi(V(t, x(t))) \leq -\psi(V_0) < 0.$$

It follows that

$$V(t, x(t)) \leq V(t_0, x_0) - \psi(V_0)(t - t_0) \to -\infty \text{ as } t \to \infty.$$

This contradicts the assumption that $V(t, x(t)) \geq 0$ and hence confirms the claim. Note that by assumption (b),

$$\lim_{t \to \infty} V(t, x(t)) = 0 \implies \lim_{t \to \infty} \phi(|x(t)|) = 0 \implies \lim_{t \to \infty} |x(t)| = 0 \implies \lim_{t \to \infty} x(t) = 0.$$

By definition, the zero solution of Eq. (E) is asymptotically stable.   □

The last theorem is about instability. Here, we use the notation $a_+ := \max\{a, 0\}$ for $a \in \mathbb{R}$. Clearly, $a_+ \geq 0$ for any $a \in \mathbb{R}$.

THEOREM 3.6.4. *Assume that*

   (a) $V(t, x) \leq \phi(|x|)$ for $(t, x) \in D$, where $\phi \in C([0, \infty), [0, \infty))$;
   (b) for any $\delta \in (0, l)$ and any $t_0 \geq 0$, there exists an $x_0 \in \mathbb{R}^n$ such that $|x_0| < \delta$ and $V(t_0, x_0) > 0$;

(c) $\dot{V}(t,x) \geq \psi(V_+(t,x))$ for $(t,x) \in D$, where $\psi \in C([0,\infty),[0,\infty))$
satisfies $\psi(0) = 0$ and $\psi(r) > 0$ for $r > 0$.

Then the zero solution of Eq. (E) is unstable.

PROOF. Without loss of generality we may assume ϕ is nondecreasing. If not, we define $\phi^*(r) = \max_{0 \leq p \leq r} \phi(p)$. Then $\phi^*(r)$ is nondecreasing and assumption (a) is satisfied with ϕ replaced by ϕ^*.

Suppose the zero solution of Eq. (E) is stable. Then for any $\epsilon > 0$ and any $t_0 \geq 0$, there exists a $\delta := \delta(\epsilon,t_0) > 0$ such that if $|x(t_0)| < \delta$, then $|x(t)| < \epsilon$ for all $t \geq t_0$. From assumption (b), there exists an $x_0 \in \mathbb{R}^n$ such that $|x_0| < \delta$ and $V(t_0, x_0) > 0$. Let $x(t)$ be a solution of Eq. (E) with $x(t_0) = x_0$. From the above we have that $|x(t)| < \epsilon$ for all $t \geq t_0$. It follows from assumption (a) and the monotone property of ϕ that

$$(3.6.3) \qquad V(t, x(t)) \leq \phi(|x(t)|) \leq \phi(\epsilon), \quad t \geq t_0.$$

On the other hand, by assumption (c) we see that $V(t, x(t))$ is nondecreasing. Hence

$$V(t, x(t)) \geq V(t_0, x_0) > 0, \quad t \geq t_0.$$

Consequently, $\psi(V_+(t,x)) = \psi(V(t,x))$ and

$$V(t, x(t)) - V(t_0, x_0) = \int_{t_0}^{t} \dot{V}(s, x(s)) \, ds \geq \int_{t_0}^{t} \psi(V(s, x(s))) \, ds$$

$$\geq \int_{t_0}^{t} \psi(V(t_0, x_0)) \, ds = \psi(V(t_0, x_0))(t - t_0) \to \infty \quad \text{as } t \to \infty.$$

This contradicts (3.6.3) and thus completes the proof. $\qquad \square$

Applications of the above theorems involve tricky and creative construction of Lyapunov functions. Here, we only give two examples for demonstration. In the following examples, $|(x, y)|_2$ stands for the 2-norm of the vector (x, y) introduced in Sect. 1.1.

EXAMPLE 3.6.1. Consider the equation

$$(3.6.4) \qquad u'' + u' + (2 + \sin t) u = 0.$$

Let $x = u$, $y = u'$. Then Eq. (3.6.4) becomes the first-order system of equations

$$(3.6.5) \qquad \begin{cases} x' = y \\ y' = -(2 + \sin t) x - y. \end{cases}$$

Let $D = \{(t,x,y) \in \mathbb{R}^3 : t \in [0,\infty), |(x,y)|_2 \leq l\}$ for some $l > 0$. Let

$$V(t,x,y) = x^2 + \frac{y^2}{(2 + \sin t)}$$

and

$$\phi_1(r) = r^2/3 \quad \text{and} \quad \phi_2(r) = r^2.$$

It is easy to see that $\phi_1(|(x,y)|_2) \leq V(t,x,y) \leq \phi_2(|(x,y)|_2)$ for $(t,x,y) \in D$; and for $i = 1, 2$, $\phi_i(0) = 0$ and $\phi_i(r) > 0$ for $r > 0$. By a simple computation we have that for $(t,x,y) \in D$,

$$\begin{aligned}
\dot{V}(t,x,y) &= 2xy - \frac{(\cos t)\,y^2}{(2+\sin t)^2} + \frac{2y\,(-(2+\sin t)x - y)}{2+\sin t} \\
&= -\frac{4 + \cos t + 2\sin t}{(2+\sin t)^2}y^2 \leq 0.
\end{aligned}$$

By Theorem 3.6.2, the zero solution $(x,y) = (0,0)$ of system (3.6.5) is uniformly stable, and hence the zero solution $u = 0$ of Eq. (3.6.4) is uniformly stable.

EXAMPLE 3.6.2. Consider the equation

(3.6.6) $u'' + a(t)\,u = 0,$

where $a \in C([0, \infty), \mathbb{R})$ such that $a(t) \leq -c$ for some $c > 0$. Let $x = u$, $y = u'$. Then Eq. (3.6.6) becomes the first-order system of equations

(3.6.7) $\begin{cases} x' = y \\ y' = -a(t)\,x. \end{cases}$

Let $D = \{(t,x,y) \in \mathbb{R}^3 : t \in [0, \infty), |(x,y)|_2 \leq l\}$ for some $l > 0$. Let

$$V(t,x,y) = xy,$$

and

$$\phi(r) = r^2/2 \quad \text{and} \quad \psi(r) = 2cr.$$

Then

$$V(t,x,y) = xy \leq \frac{1}{2}(x^2 + y^2) = \frac{1}{2}|(x,y)|_2^2 = \phi(|(x,y)|_2)$$

for $(t,x,y) \in D$, $\phi \in C([0,\infty), [0,\infty))$, and ψ satisfies $\psi(0) = 0$ and $\psi(r) > 0$ for $r > 0$. Clearly, for any $\delta \in (0, l)$ and any $t_0 \geq 0$, there exists an $x_0 \in \mathbb{R}^n$ such that $|x_0| < \delta$ and $V(t_0, x_0) > 0$. Moreover,

$$\begin{aligned}
\dot{V}(t,x,y) &= y^2 + x(-a(t)x) \geq y^2 + cx^2 \\
&\geq c^*(x^2 + y^2) \geq 2c^*|xy| = 2c^*|V(t,x,y)| \geq 2c^*V_+(t,x,y),
\end{aligned}$$

where $c^* = \min\{1, c\}$. By Theorem 3.6.4, the zero solution $(x,y) = (0,0)$ of system (3.6.7) is unstable, and hence the zero solution $u = 0$ of Eq. (3.6.6) is unstable.

Finally, we point out that the Lyapunov stability discussed in this chapter is regarded as "local stability" which characterizes the behavior of solutions near an equilibrium as $t \to \infty$. The theory for global stability has also been developed to characterize the behavior of solutions in the whole space. However, we will not include it in this book.

Exercises

3.1. Show that the zero solution of the equation $x' = (3t\sin t - t)x$ is asymptotically stable but is not uniformly stable.

3.2. Assume $A \in \mathbb{R}^{n \times n}$ and let $X(t)$ be a fundamental matrix solution of Eq. (H-c).

(a) Show that for any $t_0 \in \mathbb{R}$, $X(t - t_0)X^{-1}(0)$ is the principal matrix solution of Eq. (H-c) at t_0.

(b) Use theorem 3.2.1, instead of Lemma 3.1.3, to show that Eq. (H-c) is stable if and only if it is uniformly stable.

3.3. Use the idea in problem 2 to show that Eq. (H-p) is stable $\Longleftrightarrow$ it is uniformly stable.

3.4. Prove that for any $x, y \in \mathbb{R}^n$, $\displaystyle\lim_{h \to 0+} \frac{|x + hy| - |x|}{h}$ exists.

3.5. Prove that for $A \in \mathbb{R}^{n \times n}$ we have

$$|A|_1 = \max_j \sum_{i=1}^{n} |a_{ij}|$$

and

$$\mu_1(A) = \max_j \{a_{jj} + \sum_{i \neq j} |a_{ij}|\}, \quad -\mu_1(-A) = \min_j \{a_{jj} - \sum_{i \neq j} |a_{ij}|\}.$$

3.6. Prove the Properties (a)-(d) of the Lozinskii measures given in Sect. 3.3.

3.7. Determine the stability of the linear system (H) where $A(t)$ is defined by

(a) $A(t) = \begin{bmatrix} -\frac{1}{t} & \frac{\sin t}{t} \\ \frac{\cos t}{t} & -\frac{1}{t} \end{bmatrix}$,

(b) $A(t) = \begin{bmatrix} 1 & k\sin t \\ k\cos t & 1 \end{bmatrix}$, where $|k| \leq 1/\sqrt{2}$.

(c) $A(t) = \begin{bmatrix} \sin t - 1 & \frac{1}{t^2}\cos t \\ \sin t & \frac{1}{t^2} - 1 \end{bmatrix}$.

3.8. Determine the stability of the linear system

$$x' = (e^{-t}I + A(t))x, \quad t \geq 0,$$

where I is the $n \times n$ identity matrix and $A \in C([0, \infty), \mathbb{R}^{n \times n})$ such that $\mu_2(A(t)) \leq k < 0$ for all $t \geq 0$.

3.9. Assume $A(t) \in C(\mathbb{R}, \mathbb{R}^{n \times n})$ is ω-periodic for some $\omega > 0$, and $\lambda_1(t)$ and $\lambda_n(t)$ are the greatest and the least eigenvalues of $\frac{1}{2}(A^T + A)(t)$, respectively.

(a) Suppose that either $\int_0^\omega \lambda_1(t)dt < 0$ or $\int_0^\omega \lambda_n(t)dt > 0$. Show that Eq. (H-p) has no nontrivial periodic solutions of any period.

(b) Can you generalize the result in Part (a) using the Lozinskii measures?

3.10. (a) For different values of α, discuss the stability of the linear system

$$\begin{cases} x_1' = -x_2 \\ x_2' = x_1 + \alpha x_2. \end{cases}$$

(b) What can you say about the stability of the zero solution of the system

$$\begin{cases} x_1' = -x_2 + x_1^3 \\ x_2' = x_1 + \alpha x_2 - 2x_1 x_2^2? \end{cases}$$

(c) What can you say about the stability of the zero solution of the system

$$\begin{cases} x_1' = e^{-t}x_1^3 - x_2 \\ x_2' = \dfrac{t^2}{1+t^2}x_1 + \alpha x_2? \end{cases}$$

3.11. Use the linearization method to determine the stability of the zero solution of the system

$$(a) \begin{cases} x_1' = x_1 - 6\sin x_2 \\ x_2' = -x_1\cos x_1 + x_1 x_2, \end{cases} \qquad (b) \begin{cases} x_1' = x_1 + e^{x_2} - \cos x_2 \\ x_2' = \sin x_1 - x_2 + x_1 x_2. \end{cases}$$

3.12. Determine the stability of the following linear system:

$$\begin{cases} x_1' = -7x_1 + 10t^2(1+t^2)^{-1}x_2 \\ x_2' = -(4+t^{-1})x_1 + 5x_2. \end{cases}$$

3.13. Suppose that the linear equation (1) $x' = Ax$ is asymptotically stable, where $A \in \mathbb{R}^{n \times n}$, and $r(t,x) \in C([0,\infty) \times \mathbb{R}^n, \mathbb{R}^n)$ be such that $|r(t,x)| \leq \lambda|x|$. Show that if $\lambda > 0$ is small enough, then the zero solution of the Eq. (2) $x' = Ax + r(t,x)$ is asymptotically stable.

3.14. Determine the stability of the zero solution of the system

$$x' = A(t)x + \frac{\sqrt{|x_1 x_2|}}{(t+1)e^t} \begin{bmatrix} 1 \\ 1 \end{bmatrix},$$

where $A(t)$ is defined by Problem 7, Part (a). What can you say about the case where $A(t)$ is defined by Problem 7, Part (b)?

3.15. The nonlinear autonomous system $\begin{cases} x' = x - y \\ y' = 1 - xy \end{cases}$ has an equilibrium at $(-1,-1)$. How would you approximate the system with a linear system near $(-1,-1)$? Determine the stability of this equilibrium.

3.16. The motion of a simple pendulum with a linear damping is governed by the equation

$$u'' + \alpha u' + \beta \sin u = 0,$$

where α, β are positive constants. Use the principle of linearized stability to show that the solution $u \equiv 0$ is uniformly stable and asymptotically stable, but the solution $u \equiv \pi$ is unstable.

3.17. Let $A(t), B(t) \in C([0, \infty), \mathbb{R}^{n \times n})$ with $A(t)$ ω-periodic for some $\omega > 0$ and $\int_0^\infty |B(t)| \, dt < \infty$.

(a) Assume all solutions of the equation (1) $x' = A(t)x$ are bounded on $[0, \infty)$. Show that all solutions of the equation (2) $y' = (A(t) + B(t))y$ are bounded on $[0, \infty)$.

(b) Assume all solutions of Eq. (1) tend to 0 as $t \to \infty$. Show that the same holds for all solutions of Eq. (2).

3.18. Use the Lyapunov function method to determine the stability of the zero solution of the following equations:

(a) $\begin{cases} x' = -x + xy^2 \\ y' = -2x^2y - y^3, \end{cases}$
(b) $\begin{cases} x' = x^3 - 2y^3 \\ y' = xy^2 + x^2y + \frac{1}{2}y^3, \end{cases}$

(c) $\begin{cases} x' = -2x + y^3 \\ y' = x - y - xy^2, \end{cases}$
(d) $\begin{cases} x' = -2x + 2y + 2xy^2 \\ y' = x - y - x^2y. \end{cases}$

3.19. Show that the equilibrium $(0, 0)$ of the system $\begin{cases} x' = y^3 \\ y' = -2x + l(x, y)y \end{cases}$ is asymptotically stable if l is continuous and $l(0, 0) < 0$.

3.20. Determine the stability of the zero solution of the system $\begin{cases} x' = y + kx^3 \\ y' = -x + ky^5 \end{cases}$ for different values of k.

3.21. Consider the nonlinear system $\begin{cases} x' = y - x^3y^2 \\ y' = -x^3 - y^3x^2 \end{cases}$.

(a) Use any method you want to determine the stability of the linearized system of the nonlinear system at $(0, 0)$.

(b) For the nonlinear system, determine the stability of the equilibrium $(0, 0)$.

3.22. Consider the system

$$\begin{cases} x' = x^3(y - b) \\ y' = y^3(x - a)^5 \end{cases}, \qquad a > 0, \ b > 0.$$

Study the stability of the equilibria $(0, 0)$ and (a, b), respectively.

3.23. Use the Lyapunov function method to study the stability of the zero solution of the system $\begin{cases} x' = y + 2y^3 \\ y' = -x - 2x^3 \end{cases}$.

3.24. Use the Lyapunov function method to study the stability of the zero solution for each of the following systems:

(a) $\begin{cases} x' = y \\ y' = z \\ z' = -y \end{cases}$,
(b) $\begin{cases} x' = -x - y + z + x^3 \\ y' = x - 2y + 2z + xz \\ z' = x + 2y + z + xy \end{cases}$.

3.25. Use a Lyapunov function to determine the stability of the zero solution of the equation

$$u'' + \mu(u^2 - 1)u' + 4u^3 = 0$$

where $\mu \leq 0$.

3.26. Determine the stability of the zero solution of the equation

$$u'' - (|u| + |u'| - 1)u' + u|u| = 0.$$

3.27. Consider the equation $u'' + u' + au^3 + f(u) = 0$, where $a > 0$, $f : \mathbb{R} \to \mathbb{R}$ is continuous, and $f(u)/u^3 \to 0$ as $u \to 0$. Show that $u = 0$ is a solution of the equation and it is uniformly stable and asymptotically stable.

3.28. Use the Lyapunov function method to show that the zero solution of the equation $u'' + (\sin \dfrac{1}{t+1})u' + u^3 = 0$ is stable.

3.29. Use the Lyapunov function method to show that the zero solution of the equation $u'' + (1 - 2e^{-t})u' + \sin u = 0$ is uniformly stable.

3.30. Determine the stability of the zero solution of the following systems:

$$(a) \begin{cases} x' = x(r^2 - 1) - g(t)\,y \\ y' = g(t)\,x + y(r^2 - 1) \end{cases}, \qquad (b) \begin{cases} x' = x(1 - r^2) - g(t)y \\ y' = g(t)x + y(1 - r^2) \end{cases};$$

where $r^2 = x^2 + y^2$ and $g \in C([0, \infty), \mathbb{R})$.

CHAPTER 4

Dynamical Systems and Planar Autonomous Equations

4.1. General Knowledge of Autonomous Equations

We will first discuss some basic properties of general autonomous equations. Consider the equation

(A) $$x' = f(x),$$

where $f : \mathbb{R}^n \to \mathbb{R}^n$, $n \geq 1$, is locally Lipschitz. Here, x' represents the derivative of x with respect to the independent variable t, and hence every solution of Eq. (A) is a function of t. By Remark 1.2.1 we see that f is locally Lipschitz on $\mathbb{R}^n$ implies that f is continuous on $\mathbb{R}^n$. Therefore by Theorem 1.3.1, any IVP associated with Eq. (A) has a unique solution. We note that for the case with $n = 1$, we can always solve Eq. (A) for the general solution using the method of separating variables; in the following, we only study the case with $n \geq 2$.

Let $x(t)$ be a solution of Eq. (A) defined on its maximal interval of existence (α, β). Recall from Definition 3.5.3 that the set

$$\{(t, x(t)) : t \in (\alpha, \beta)\}$$

in $\mathbb{R}^{n+1}$ is called a solution curve, and the projection of this solution curve into $\mathbb{R}^n$

$$\{x(t) : t \in (\alpha, \beta)\}$$

is called an orbit or trajectory. The space $\mathbb{R}^n$, where all orbits stay, is called the phase space. Since the phase space $\mathbb{R}^n$ has one dimension less than $\mathbb{R}^{n+1}$, where all solution curves stay, we often use orbits, rather than solution curves, to study autonomous equations.

We also recall from Lemma 3.1.2 that if $x(t)$ is a solution of Eq. (A) on (α, β), then for any $c \in \mathbb{R}$, $x(t - c)$ is a solution of Eq. (A) on $(\alpha + c, \beta + c)$.

EXAMPLE 4.1.1. Consider the system of equations $x'_1 = x_2$, $x'_2 = -x_1$. We observe that

(i) $x_1 = 0$, $x_2 = 0$ is a solution on $(-\infty, \infty)$ whose solution curve is the t-axis in $\mathbb{R}^3$ and whose orbit is the origin in $\mathbb{R}^2$.

(ii) $x_1 = \sin t$, $x_2 = \cos t$ is a solution on $(-\infty, \infty)$ whose solution curve is a helix in $\mathbb{R}^3$ and whose orbit is the unit circle in $\mathbb{R}^2$.

© Springer International Publishing Switzerland 2014   101
Q. Kong, *A Short Course in Ordinary Differential Equations*, Universitext,
DOI 10.1007/978-3-319-11239-8_4

(iii) For any $\theta \in [0, \pi)$, $x_1 = \sin(t - \theta)$, $x_2 = \cos(t - \theta)$ is also a solution on $(-\infty, \infty)$. Note that the solution curve is a shifted helix from the one in Part (ii), but the orbit is exactly the same unit circle as in Part (ii).

For the shape of a helix, see Fig. 3.7.

REMARK 4.1.1. In general, if $x(t)$ is a solution of Eq. (A) and $c \in \mathbb{R}$, then the solution curve for $x(t-c)$ is a shift of the solution curve for $x(t)$, but $x(t)$ and $x(t-c)$ have exactly the same orbit. Therefore, one orbit represents an infinite number of solution curves, all of which can be obtained by shifting one of them in the direction of the t-axis.

Now we introduce some further properties of orbits of Eq. (A).

LEMMA 4.1.1. *There exists a unique orbit of Eq.* (A) *through each point in the phase space.*

PROOF. For any $x_0 \in \mathbb{R}^n$ and $t_0 \in \mathbb{R}$, let $x(t)$ be the solution of Eq. (A) satisfying $x(t_0) = x_0$. Then the orbit for $x(t)$ passes through x_0.

Let Γ_1 and Γ_2 be orbits for solutions $x_1(t)$ and $x_2(t)$ of Eq. (A), respectively, both pass through x_0. This means that there exist $t_1, t_2 \in \mathbb{R}$ such that $x_1(t_1) = x_0$ and $x_2(t_2) = x_0$. Let $t_3 = t_1 - t_2$ and $x_3(t) = x_2(t - t_3)$. Then $x_3(t)$ is a solution of Eq. (A) satisfying $x_3(t_1) = x_2(t_2) = x_1(t_1)$. By the uniqueness of IVPs, $x_3(t) \equiv x_1(t)$ on $\mathbb{R}$. Since $x_3(t)$ has the same orbit as $x_2(t)$, we see that Γ_1 and Γ_2 are the same. $\square$

From Lemma 4.1.1 we see that the orbits for any two solutions of Eq. (A) are either identically the same or do not intersect each other.

COROLLARY 4.1.1. *(a) Let $x = a$ be an equilibrium of Eq.* (A). *Assume a solution $x(t)$ of Eq.* (A) *satisfies $x(t) \neq a$ and*

$$\lim_{t \to \alpha+} x(t) = a \quad [\lim_{t \to \beta-} x(t) = a].$$

Then $\alpha = -\infty$ $[\beta = \infty]$.

(b) Assume $x(t)$ is a solution of Eq. (A) *satisfying*

$$\lim_{t \to -\infty} x(t) = a \quad or \quad \lim_{t \to \infty} x(t) = a.$$

Then $x = a$ is an equilibrium of Eq. (A).

PROOF. (a) Assume $\lim_{t \to \alpha+} x(t) = a$ with $\alpha \in (-\infty, \infty)$. Then the orbits for $x(t)$ and $x = a$ both pass through the point a in the phase space. This contradicts Lemma 4.1.1 and hence shows that $\alpha = -\infty$. Similarly for the case when $\lim_{t \to \beta-} x(t) = a$.

(b) Since $x(t)$ is a solution of Eq. (A), we have that $x'(t) \equiv f(x(t))$. Assume $\lim_{t \to -\infty} x(t) = a$. Taking limits on both sides as $t \to -\infty$ we get that $\lim_{t \to -\infty} x'(t) = f(a)$. If $x = a$ is not an equilibrium of Eq. (A), then there exists an $i \in \{1, \ldots, n\}$ such that $f_i(a) \neq 0$, say $f_i(a) > 0$. It follows that $x_i'(t) > f_i(a)/2$ for t sufficiently small. This implies that $\lim_{t\to-\infty} x_i(t) = -\infty$, contradicting the assumption. Similarly for the case when $\lim_{t\to\infty} x(t) = a$.

□

Corollary 4.1.1 shows that no orbit can "pass" an equilibrium at a finite time.

To characterize the orbits for periodic solutions, we need the following definition.

DEFINITION 4.1.1. A curve C in $\mathbb{R}^n$ is said to be a closed curve if it is the graph of a non-constant function $x \in C([a, b], \mathbb{R}^n)$ such that $x(a) = x(b)$. Furthermore, C is said to be a simple closed curve if the function x also satisfies $x(t_1) \neq x(t_2)$ for $a \leq t_1 < t_2 < b$.
A simple closed curve in $\mathbb{R}^2$ is also called a Jordan curve.

Intuitively, a simple closed curve is a non-self-intersecting continuous loop in space. As the well known Jordan curve theorem states, a Jordan curve dissects the whole space $\mathbb{R}^2$ into two disjoint regions: the interior and the exterior. This result may seem obvious at first, but it is rather difficult to prove. The interested reader is referred to A. F. Filippov [15] for an elementary proof of the Jordan curve theorem. We note that a simple closed curve in $\mathbb{R}^n$ does not have this property for $n \geq 3$.

LEMMA 4.1.2. Let $x(t)$ be a nontrivial solution of Eq. (A) on $\mathbb{R}$. Then $x(t)$ is a periodic solution if and only if the orbit Γ for $x(t)$ is a simple closed curve.

PROOF. Let $x(t)$ be a periodic solution of Eq. (A) with a period ω. Then for any $t \in \mathbb{R}$, $x(t+\omega) = x(t)$ and hence Γ is a closed curve. We claim that for any $0 \leq t_1 < t_2 < \omega$, $x(t_1) \neq x(t_2)$. Otherwise, there exist $0 \leq t_1 < t_2 < \omega$ such that $x(t_1) = x(t_2)$. Note that $x_1(t) := x(t + t_2 - t_1)$ is also a solution of Eq. (A) and satisfies the IC $x_1(t_1) = x(t_2) = x(t_1)$. By the uniqueness of solutions of IVPs, we have that $x_1(t) = x(t)$ or $x(t + t_2 - t_1) = x(t)$ for any $t \in \mathbb{R}$. This means that the largest possibility for the period of $x(t)$ is $t_2 - t_1$ which contradicts the assumption that $t_2 - t_1 < \omega$. Therefore, $\Gamma = \{x(t) : t \in [0, \omega]\}$ is a simple closed curve.

Let Γ be a simple closed curve. Then there exist a $t_0 \in \mathbb{R}$ and a least $\omega > 0$ such that $x(t_0) = x(t_0 + \omega) \in \Gamma$. Note that $x(t + \omega)$ is also a solution of Eq. (A). Similar to the above, we have that $x(t) = x(t+\omega)$ for any $t \in \mathbb{R}$. Since $x(t_0) \neq x(t_0 + t)$ for any $t \in (0, \omega)$, $x(t)$ is ω-periodic.   □

Now we present a property of the solutions of the autonomous equation (A) which is useful in dealing with the orbits, see Fig. 4.1 for an illustration.

LEMMA 4.1.3. For $x_0 \in \mathbb{R}^n$, denote by $x(t) := \phi(t; x_0)$ the solution of Eq. (A) satisfying the IC $x(0) = x_0$. Assume its maximal interval of existence is $\mathbb{R}$. Then for any $t_1, t_2 \in \mathbb{R}$ we have

(4.1.1) $$\phi(t_2; \phi(t_1; x_0)) = \phi(t_2 + t_1; x_0).$$

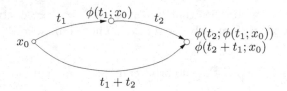

FIGURE 4.1. Property of solutions of Eq. (A)

PROOF. Let $x_1(t) = \phi(t; \phi(t_1; x_0))$ and $x_2(t) = \phi(t + t_1; x_0)$. By Lemma 3.1.2, both $x_1(t)$ and $x_2(t)$ are solutions of Eq. (A). Note that

$$x_1(0) = \phi(t_1; x_0) = x_2(0).$$

By the uniqueness of solutions of IVPs, we have $x_1(t) \equiv x_2(t)$, i.e.,

$$\phi(t; \phi(t_1; x_0)) = \phi(t + t_1; x_0).$$

Let $t = t_2$ in the above, we obtain (4.1.1). $\qquad\square$

COROLLARY 4.1.2. (a) Assume Γ is a nontrivial non-closed orbit of Eq. (A). Then Γ does not intersect itself.

(b) Assume Γ is a closed orbit of Eq. (A). Then Γ is a simple closed curve.

PROOF. We only prove Part (a). The proof for Part (b) is left as an exercise, see Exercise 4.2.

Assume the contrary. Then orbit Γ intersects itself at a point $x_0 \in \Gamma$. Denote by $x(t) := \phi(t; x_0)$ the solution of Eq. (A) satisfying $x(0) = x_0$. Clearly, $x(t)$ has Γ as its orbit. This implies that there exists a $t_1 \neq 0$ such that $\phi(t_1; x_0) = x_0$. By Lemma 4.1.3 we have that for any $t \in \mathbb{R}$

$$\phi(t; x_0) = \phi(t; \phi(t_1; x_0)) = \phi(t + t_1, x_0).$$

Hence $x(t)$ is ω-periodic for some $\omega \in (0, |t_1|]$. This contradicts the assumption that Γ is non-closed orbit and lemma 4.1.2. $\qquad\square$

To give a geometric sketch of the orbits for Eq. (A), we observe that the right-hand side function f of Eq. (A) determines a vector field. This vector field is called the vector field for Eq. (A). It is easy to see that this vector field uniquely determines the positive direction of the orbits at each regular point, i.e., a point $x \in \mathbb{R}^n$ such that $f(x) \neq 0$. However, there is no fixed direction of the vector field at a point $x \in \mathbb{R}^n$ such that $f(x) = 0$. Such a point is called a singular point of the vector field. Clearly, a singular point of a vector field is an equilibrium of the corresponding autonomous differential equation.

Using the elimination method, we are able to obtain a system of equations for the orbits from the component form of Eq. (A):

(4.1.2) $$\frac{dx_1}{f_1(x)} = \frac{dx_2}{f_2(x)} = \cdots = \frac{dx_n}{f_n(x)}.$$

Note that Eq. (4.1.2) is an $(n-1)$-dimensional system of equations and hence is easier to deal with than Eq. (A). In particular, when $n = 2$, the equation for orbits becomes the scalar equation

(4.1.3) $$\frac{dx_2}{dx_1} = \frac{f_2(x_1, x_2)}{f_1(x_1, x_2)} \quad \text{or} \quad \frac{dx_1}{dx_2} = \frac{f_1(x_1, x_2)}{f_2(x_1, x_2)}.$$

However, when we solve (4.1.2) or its special form (4.1.3) to obtain orbits, some solutions may be neglected due to the fact that all or some f_i's are moved to the denominators. Therefore, additional care should be taken for the case where $f_i(x) = 0$ for $i \in \{1, \ldots, n\}$.

The geometric representation of the topological structure of orbits with the directions given by the vector field is called a phase portrait.

EXAMPLE 4.1.2. Consider the system of equations

(4.1.4) $$\begin{cases} x_1' = 2x_1 x_2, \\ x_2' = -x_1^2 + x_2^2. \end{cases}$$

Clearly, $(0, 0)$ is the only equilibrium. Note that

$$x_1' = 0 \iff x_1 = 0 \text{ or } x_2 = 0 \quad \text{and} \quad x_2' = 0 \iff x_2 = \pm x_1.$$

Then the $x_1 x_2$-plane is divided by the x_1-axis, x_2-axis, and the two lines $x_2 = \pm x_1$ into eight regions. In each region, x_1' and x_2' have fixed signs, and hence the directions of the orbits can be determined by the vector field $f(x_1, x_2) = (2x_1 x_2, -x_1^2 + x_2^2)^T$, as shown in Fig. 4.2.

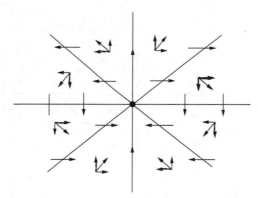

FIGURE 4.2. Vector field for Example 4.1.2

The equation for orbits is

(4.1.5) $$\frac{dx_2}{dx_1} = \frac{-x_1^2 + x_2^2}{2x_1 x_2} = -\frac{x_1}{2x_2} + \frac{x_2}{2x_1}$$

which is a homogeneous equation. Let $v = x_2/x_1$, i.e., $x_2 = x_1 v$. Then the above equation is transformed to the separable equation for v

$$\frac{2v}{v^2 + 1} dv = -\frac{dx_1}{x_1}$$

whose general solution can be simplified to $x_1(v^2 + 1) = c$, where c is an arbitrary constant. Therefore, the general solution for Eq. (4.1.5) is $x_1^2 + x_2^2 = cx_1$, i.e.,

$$\left(x_1 - \frac{c}{2}\right)^2 + x_2^2 = \frac{c^2}{4}.$$

It is easy to see that for each $c \in \mathbb{R}$, the orbit is a circle centered at $(c/2, 0)$ with radius $|c|/2$, see Fig. 4.3. From system (4.1.1) we also note that the line $x_1 = 0$ contains orbits on the two sides of the equilibrium $(0, 0)$. These orbits were neglected in the above solution process. However, the line $x_2 = 0$ does not contain orbits since it intersects the circles. The arrows for the directions of the orbits are determined by the vector field given by Fig. 4.2.

4.2. Rudiments of Dynamical Systems

The theory of dynamical systems originates from the qualitative study of differential equations. It is a means of describing how the global behavior of a structure develops over the course of time. In this section, we introduce basic knowledge of dynamical systems related to differential equations.

For any $x_0 \in \mathbb{R}^n$, denote by $x(t) := \phi(t; x_0)$ the solution of Eq. (A) satisfying the IC $x(0) = x_0$. Assume its maximal interval of existence is $\mathbb{R}$. Then from the definition and Lemma 4.1.3, $\phi(t; x_0)$ satisfies the following properties:

   (i) $\phi(0; x_0) = x_0$ for any $x_0 \in \mathbb{R}^n$,

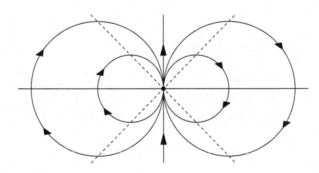

FIGURE 4.3. Phase portrait for Example 4.1.2

   (ii) $\phi(t_2; \phi(t_1; x_0)) = \phi(t_2 + t_1; x_0)$ for any $x_0 \in \mathbb{R}^n$ and $t_1, t_2 \in \mathbb{R}$.

Traditionally, the solution $\phi(t; x_0)$ is considered as a function of t, while the initial point x_0 is regarded as a parameter. It seams that H. Poincaré was the first person to treat $\phi(t; x_0)$ in a different point of view: for any given $t \in \mathbb{R}$, define a mapping $\phi_t : \mathbb{R}^n \to \mathbb{R}^n$ by

$$\phi_t(x) = \phi(t; x), \quad x \in \mathbb{R}^n.$$

In this way, $\phi_t(x)$ is considered as a function of x, while t is regarded as a parameter. Obviously, this mapping satisfies the following properties:

(I) $\phi_0 = I$, the identity operator on $\mathbb{R}^n$;

(II) $\phi_{t_2}\phi_{t_1} = \phi_{t_2+t_1}$ for any t_1, $t_2 \in \mathbb{R}$.

We call the one-parameter family of such mappings $\Phi := \{\phi_t : t \in \mathbb{R}\}$ a dynamical system or a flow.

Note that the above dynamical system is induced by Eq. (A). Now we present the general concept of dynamical systems over $\mathbb{R}^n$.

DEFINITION 4.2.1. Let D be an open subset of $\mathbb{R}^n$. For each $t \in \mathbb{R}$, let $\phi_t : D \to D$ be a transformation satisfying assumptions (I) and (II). Define $\Phi := \{\phi_t : t \in \mathbb{R}\}$. Then Φ is said to be a dynamical system or a flow on D. Furthermore, if $\phi_t(x)$ is C $[C^i]$ in t and x [for some $i \in \mathbb{N}$], then Φ is said to be a C $[C^i]$ dynamical system or flow on D.

From Theorems 1.5.2 and 1.5.3, if $f : \mathbb{R}^n \to \mathbb{R}^n$ is locally Lipschitz, then the dynamical system induced by Eq. (A) is a C dynamical system; and if $f \in C^1(\mathbb{R}^n, \mathbb{R}^n)$, then the dynamical system induced by Eq. (A) is a C^1 dynamical system.

It is easy to see that with the composition operation as the multiplication, a dynamical system is a commutative transformation group of one-parameter. In fact, assumptions (I) and (II) show that ϕ_0 is the unit element and Φ is closed under multiplication. Moreover, since

$$\phi_{-t}\phi_t = \phi_t\phi_{-t} = \phi_0,$$

$\phi_t^{-1} := \phi_{-t}$ is the inverse of ϕ_t for any $t \in \mathbb{R}$. The commutative property follows from the observation that for any t_1, $t_2 \in \mathbb{R}$

$$\phi_{t_2}\phi_{t_1} = \phi_{t_2+t_1} = \phi_{t_1+t_2} = \phi_{t_1}\phi_{t_2}.$$

Although dynamical systems are defined in a general way, each C^1 dynamical system has a unique representation by an autonomous differential equation, as shown in the lemma below.

LEMMA 4.2.1. Assume $\phi_t(x)$ is C^1 in t for all $t \in \mathbb{R}$ and $x \in D \subset \mathbb{R}^n$, and $[d\phi_t(x)/dt]_{t=0}$ is locally Lipschitz on D. Then there exists a unique vector field $f : D \to \mathbb{R}^n$ such that $\Phi := \{\phi_t : t \in \mathbb{R}\}$ is the dynamical system induced by Eq. (A) with this f.

PROOF. For $x \in D$ define

$$f(x) = \left.\frac{d\phi_t(x)}{dt}\right|_{t=0} = \lim_{h \to 0} \frac{\phi_h(x) - \phi_0(x)}{h}.$$

By the assumption, $f : D \to \mathbb{R}^n$ is a locally Lipschitz vector field which determines an autonomous equation $x' = f(x)$. Since

$$\frac{d\phi_t(x)}{dt} = \lim_{h \to 0} \frac{\phi_{t+h}(x) - \phi_t(x)}{h} = \lim_{h \to 0} \frac{\phi_h(\phi_t(x)) - \phi_0(\phi_t(x))}{h} = f(\phi_t(x)),$$

$\phi_t(x)$ is a solution of the equation. It also satisfies the IC $\phi_0(x) = x$. Clearly, $\phi_t(x)$ is the only solution of the IVP. Therefore, Φ is the dynamical systems induced by Eq. (A).

Assume there exists another vector field $g : D \to \mathbb{R}^n$ such that the equation $x' = g(x)$ induces Φ. Then

$$\frac{d\phi_t(x)}{dt} = f(\phi_t(x)) \quad \text{and} \quad \frac{d\phi_t(x)}{dt} = g(\phi_t(x)) \quad \text{for any } t \in \mathbb{R} \text{ and } x \in D.$$

This implies that $f(\phi_t(x)) = g(\phi_t(x))$ for all $t \in \mathbb{R}$ and $x \in D$. In particular, letting $t = 0$ and using the fact that $\phi_0(x) = x$, we have $f(x) \equiv g(x)$ on D. This shows that the autonomous equation that induces the C^1 flow Φ is unique. $\qquad\square$

In the sequel, we will use the term "flow" instead of "dynamical system" for the dynamical system induced by Eq. (A) due to its natural and intuitive meaning from physics and engineering. We first introduce several definitions and notation for the flow $\Phi := \{\phi_t : t \in \mathbb{R}\}$ on D induced by Eq. (A).

DEFINITION 4.2.2. (a) For a fixed $x \in D$, the set $\{\phi_t(x) : t \in \mathbb{R}\}$ is called the orbit of Φ passing through x.

(b) If $x \in D$ satisfies $\phi_t(x) = x$ for any $t \in \mathbb{R}$, then x is called an equilibrium or a fixed point of Φ.

(c) If $x \in D$ satisfies $\phi_\omega(x) = x$ for some but not all $\omega > 0$, then x is called a periodic point of Φ; the smallest such $\omega > 0$ is called the period of x; and the orbit passing through x is called a closed orbit of Φ.

Sometimes, we may write the orbit passing through a point $x \in D$ as $\Gamma := \{\phi(t, x) : t \in \mathbb{R}\}$ to avoid complicated subscripts, and when we address an orbit without emphasizing the initial point x, we may denote it by $\Gamma := \{\phi(t) : t \in \mathbb{R}\}$. Since the flow Φ is induced by Eq. (A), an orbit Γ of Φ may be called (in fact, it is) an orbit of Eq. (A).

REMARK 4.2.1. (i) If x is an equilibrium of Φ, then $f(x) = 0$ and hence x is an equilibrium of Eq. (A) and a singular point of the vector field f.

(ii) If x is a periodic point of Φ with a period ω, then every point y on the orbit $\Gamma := \{\phi_t(x) : t \in \mathbb{R}\}$ passing through x is a periodic point of Φ with the same period ω. In fact, there exists an $s \in \mathbb{R}$ such that $y = \phi_s(x)$. Hence

$$\phi_\omega(y) = \phi_\omega(\phi_s(x)) = \phi_{\omega+s}(x) = \phi_{s+\omega}(x) = \phi_s(\phi_\omega(x)) = \phi_s(x) = y.$$

DEFINITION 4.2.3. A set $E \subset D$ is said to be positively [negatively] invariant for the flow Φ or for Eq. (A) if it satisfies $\phi_t(x) \in E$ for all $t \geq 0$ [$t \leq 0$] whenever $x \in E$; It is said to be invariant for the flow Φ or for Eq. (A) if it is both positively invariant and negatively invariant for the flow Φ or for Eq. (A).

It is easy to see that every orbit and any union of orbits are invariant sets for Φ, and an invariant set must consist of entire orbits. Now we present the concept of limit sets of orbits.

DEFINITION 4.2.4. Let $\Gamma := \{\phi(t) : t \in \mathbb{R}\}$ be an orbit of a flow Φ. Then the sets

$$\Gamma^+ = \{\phi(t) : t \in [0, \infty)\} \quad \text{and} \quad \Gamma^- = \{\phi(t) : t \in (-\infty, 0]\}$$

are called the positive and negative semi-orbits of Γ, respectively; and the sets

$$\Omega(\Gamma^+) = \{x^* \in D : \text{ there exists a sequence } t_i \to \infty \text{ such that } \phi(t_i) \to x^*\}$$

and

$$A(\Gamma^-) = \{x_* \in D : \text{ there exists a sequence } t_i \to -\infty \text{ such that } \phi(t_i) \to x_*\}$$

are called the ω-limit set and the α-limit set of Γ, respectively. The set

$$L(\Gamma) = \Omega(\Gamma^+) \cup A(\Gamma^-)$$

is called the limit set of Γ.

EXAMPLE 4.2.1. Let Γ be an orbit in $\mathbb{R}^2$.

(i) If Γ is an equilibrium or a closed orbit, then $\Gamma^+ = \Gamma^- = \Gamma$ and hence $\Omega(\Gamma^+) = A(\Gamma^-) = L(\Gamma) = \Gamma$.

(ii) If Γ approaches $(0,0)$ as $t \to \infty$ and approaches infinity as $t \to -\infty$, then $\Omega(\Gamma^+) = \{(0,0)\}$, $A(\Gamma^-) = \emptyset$, and hence $L(\Gamma) = \{(0,0)\}$.

(iii) Assume $(0,0)$ is contained in the interior of a closed orbit Γ_0. If Γ is in the interior of Γ_0 approaching $(0,0)$ as $t \to \infty$ and approaching Γ_0 as $t \to -\infty$, as shown in Fig. 4.4, then $\Omega(\Gamma^+) = \{(0,0)\}$, $A(\Gamma^-) = \Gamma_0$, and hence $L(\Gamma) = \{(0,0)\} \cup \Gamma_0$.

REMARK 4.2.2. For the ω-limit set and the α-limit set of the orbit Γ defined in Definition 4.2.4, we have

(4.2.1) $\qquad \Omega(\Gamma^+) = \cap_{s \geq 0} \overline{\cup_{t \geq s} \phi(t)} \quad \text{and} \quad A(\Gamma^-) = \cap_{s \leq 0} \overline{\cup_{t \leq s} \phi(t)}.$

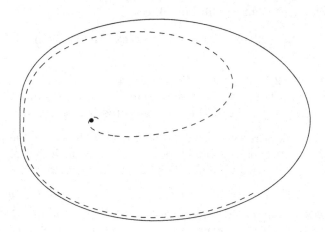

FIGURE 4.4. Limit set

Consequently,

$$(4.2.2) \qquad \Omega(\Gamma^+) \subset \overline{\Gamma^+}, \quad A(\Gamma^-) \subset \overline{\Gamma^-}, \quad \text{and} \quad L(\Gamma) \subset \overline{\Gamma}.$$

The proof is left as an exercise, see Exercise 4.9.

In the following we introduce some properties satisfied by ω-limit sets and α-limit sets of orbits.

LEMMA 4.2.2. *Let* $\Gamma := \{\phi(t) : t \in \mathbb{R}\}$ *be an orbit of a flow* Φ. *Then the* ω-*limit set* $\Omega(\Gamma^+)$ *and the* α-*limit set* $A(\Gamma^-)$ *of* Γ *are both closed and invariant for* Φ.

PROOF. The closedness of $\Omega(\Gamma^+)$ and $A(\Gamma^-)$ follows from (4.2.1).

Now we show that $\Omega(\Gamma^+)$ is an invariant set. Let $\phi(t) := \phi(t, x_0)$ for some $x_0 \in \mathbb{R}^n$. Then we need to show that for any $x^* \in \Omega(\Gamma^+)$, $\phi(t, x^*) \in \Omega(\Gamma^+)$ for all $t \in \mathbb{R}$. In fact, by the definition of $\Omega(\Gamma^+)$, there exists a sequence $t_i \to \infty$ such that $\phi(t_i, x_0) \to x^*$. For any fixed $t \in \mathbb{R}$, by the continuous dependence of solutions of IVPs on ICs, we see that

$$\phi(t + t_i, x_0) = \phi(t, \phi(t_i, x_0)) \to \phi(t, x^*).$$

Note that $\{s_i := t + t_i\}_{i=0}^{\infty}$ is a sequence satisfying $s_i \to \infty$ and $\phi(s_i, x_0) \to \phi(t, x^*)$. Therefore, $\phi(t, x^*) \in \Omega(\Gamma^+)$.

Similarly, we can prove that $A(\Gamma^-)$ is an invariant set of Φ. $\qquad \square$

As a consequence of Lemma 4.2.2 we have the result below.

COROLLARY 4.2.1. *Let* $\Gamma := \{\phi(t) : t \in \mathbb{R}\}$ *be an orbit of a flow* Φ. *Assume* $\Omega(\Gamma^+) = \{x^*\}$ *[*$A(\Gamma^-) = \{x^*\}$*] for some* $x^* \in \mathbb{R}^n$. *Then* x^* *is an equilibrium of* Φ.

PROOF. Without loss of generality we assume $\Omega(\Gamma^+) = \{x^*\}$. Since $x^* \in \Omega(\Gamma^+)$, by Lemma 4.2.2, $\phi_t(x^*) \in \Omega(\Gamma^+)$ for all $t \in \mathbb{R}$, i.e., $\phi_t(x^*) \equiv x^*$ on $\mathbb{R}$. Therefore, x^* is an equilibrium of Φ. $\qquad \square$

To present the next result, we need to recall the following definition for connected sets.

DEFINITION 4.2.5. A set $E \in \mathbb{R}^n$ is said to be disconnected if there are two disjoint open sets O_j, $j = 1, 2$, such that $E \subset O_1 \cup O_2$ and $E \cap O_j \neq \emptyset$ for $j = 1, 2$. A set $E \in \mathbb{R}^n$ is said to be connected if it is not disconnected.

REMARK 4.2.3. It is easy to show that if $E \in \mathbb{R}^n$ is a disconnected closed set, then there are two disjoint closed sets E_j, $j = 1, 2$, such that $E = E_1 \cup E_2$. In fact, $E_j = E \cap O_j$ with O_j defined as in Definition 4.2.5.

LEMMA 4.2.3. *Let* $\Gamma := \{\phi(t) : t \in \mathbb{R}\}$ *be an orbit of a flow* Φ.

(a) *Assume that the positive semi-orbit* Γ^+ *is bounded. Then the* ω-*limit set* $\Omega(\Gamma^+)$ *is a nonempty connected compact set. Moreover, the distance* $d(\phi(t), \Omega(\Gamma^+)) \to 0$ *as* $t \to \infty$.

(b) *Assume that the negative semi-orbit Γ^- is bounded. Then the α-limit set $A(\Gamma^-)$ is a nonempty connected compact set. Moreover, the distance $d(\phi(t), A(\Gamma^-)) \to 0$ as $t \to -\infty$.*

PROOF. We only prove Part (a). The proof for Part (b) is similar and hence is omitted.

Since Γ^+ is bounded, so is $\overline{\Gamma^+}$. By (4.2.2), $\Omega(\Gamma^+)$ is bounded. We also know from Lemma 4.2.2 that $\Omega(\Gamma^+)$ is closed. Thus, it is compact.

Now we show that $\Omega(\Gamma^+)$ is nonempty. Choose a sequence $t_i \to \infty$. Since $\Gamma^+ := \{\phi(t) : t \geq 0\}$ is a bounded set, the sequence $\{\phi(t_i)\}_{i=1}^\infty$ is bounded. Hence it has a convergent subsequence $\{\phi(t_{i_k})\}_{k=1}^\infty$. Assume $\lim_{k\to\infty} \phi(t_{i_k}) = x^*$. Since $t_{i_k} \to \infty$, $x^* \in \Omega(\Gamma^+)$.

Then we prove that $\Omega(\Gamma^+)$ is a connected set. Assume the contrary. By Remark 4.2.3, there are two disjoint bounded closed sets E_j, $j = 1, 2$, such that $\Omega(\Gamma^+) = E_1 \cup E_2$. It is easy to see (and we leave it as Exercise 4.11 to show) that $r := d(E_1, E_2) > 0$. For $j = 1, 2$, let O_j be the $(r/3)$-neighborhood of E_j. Then O_1 and O_2 are disjoint. Since all points in E_j are ω-limit points for $j = 1, 2$, there are two sequences t_i, $s_i \to \infty$, $t_i < s_i < t_{i+1}$, such that $\phi(t_i) \in O_1$ and $\phi(s_i) \in O_2$. By the continuity of $\phi(t)$, for each $i \in \mathbb{N}$, there exists a $\tau_i \in (t_i, s_i)$ such that $\phi(\tau_i) \notin O_1 \cup O_2$. This implies that

$$(4.2.3) \qquad d(\phi(\tau_i), E_j) \geq r/3 \quad \text{for } i \in \mathbb{N} \text{ and } j = 1, 2.$$

Note that the set $\{\phi(\tau_i) : i \in \mathbb{N}\} \subset \Gamma^+$ and hence is bounded. Thus, the sequence $\{\phi(\tau_i)\}_{i=1}^\infty$ has a convergent subsequence $\{\phi(\tau_{i_k})\}_{k=1}^\infty$. Assume $\lim_{k\to\infty} \phi(\tau_{i_k}) = x_*$. Since $i_k \to \infty$, $x_* \in \Omega(\Gamma^+)$, it follows that either $x_* \in E_1$ or $x_* \in E_2$. This contradicts (4.2.3).

Finally, we prove that $d(\phi(t), \Omega(\Gamma^+)) \to 0$ as $t \to \infty$. Assume the contrary. Then there exists a sequence $t_i \to \infty$ such that

$$(4.2.4) \qquad d(\phi(t_i), \Omega(\Gamma^+)) \geq \delta \quad \text{for some } \delta > 0 \text{ and all } i \in \mathbb{N}.$$

By the same argument, as in the above, we see that the sequence $\{\phi(t_i)\}_{i=1}^\infty$ has a subsequence which converges to a point in $\Omega(\Gamma^+)$. This contradicts (4.2.4) and hence completes the proof. □

REMARK 4.2.4. We point out that in Lemma 4.2.3, the connectedness of the set $\Omega(\Gamma^+)$ would not be guaranteed if the assumption that Γ^+ is bounded were removed. This can be seen from the system

$$(4.2.5) \qquad \begin{cases} x_1' = (x_1^2 - 1)(x_2 - x_1) \\ x_2' = x_1. \end{cases}$$

By drawing the phase portrait for system (4.2.5) we will find that any orbit starting from a point $x_0 := (x_{01}, x_{02})$ with $x_{01} \in (-1, 1)$ will spiral out and have the two lines $x_1 = \pm 1$ as its ω-limit set, which is obviously disconnected. Moreover, the conclusion that $d(\phi(t), \Omega(\Gamma^+)) \to 0$ as $t \to \infty$ fails to hold. We leave the details for drawing the phase portrait as an exercise, see Exercise 4.12.

Before we end this section, let us use the knowledge of dynamical systems to give a proof of Theorem 3.5.2 on the asymptotic stability by the Lyapunov function method. Recall that Theorem 3.5.2 reads as follows:

THEOREM. *Let $D = \{x \in \mathbb{R}^n : |x| \leq l\}$ for some $l > 0$ and $V \in C^1(D, \mathbb{R})$. Assume that $V(x)$ is positive definite and $\dot{V}(x)$ is negative semi-definite. Moreover, if the set $D_0 := \{x \in D : \dot{V}(x) = 0\}$ does not contain any nontrivial orbit of Eq. (A). Then the zero solution of Eq. (A) is uniformly stable and asymptotically stable.*

PROOF. Since $V(x)$ is positive definite and $\dot{V}(x)$ is negative semi-definite, by Theorem 3.5.1, the zero solution of Eq. (A) is uniformly stable. This means that there exists a $\delta > 0$ such that if $|x_0| < \delta$, then the positive semi-orbit Γ^+ passing through x_0 remains in D and hence is bounded. By Lemma 4.2.3, Part (a), the ω-limit set $\Omega(\Gamma^+)$ is nonempty.

Denote $\Gamma^+ := \{\phi(t) : t \geq 0\}$. We claim that $\Omega(\Gamma^+) = \{0\}$. In fact, $\dot{V}(x)$ is negative semi-definite implies that $V(\phi(t))$ is nonincreasing. Since $V(\phi(t)) \geq 0$, we have $\lim_{t \to \infty} V(\phi(t)) = c \geq 0$. Thus,

$$\lim_{i \to \infty} V(\phi(t_i)) = c \quad \text{for any sequence } t_i \to \infty.$$

This implies that $V(x) = c$ for all points $x \in \Omega(\Gamma^+)$. Assume there is a point $x^* \neq 0$ such that $x^* \in \Omega(\Gamma^+)$. Then $V(x^*) > 0$ since V is positive definite. By Lemma 4.2.2, the whole orbit Γ^* passing through x^* must be contained in $\Omega(\Gamma^*)$. Thus, along Γ^* we have $\dot{V}(x) = 0$ which means that $\Gamma^* \subset D_0$. This contradicts the assumption that D_0 does not contain any nontrivial orbits. As a result, $\Omega(\Gamma^+) = \{0\}$.

By Lemma 4.2.3, Part (a) again, we have $d(\phi(t), 0) \to 0$ as $t \to \infty$, i.e., $\lim_{t \to \infty} \phi(t) = 0$. Therefore, the zero solution of Eq. (A) is asymptotically stable. This completes the proof. $\qquad\square$

4.3. Linear Autonomous Systems in the Plane

In this section, we study the homogeneous linear autonomous system of dimension $n = 2$

$$(4.3.1) \qquad \begin{cases} x_1' = ax_1 + bx_2 \\ x_2' = cx_1 + dx_2 \end{cases} \quad \text{or} \quad x' = Ax,$$

where $x = \begin{bmatrix} x_1 \\ x_2 \end{bmatrix}$ and $A = \begin{bmatrix} a & b \\ c & d \end{bmatrix}$. It is known from Sect. 3.2 that the stability of system (4.3.1) is determined by the real parts of eigenvalues λ_1 and λ_2 of A. Let

$$p = a + d, \quad q = ad - bc, \quad \text{and} \quad \Delta = p^2 - 4q.$$

Since

$$|A - \lambda I| = \begin{vmatrix} a - \lambda & b \\ c & d - \lambda \end{vmatrix} = \lambda^2 - (a + d)\lambda + ad - bc = \lambda^2 - p\lambda + q,$$

we obtain that

$$\lambda_{1,2} = \frac{1}{2}(p \pm \sqrt{p^2 - 4q}) = \frac{1}{2}(p \pm \sqrt{\Delta}).$$

From this we see that

(i) λ_1 and λ_2 are distinct real eigenvalues of $A \Leftrightarrow \Delta > 0$,
(ii) λ_1 and λ_2 are equal real eigenvalues of $A \Leftrightarrow \Delta = 0$, and
(iii) λ_1 and λ_2 are complex conjugate eigenvalues of $A \Leftrightarrow \Delta < 0$.

Then we use the values of p and q to determine the signs of $\operatorname{Re}\lambda_i$, $i = 1, 2$, and hence determine the stability of system (4.3.1).

(a) When $q < 0$, then $\lambda_1 < 0 < \lambda_2$, and hence system (4.3.1) is unstable.
(b) When $q > 0$, then

$$\begin{cases} p > 0 \Rightarrow \operatorname{Re}\lambda_{1,2} > 0, & \text{and hence (4.3.1) is unstable;} \\ p < 0 \Rightarrow \operatorname{Re}\lambda_{1,2} < 0, & \text{and (4.3.1) is asymptotically stable;} \\ p = 0 \Rightarrow \lambda_{1,2} = \pm\sqrt{q}\,\mathrm{i}, & \text{and (4.3.1) is stable but not asymptotically stable.} \end{cases}$$

(c) When $q = 0$, then

$$\begin{cases} p > 0 \Rightarrow \lambda_2 > \lambda_1 = 0, \text{and hence (4.3.1) is unstable;} \\ p < 0 \Rightarrow \lambda_2 < \lambda_1 = 0, \text{and hence (4.3.1) is stable but not asymptotically stable;} \\ p = 0 \Rightarrow \lambda_2 = \lambda_1 = 0, \text{and (4.3.1) may be stable or unstable.} \end{cases}$$

For the last case, where $p = q = 0$, system (4.3.1) is stable if $A = \begin{bmatrix} 0 & 0 \\ 0 & 0 \end{bmatrix}$; and is unstable if not all entries of A are 0, in this case, the Jordan canonical form of A is $J = \begin{bmatrix} 0 & 1 \\ 0 & 0 \end{bmatrix}$.

We observe that in case (c), the equilibrium $(0, 0)$ is not isolated. In fact, when $q = 0$, the equation $Ax = 0$ has an infinite number of solutions which include either all points on the line $ax_1 + bx_2 = 0$ or all points in $\mathbb{R}^2$.

The stability results for all values of p and q are summarized in the chart in Fig. 4.5, where (S), (US), (AS), and (NS) stand for stable, uniformly stable, asymptotically stable, and unstable, respectively.

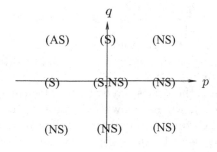

FIGURE 4.5. Stability chart

We now further study the phase portrait for system (4.3.1) near the equilibrium $(0,0)$, i.e., the geometric representation of the topological structure of orbits in a neighborhood of $(0,0)$. We only consider the case when $(0,0)$ is an isolated equilibrium, in other words, when $q \neq 0$.

Recall that λ_1 and λ_2 may be real-valued or complex-valued. When they are real-valued, the Jordan canonical form J of A is either $\begin{bmatrix} \lambda_1 & \\ & \lambda_2 \end{bmatrix}$, or $\begin{bmatrix} \lambda & 1 \\ & \lambda \end{bmatrix}$ if $\lambda_1 = \lambda_2 = \lambda$ and they are not simple eigenvalues. When they are complex-valued, i.e., $\lambda_{1,2} = \alpha \pm i\beta$, then $J = \begin{bmatrix} \alpha - i\beta & \\ & \alpha + i\beta \end{bmatrix}$. However, since $\begin{bmatrix} \alpha - i\beta & \\ & \alpha + i\beta \end{bmatrix} \sim \begin{bmatrix} \alpha & \beta \\ -\beta & \alpha \end{bmatrix}$, we would rather let $J = \begin{bmatrix} \alpha & \beta \\ -\beta & \alpha \end{bmatrix}$ to make it real-valued. In all the cases, there exists a $T \in \mathbb{R}^{n \times n}$ such that $TAT^{-1} = J$. For any solution $x(t)$ of system (4.3.1), let $y(t) = Tx(t)$. Then $y(t)$ satisfies the system

$$(4.3.2) \qquad\qquad y' = Jy.$$

Clearly, systems (4.3.1) and (4.3.2) have the same topological structure of orbits. Therefore, we will use system (4.3.2) to investigate the topological structure of system (4.3.1).

Case 1. $\lambda_{1,2}$ are distinct real eigenvalues of the same sign.

(i) Assume $\Delta > 0$, $q > 0$, and $p < 0$. In this case, $\lambda_2 < \lambda_1 < 0$. Hence the equilibrium $(0,0)$ is asymptotically stable. Note that system (4.3.2) becomes

$$(4.3.3) \qquad\qquad \begin{bmatrix} y_1 \\ y_2 \end{bmatrix}' = \begin{bmatrix} \lambda_1 & \\ & \lambda_2 \end{bmatrix} \begin{bmatrix} y_1 \\ y_2 \end{bmatrix},$$

and the general solution $y_1 = c_1 e^{\lambda_1 t}$, $y_2 = c_2 e^{\lambda_2 t}$ satisfies that for $i = 1, 2$, $y_i(t) \to 0 \, [\pm\infty]$ as $t \to \infty \, [-\infty]$.

Moreover, by taking $c_1 = 0$ and $c_2 = 0$, respectively, we see that there are orbits on both sides of the origin on the y_2-axis and on the y_1-axis. For $c_1, c_2 \neq 0$,

$$\frac{y_2(t)}{y_1(t)} = \frac{c_2}{c_1} e^{(\lambda_2 - \lambda_1)t} \to 0 \, [\pm\infty] \quad \text{as } t \to \infty \, [-\infty].$$

This implies that for orbits not on the y_1-axis and y_2-axis, the slopes tend to 0 as the orbits approach $(0,0)$ and tend to $\pm\infty$ as the orbits approach infinity.

Therefore, the orbits have the topological structure as shown in Fig. 4.6, (a), where arrows indicate the directions of orbits as $t \to \infty$. The equilibrium $(0,0)$ is called a stable improper-node which means that all orbits approach $(0,0)$ in one pair of opposite directions as $t \to \infty$ except two which approach $(0,0)$ in a different pair of opposite directions as $t \to \infty$.

(ii) Assume $\Delta > 0$, $q > 0$, and $p > 0$. In this case, $\lambda_2 > \lambda_1 > 0$. Hence the equilibrium $(0,0)$ is unstable. Note that system (4.3.2) is the same as system (4.3.3) except that the signs of λ_1 and λ_2 change. Therefore, the topological structure of orbits is essentially the same as in Part (i) except that the directions of orbits are reversed, as shown in Fig. 4.6, (b). The equilibrium $(0,0)$ is then called an unstable improper-node.

Case 2. $\lambda_{1,2}$ are equal real eigenvalues.

This is the case when $\Delta = 0$. Thus $\lambda_{1,2} = p/2$. We only consider $p \neq 0$ since otherwise, $p = q = 0$ which implies that $(0,0)$ is not isolated. We know that the equilibrium $(0,0)$ is asymptotically stable for $p < 0$ and unstable for $p > 0$. In this case, $J = \begin{bmatrix} p/2 & k \\ & p/2 \end{bmatrix}$ for $k = 0$ or 1.

a

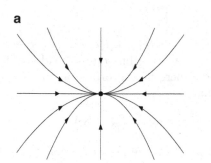

b

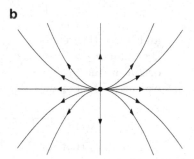

FIGURE 4.6. Improper-node. (a) Stable improper-node. (b) Unstable improper-node

(i) Assume $k = 0$. Note that system (4.3.2) becomes

$$\begin{bmatrix} y_1 \\ y_2 \end{bmatrix}' = \begin{bmatrix} p/2 & \\ & p/2 \end{bmatrix} \begin{bmatrix} y_1 \\ y_2 \end{bmatrix},$$

and hence the general solution is $y_1 = c_1 e^{pt/2}$, $y_2 = c_2 e^{pt/2}$. By taking $c_1 = 0$ and $c_2 = 0$, respectively, we see that there are orbits on both sides of the origin on the y_2-axis and on the y_1-axis. For $c_1, c_2 \neq 0$, we have $y_2(t) = (c_2/c_1)y_1(t)$ which shows that there are orbits on both sides of the origin on straight lines passing through the origin with any slopes.

Therefore, the orbits have the topological structure as shown in Fig. 4.7, where Part (a) is for the asymptotically stable case with $p < 0$ and Part (b) is for the unstable case with $p > 0$. The equilibrium $(0,0)$ is then called a proper-node, stable or unstable, which means that there are orbits approaching $(0,0)$, or going away from $(0,0)$, in any direction as $t \to \infty$.

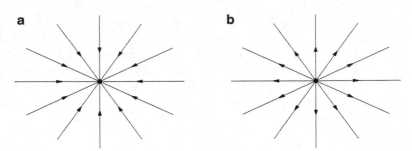

FIGURE 4.7. Proper-node. (**a**) Stable proper-node. (**b**) Unstable proper-node

(ii) Assume $k = 1$. Note that system (4.3.2) becomes

$$\begin{bmatrix} y_1 \\ y_2 \end{bmatrix}' = \begin{bmatrix} p/2 & 1 \\ & p/2 \end{bmatrix} \begin{bmatrix} y_1 \\ y_2 \end{bmatrix},$$

and hence the general solution is $y_1 = (c_1 + c_2 t)e^{pt/2}$, $y_2 = c_2 e^{pt/2}$. By taking $c_2 = 0$, we see that there are orbits on both sides of the origin on the y_1-axis. However, there are no orbits on the y_2-axis except $(0,0)$. For $c_2 \neq 0$, we have

$$\frac{y_1(t)}{y_2(t)} = \frac{c_1}{c_2} + t \to \infty\,[-\infty] \quad \text{as } t \to \infty\,[-\infty].$$

This implies that for the orbits not on the y_1-axis, the slopes tend to 0 as the orbits approach $(0,0)$ and as the orbits approach infinity.

Therefore, the orbits have the topological structure as shown in Fig. 4.8, where Part (a) is for the asymptotically stable case with $p < 0$ and Part (b) is for the unstable case with $p > 0$. The equilibrium $(0,0)$ is called a degenerate-node, stable or unstable, which means that all orbits approach $(0,0)$, or go away from $(0,0)$, in only one pair of opposite directions as $t \to \infty$.

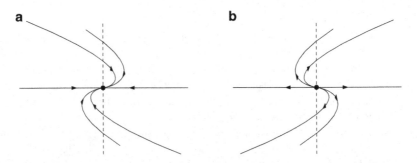

FIGURE 4.8. Degenerate-node. (**a**) Stable proper-node. (**b**) Unstable proper-node

Case 3. $\lambda_{1,2}$ are real eigenvalues of the opposite signs.

This is the case when $q < 0$. Thus $\lambda_1 < 0 < \lambda_2$. Hence the equilibrium $(0,0)$ is unstable. Note that system (4.3.2) becomes

$$\begin{bmatrix} y_1 \\ y_2 \end{bmatrix}' = \begin{bmatrix} \lambda_1 & \\ & \lambda_2 \end{bmatrix} \begin{bmatrix} y_1 \\ y_2 \end{bmatrix},$$

and the general solution $y_1 = c_1 e^{\lambda_1 t}$, $y_2 = c_2 e^{\lambda_2 t}$ satisfies

$$y_1(t) \to 0\,[\pm\infty] \quad \text{and} \quad y_2(t) \to \pm\infty\,[0] \quad \text{as } t \to \infty\,[-\infty].$$

By taking $c_1 = 0$ and $c_2 = 0$, respectively, we see that there are orbits on both sides of the origin on the y_2-axis going away from $(0,0)$ and there are orbits on both sides of the origin on the y_1-axis approaching $(0,0)$, as $t \to \infty$. By looking at the solutions with $c_1, c_2 \neq 0$, we see that the orbits not on the y_1-axis or y_2-axis will go away from $(0,0)$ as $t \to \infty$ and as $t \to -\infty$.

Therefore, the orbits have the topological structure as shown in Fig. 4.9. The equilibrium $(0,0)$ is then called a saddle-point. Note that a saddle-point is always unstable.

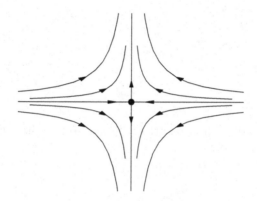

FIGURE 4.9. Saddle-point

Case 4. $\lambda_{1,2}$ are complex conjugate eigenvalues.

This is the case when $\Delta < 0$. Thus $\lambda_{1,2} = \alpha \pm i\beta$ with $\alpha, \beta \in \mathbb{R}$ such that $\beta \neq 0$. Hence the equilibrium $(0,0)$ is asymptotically stable for $\alpha < 0$, stable for $\alpha = 0$, and unstable for $\alpha > 0$. Note that system (4.3.2) becomes

$$(4.3.4) \qquad \begin{bmatrix} y_1 \\ y_2 \end{bmatrix}' = \begin{bmatrix} \alpha & \beta \\ -\beta & \alpha \end{bmatrix} \begin{bmatrix} y_1 \\ y_2 \end{bmatrix}.$$

We will change system (4.3.4) into a "polar-coordinate form" by letting

$$y_1 = r\cos\theta \quad \text{and} \quad y_2 = r\sin\theta.$$

Hence

$$r^2 = y_1^2 + y_2^2 \quad \text{and} \quad \tan\theta = \frac{y_2}{y_1}.$$

It follows that
$$rr' = y_1 y_1' + y_2 y_2' = \alpha r^2 \quad \text{and} \quad r^2 \theta' = y_1 y_2' - y_1' y_2 = -\beta r^2$$
which are simplified to
$$r' = \alpha r \quad \text{and} \quad \theta' = -\beta.$$
Then the general solution is $r = r_0 e^{\alpha t}$, $\theta = \theta_0 - \beta t$ with arbitrary $r_0 \geq 0$ and $\theta_0 \in \mathbb{R}$.

Therefore, for $\beta < 0$, the orbits have the topological structure as shown in Fig. 4.10. The equilibrium $(0,0)$ in Part (a), where $\alpha < 0$, is called a stable spiral-point; the one in Part (b), where $\alpha > 0$, is called an unstable spiral-point. The one in Fig. 4.11 is for $\alpha = 0$ and is called a center. Note that a center is stable but not asymptotically stable. The topological structure for $\beta > 0$ is similar.

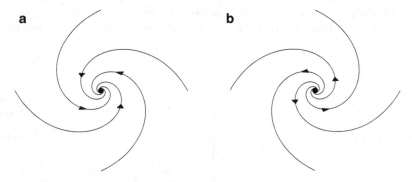

FIGURE 4.10. Spiral-point. (a) Stable spiral-point. (b) Unstable spiral-point

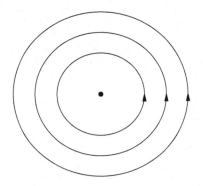

FIGURE 4.11. Center

Now, we have completed the discussions on the global phase portraits of Eq. (4.3.1) for all values of p and q except $q = 0$. The results on the types and stabilities of equilibria are summarized in the chart in Fig. 4.12, where (S), (US), (AS), and (NS) stand for stable, uniformly stable, asymptotically stable, and unstable, respectively.

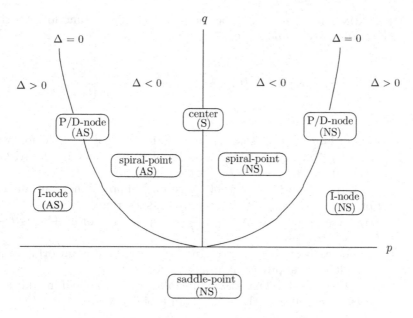

FIGURE 4.12. Equilibrium classification chart

4.4. Autonomous Perturbations of Linear Systems in the Plane

We then investigate how the phase portrait for the linear system (4.3.1) is affected by higher-order perturbations. Consider the nonlinear autonomous system

$$(4.4.1) \qquad \begin{cases} x_1' = ax_1 + bx_2 + f_1(x_1, x_2) \\ x_2' = cx_1 + dx_2 + f_2(x_1, x_2) \end{cases} \quad \text{or} \quad x' = Ax + f(x),$$

where $x = \begin{bmatrix} x_1 \\ x_2 \end{bmatrix}$, $A = \begin{bmatrix} a & b \\ c & d \end{bmatrix}$, and $f = \begin{bmatrix} f_1 \\ f_2 \end{bmatrix}$. Here we assume that $ad - bc \neq 0$ and for $i = 1, 2$, $f_i : \mathbb{R}^2 \to \mathbb{R}$ is continuous in a neighborhood of $(0, 0)$ satisfying that $f_i(0, 0) = 0$ and $f_i(x_1, x_2) = o(\sqrt{x_1^2 + x_2^2})$ as $(x_1, x_2) \to (0, 0)$. Therefore, $(0, 0)$ is an isolated equilibrium for both systems (4.3.1) and (4.4.1).

Definitely, for Eq. (4.4.1), the global topological structure for orbits will be affected by the nonlinear perturbation terms. However, we may expect that the local topological structure for orbits near equilibrium $(0, 0)$ remains the same if $(0, 0)$ belongs to certain types of equilibria. To see the phase portrait of system (4.4.1) near $(0, 0)$, we need to extend the classification of equilibria for linear systems to nonlinear systems.

DEFINITION 4.4.1. Let r and θ be the polar-coordinate functions for the Cartesian coordinates x_1 and x_2, i.e., $r \in C(\mathbb{R}, [0, \infty))$ and $\theta \in C(\mathbb{R}, \mathbb{R})$ such that

$$(4.4.2) \qquad r^2(t) = x_1^2(t) + x_2^2(t) \quad \text{and} \quad \tan \theta(t) = \frac{x_2(t)}{x_1(t)}.$$

Then

(a) Equilibrium $(0,0)$ is said to be a stable [unstable] node for system (4.4.1) if in a neighborhood $\mathcal{N}$ of $(0,0)$, every orbit satisfies $r(t) \to 0$ and $\theta(t) \to c$ as $t \to \infty[-\infty]$ for some $c \in [0, 2\pi)$.

$(0,0)$ is said to be a stable [unstable] improper-node if in addition, there exist $c_1, c_2 \in [0, \pi)$, $c_1 \neq c_2$, such that

(i) for each orbit in $\mathcal{N}$, $\theta(t) \to c_1, c_1 + \pi, c_2$, or $c_2 + \pi$ as $t \to \infty[-\infty]$;

(ii) for each of $c_1, c_1 + \pi, c_2$, and $c_2 + \pi$, there is an orbit in $\mathcal{N}$ with $\theta(t)$ approaching it as $t \to \infty[-\infty]$.

$(0,0)$ is said to be a stable [unstable] proper-node if in addition, for every $c \in [0, 2\pi)$, there exists an orbit in $\mathcal{N}$ such that $\theta(t) \to c$ as $t \to \infty[-\infty]$.

$(0,0)$ is said to be a stable [unstable] degenerate-node if in addition, there exists a $c \in [0, \pi)$ such that

(i) for each orbit in $\mathcal{N}$, $\theta(t) \to c$ or $c + \pi$ as $t \to \infty[-\infty]$;

(ii) for each of c and $c + \pi$, there is an orbit in $\mathcal{N}$ with $\theta(t)$ approaching it as $t \to \infty[-\infty]$.

(b) Equilibrium $(0,0)$ is said to be a stable [unstable] spiral-point for system (4.4.1) if every orbit in a neighborhood of $(0,0)$ satisfies $r(t) \to 0$ and $\theta(t) \to \infty$ or $-\infty$ as $t \to \infty[-\infty]$.

(c) Equilibrium $(0,0)$ is said to be a center for system (4.4.1) if every neighborhood of $(0,0)$ contains a closed orbit.

(d) Equilibrium $(0,0)$ is said to be a saddle-point for system (4.4.1) if there exist $c_1, c_2 \in [0, \pi)$, $c_1 \neq c_2$, such that in a neighborhood $\mathcal{N}$ of $(0,0)$,

(i) there is a unique orbit satisfying $r(t) \to 0$ and $\theta(t) \to c_1$ and a unique orbit satisfying $r(t) \to 0$ and $\theta(t) \to c_1 + \pi$ as $t \to \infty$;

(ii) there is a unique orbit satisfying $r(t) \to 0$ and $\theta(t) \to c_2$ and a unique orbit satisfying $r(t) \to 0$ and $\theta(t) \to c_2 + \pi$ as $t \to -\infty$;

(iii) all other orbits tend away from $(0,0)$ as $t \to \infty$ and as $t \to -\infty$.

From the definition we see that a center is stable and a saddle-point is unstable. From Example 4.6.3 in Sect. 4.6, we will see that unlike those of the linear system (4.3.1), the orbits of the nonlinear system (4.4.1) surrounding a center may not always be closed orbits.

Now we are ready to show that for certain cases, the phase portrait for system (4.3.1) near the equilibrium $(0,0)$ is preserved by system (4.4.1).

THEOREM 4.4.1. *(a) If the equilibrium $(0,0)$ is a spiral-point for system $(4.3.1)$, then it is a spiral-point for system $(4.4.1)$;*

(b) If the equilibrium $(0,0)$ is an improper-node for system $(4.3.1)$, then it is an improper-node for system $(4.4.1)$;

(c) If the equilibrium $(0,0)$ is a saddle-point for system $(4.3.1)$, then it is a saddle-point for system $(4.4.1)$.

In all the above cases, the stability property of the equilibrium $(0,0)$ is also preserved.

PROOF. (a) Without loss of generality, we may let $(0,0)$ be a stable spiral-point for system $(4.3.1)$. The proof for an unstable spiral-point case is similar. As shown in Case 4 of Sect. 4.3, by taking a linear transformation if necessary, we may assume that the coefficient matrix $A = \begin{bmatrix} \alpha & \beta \\ -\beta & \alpha \end{bmatrix}$, where $\alpha, \beta \in \mathbb{R}$ such that $\alpha < 0$ and $\beta \neq 0$. Hence system $(4.4.1)$ becomes

$$(4.4.3) \qquad \begin{cases} x_1' = \alpha x_1 + \beta x_2 + f_1(x_1, x_2) \\ x_2' = -\beta x_1 + \alpha x_2 + f_2(x_1, x_2), \end{cases}$$

where for $i = 1, 2$, $f_i : \mathbb{R}^2 \to \mathbb{R}$ is continuous in a neighborhood of $(0,0)$ satisfying $f_i(0,0) = 0$ and $f_i(x_1, x_2) = o(\sqrt{x_1^2 + x_2^2})$ as $(x_1, x_2) \to (0,0)$.

Let $r(t)$ and $\theta(t)$ be defined by $(4.4.2)$. Note that $\operatorname{Re} \lambda_i(A) = \alpha < 0$ for $i = 1, 2$. By Theorem 3.4.1, the zero solution $(0,0)$ of system $(4.4.3)$ is asymptotically stable, which implies that $r(t) \to 0$ as $t \to \infty$. From $(4.4.2)$ we have

$$
\begin{aligned}
r^2 \theta' &= x_1 x_2' - x_1' x_2 \\
&= x_1(-\beta x_1 + \alpha x_2 + f_2(x_1, x_2)) - x_2(\alpha x_1 + \beta x_2 + f_1(x_1, x_2)) \\
&= -\beta r^2 + o(r^2) \quad \text{as } t \to \infty
\end{aligned}
$$

which is simplified to $\theta' = -\beta + o(1)$ as $t \to \infty$. Thus $\theta(t) = -\beta t + o(t) \to \infty$ or $-\infty$ as $t \to \infty$. Therefore, $(0,0)$ is a stable spiral-point for system $(4.4.1)$.

The proofs for Parts (b) and (c) are quite involved and lengthy and hence are omitted. The interested reader is referred to [11, Chapter 15] or [18, Section 5.3] for the details. $\qquad \square$

REMARK 4.4.1. We point out that when the equilibrium $(0,0)$ is a proper-node, degenerate-node, or a center, the topological structure of orbits of system $(4.3.1)$ near $(0,0)$ is generally not preserved by the perturbed system $(4.4.1)$. In fact, we have the following:

(i) If the equilibrium $(0,0)$ is a proper-node or degenerate-node for system $(4.3.1)$, then it may be any kind of node or a spiral-point for system $(4.4.1)$ with the stability property preserved;

(ii) If the equilibrium $(0,0)$ is a center for system $(4.3.1)$, then it may be a center for system $(4.4.1)$, or a spiral-point for system $(4.4.1)$, either asymptotically stable or unstable.

EXAMPLE 4.4.1. Let $r(t)$ and $\theta(t)$ be defined by (4.4.2). Consider the system

$$(4.4.4) \qquad \begin{cases} x_1' = -x_1 - x_2(\ln r)^{-1} \\ x_2' = -x_2 + x_1(\ln r)^{-1}. \end{cases}$$

Although the right-hand functions are not defined at $(0,0)$, we may extend them to $(0,0)$ by continuity. Thus, $(0,0)$ is an equilibrium. Note that the coefficient matrix of the linearized system is $A = \begin{bmatrix} -1 & 0 \\ 0 & -1 \end{bmatrix}$ and the nonlinear terms $x_i(\ln r)^{-1} = o(r)$ as $r \to 0$ for $i = 1, 2$. By Case 2, (i) of Sect. 4.3, $(0,0)$ is a stable proper-node for the linearized system.

However, from (4.4.2) we have

$$rr' = x_1 x_1' + x_2 x_2' = -r^2 \quad \text{and} \quad r^2\theta' = x_1 x_2' - x_1' x_2 = r^2(\ln r)^{-1}$$

which are simplified to

$$r' = -r \quad \text{and} \quad \theta' = (\ln r)^{-1}.$$

By solving this system we obtain that

$$r(t) = r_0 e^{-t} \to 0 \quad \text{and} \quad \theta(t) = \theta_0 - \ln(t - \ln r_0) \to -\infty \quad \text{as } t \to \infty.$$

This shows that $(0,0)$ is a stable spiral-point for system (4.4.4).

EXAMPLE 4.4.2. Let $r(t)$ and $\theta(t)$ be defined by (4.4.2). Consider the system

$$(4.4.5) \qquad \begin{cases} x_1' = x_2 - x_1 r^2 \\ x_2' = -x_1 - x_2 r^2. \end{cases}$$

Note that the coefficient matrix of the linearized system is $A = \begin{bmatrix} 0 & 1 \\ -1 & 0 \end{bmatrix}$ and hence by Case 4 of Sect. 4.3, $(0,0)$ is a center for the linearized system.

However, from (4.4.2) we have

$$rr' = x_1 x_1' + x_2 x_2' = -r^4 \quad \text{and} \quad r^2\theta' = x_1 x_2' - x_1' x_2 = -r^2$$

which are simplified to

$$r' = -r^3 \quad \text{and} \quad \theta' = -1.$$

By solving this system we obtain that

$$r(t) = \frac{1}{\sqrt{2t + 1/r_0^2}} \to 0 \quad \text{and} \quad \theta(t) = -t + \theta_0 \to -\infty \quad \text{as } t \to \infty.$$

This shows that $(0,0)$ is a stable spiral-point for system (4.4.5).

In the same way, we can show that the equilibrium $(0,0)$ is an unstable spiral-point for the system

$$\begin{cases} x_1' = x_2 + x_1 r^2 \\ x_2' = -x_1 + x_2 r^2. \end{cases}$$

Then, we introduce two results on the preservation of proper-nodes, degenerate-nodes, and centers with stronger assumptions on the perturbation functions f_1 and f_2.

THEOREM 4.4.2. *In addition to the assumptions for system* (4.4.1), *we assume that for* $i = 1, 2$, $f_i(x_1, x_2) = o((x_1^2 + x_2^2)^{\delta/2})$ *as* $(x_1, x_2) \to (0, 0)$ *for some* $\delta > 1$. *If the equilibrium* $(0, 0)$ *is a proper-node [degenerate-node] for system* (4.3.1), *then it is a proper-node [degenerate-node] for system* (4.4.1).

We refer the interested reader to Theorem 3.1 and its corollary in [11, Chapter 15] for the proof. We point out that system (4.4.4) in Example 4.4.2 does not satisfy the assumption of Theorem 4.4.2.

THEOREM 4.4.3. *Let* $(0, 0)$ *be a center for system* (4.3.1). *By taking a linear transformation if necessary, we may assume* $a = d = 0$ *and* $bc < 0$ *in* (4.3.1). *If either*

$$f_1(-x_1, x_2) \equiv f_1(x_1, x_2) \quad and \quad f_2(-x_1, x_2) \equiv -f_2(x_1, x_2)$$

or

$$f_1(x_1, -x_2) \equiv -f_1(x_1, x_2) \quad and \quad f_2(x_1, -x_2) \equiv f_2(x_1, x_2),$$

then $(0, 0)$ *is a center for system* (4.4.1).

We observe that if an orbit is symmetric about the x_1-axis or the x_2-axis, then the orbit must be a closed orbit. Then the proof is based on Remark 4.4.1, Part (b) and the symmetric properties of the orbits determined by the assumptions. We leave it as an exercise, see Exercise 4.17.

Before we conclude this section, we point out that all the equilibria of Eq. (4.4.1) discussed above are under the assumption that $ad - bc \neq 0$. Such equilibria are called elementary equilibria. We observe that all elementary equilibria are isolated. However, it is not always true that the equilibria of Eq. (4.4.1) are elementary, even if they are isolated. For example, $(0, 0)$ is the only equilibrium of the system

$$\begin{cases} x_1' = x_1^2 - x_2^2 \\ x_2' = x_1 x_2, \end{cases}$$

but it is not elementary. In this case, we have $a = b = c = d = 0$.

4.5. Poincaré–Bendixson Theorem

Starting from this section, we study the general nonlinear autonomous system in the plane

$$\text{(A-2)} \qquad \begin{cases} x_1' = f_1(x_1, x_2) \\ x_2' = f_2(x_1, x_2) \end{cases} \quad \text{or} \quad x' = f(x),$$

where $x = \begin{bmatrix} x_1 \\ x_2 \end{bmatrix}$, and $f = \begin{bmatrix} f_1 \\ f_2 \end{bmatrix} \in C^1(\mathbb{R}^2, \mathbb{R}^2)$.

By definition, we know that if a positive semi-orbit Γ^+ of system (A-2) is unbounded, then the ω-limit set $\Omega(\Gamma^+)$ may be an empty set. However, when

Γ^+ is bounded, then $\Omega(\Gamma^+)$ is guaranteed to be nonempty. The same holds for the α-limit set $A(\Gamma^-)$ for a negative semi-orbit Γ^- of system (A-2). Now, we investigate the structure of $\Omega(\Gamma^+)$ and $A(\Gamma^-)$ when Γ^+ and Γ^- are bounded, respectively. The answer is given by the famous Poincaré–Bendixson theorem below.

THEOREM 4.5.1 (Poincaré–Bendixson). *Let $x(t)$ be a solution of system* (A-2) *with Γ its orbit. Assume Γ^+ is bounded and $\Omega(\Gamma^+)$ contains only regular points. Then either*

    (a) Γ is a closed orbit, or
    (b) $\Omega(\Gamma^+)$ is a closed orbit.

The same conclusion holds when Γ^+ and $\Omega(\Gamma^+)$ are replaced by Γ^- and $A(\Gamma^-)$, respectively.

To prove Theorem 4.5.1, we introduce the following concept.

DEFINITION 4.5.1. A finite closed line segment L in the x_1x_2-plane is called a transversal for system (A-2) if it consists of only regular points and the direction field determined by f is not tangent to L at any point of L.

REMARK 4.5.1. From Definition 4.5.1 we see directly that

    (i) Every regular point $x \in \mathbb{R}^2$ is an interior point of some transversal, which may have any direction except that of $f(x)$.
    (ii) Every orbit which meets a transversal must cross it, and all such orbits cross it from the same side to the other. Otherwise, there exists a point $x^* \in L$ where $f(x)$ reverses its direction. As a result, either $f(x^*) = 0$ which means that x^* is a singular point, or f is tangent to L at x^*. This contradicts the definition of transversal.

For further discussions on the properties of transversals, we need to use the implicit function theorem stated below.

IMPLICIT FUNCTION THEOREM. *Let $g \in C^1(D, \mathbb{R})$ with D an open subset of $\mathbb{R} \times \mathbb{R}^2$ and $(t_0, x_0) \in D$. Assume that*

$$g(t_0, x_0) = 0 \quad and \quad \frac{\partial g}{\partial t}(t_0, x_0) \neq 0.$$

Then there is a neighborhood $D_t \times D_x \subset D$ of (t_0, x_0) with $D_t \subset \mathbb{R}$ and $D_x \subset \mathbb{R}^2$ and a unique function $h \in C^1(D_x, D_t)$ such that

$$g(h(x), x) \equiv 0 \quad and \quad h(x_0) = t_0.$$

The next lemma for a property of transversals is derived using this implicit function theorem and is illustrated by Fig. 4.13.

LEMMA 4.5.1. *Let x_0 be an interior point of a transversal L. Then for any $\epsilon > 0$, there is a $\delta > 0$ such that every orbit passing through a point in the δ-neighborhood of x_0 at $t = t_0$ crosses L at some t with $|t - t_0| < \epsilon$.*

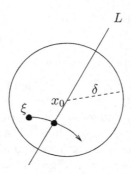

FIGURE 4.13. Transversal property

PROOF. Assume the points $x = (x_1, x_2)$ on the transversal L satisfy the equation $ax_1 + bx_2 + c = 0$. Then L is perpendicular to the vector (a, b). Since x_0 is a regular point, by the continuity of f, there is a neighborhood $\mathcal{N}$ of x_0 which contains only regular points. For any $\xi \in \mathcal{N}$, let $\phi(t; \xi) := (\phi_1, \phi_2)(t; \xi)$ be the solution of system (A-2) satisfying $\phi(t_0, \xi) = \xi$. By Theorem 1.5.1, $\phi(t; \xi)$ is continuous in (t, ξ). Define a function

$$g(t, \xi) = a\phi_1(t; \xi) + b\phi_2(t; \xi) + c.$$

Then $\phi(t; \xi)$ crosses the transversal L at some time t^* if and only if $g(t^*, \xi) = 0$. Note that $x_0 \in L$ and $\phi(t_0, x_0) = x_0$, we have that $g(t_0, x_0) = 0$. We claim that $\partial g / \partial t(t_0, x_0) \neq 0$. Otherwise,

$$\frac{\partial g}{\partial t}(t_0, x_0) = a\frac{\partial \phi_1}{\partial t}(t_0, x_0) + b\frac{\partial \phi_2}{\partial t}(t_0, x_0) = 0$$

which implies that

$$\left(\frac{\partial \phi_1}{\partial t}, \frac{\partial \phi_2}{\partial t}\right)(t_0, x_0) \perp (a, b).$$

This means that the vector field determined by f is tangent to L at x_0, and hence contradicts the definition of transversal. By the implicit function theorem, when the neighborhood $\mathcal{N}$ is small enough, there exists a unique C^1 function $t = h(\xi)$ defined in $\mathcal{N}$ such that

$$g(h(\xi), \xi) \equiv 0 \text{ in } \mathcal{N} \quad \text{and} \quad h(x_0) = t_0.$$

Hence $h(\xi)$ is the time for $\phi(t; \xi)$ to cross the transversal L. Therefore, the conclusion of the lemma follows from the continuity of the function h. $\qquad \square$

Then we introduce a result on the order of the points where an orbit crosses a transversal.

LEMMA 4.5.2. *Let Γ be an orbit and L a transversal.*

(a) *Assume Γ and L intersect at a sequence of points $S := \{x_1, x_2, \ldots\}$ (finite or infinite), which are ordered as t increases or decreases. If any two points in S are different, then all points in S are distinct, and the sequence is monotone on L.*

(b) *Assume Γ is a closed orbit. Then Γ and L can only intersect at one point.*

PROOF. Let Γ be the orbit of a solution $\phi(t)$ of system (A-2).

(a) Let $x_i = \phi(t_i)$, $i = 1, 2, \ldots$. Without loss of generality assume $t_1 < t_2 < \cdots$ and $x_1 \neq x_2$. Denote by $J_1 := \{\phi(t) : t_1 \leq t \leq t_2\}$ the arc of Γ between x_1 and x_2 and by J_2 the segment of L between x_1 and x_2, and let $J = J_1 \cup J_2$. Then J is a Jordan curve. This means that J separates the plane $\mathbb{R}^2$ into two regions: the interior G_1 and the exterior G_2 as shown in Fig. 4.14. Without loss of generality we may assume that J has the shape as in Fig. 4.14, Part (a).

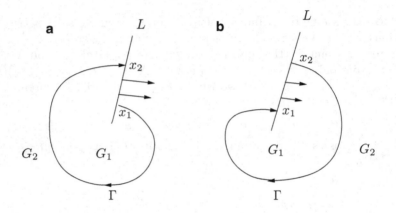

FIGURE 4.14. Jordan curve

We know from Corollary 4.1.2 that Γ does not intersect itself and from Remark 4.5.1 that Γ crosses L from the same side at all x_i, $i = 1, 2 \ldots$. As a result, Γ cannot enter G_1 either from J_1 or from J_2 for $t > t_2$. Therefore, x_3 must be outside G_1 which implies that $x_3 \neq x_i$, $i = 1, 2$, and x_1, x_2, x_3 remain the same order on L. By induction, we see that x_i, $i = 1, 2, \ldots$ are distinct and remain the same order on L.

(b) Assume Γ and L intersect at two different point x_1 and x_2. By Part (a), if Γ intersects L later at a point x_3, then x_1, x_2, x_3 must remain the same order on L, which implies that Γ can never return to x_1 at any later time. This shows that Γ is not a closed orbit.

□

The next two lemmas discuss some further properties of the limit sets of orbits. To avoid redundancy, we only present the results for the ω-limit sets. The same also hold for the α-limits.

LEMMA 4.5.3. *Let Γ be an orbit and L a transversal.*

(a) *Assume $\Omega(\Gamma^+)$ intersects L at a point x^*. Then there exists a sequence of points $\{x_i\}_{i=1}^\infty \subset \Gamma \cap L$ such that $x_i \to x^*$.*

(b) *$\Omega(\Gamma^+)$ intersects L at no more than one point.*

PROOF. (a) Let Γ be the orbit of a solution $\phi(t)$ of system (A-2). Since $x^* \in \Omega(\Gamma^+)$, there exists a sequence $t_i \to \infty$ such that $y_i := \phi(t_i) \to x^*$ as $i \to \infty$. By Lemma 4.5.1, for $i \in \mathbb{N}$ sufficiently large, the orbit passing through y_i at t_i will cross L at a time $\bar{t}_i$ which satisfies $|\bar{t}_i - t_i| \to 0$ as $i \to \infty$. Let $x_i = \phi(\bar{t}_i)$. By the continuity of the solution $\phi(t)$ we have

$$
\begin{aligned}
\lim_{i\to\infty} x_i &= \lim_{i\to\infty} [y_i + (x_i - y_i)] \\
&= \lim_{i\to\infty} [y_i + \phi(\bar{t}_i) - \phi(t_i)] = \lim_{i\to\infty} y_i = x^*.
\end{aligned}
$$

(b) Assume $\Omega(\Gamma^+)$ intersects L at two points x_1^* and x_2^*. By the above, there exist two sequences

$$
\{\bar{x}_i = \phi(\bar{t}_i)\}_{i=1}^\infty \quad \text{and} \quad \{\tilde{x}_i = \phi(\tilde{t}_i)\}_{i=1}^\infty
$$

in L such that $\bar{t}_i \to \infty$, $\tilde{t}_i \to \infty$, $\bar{x}_i \to x_1^*$, and $\tilde{x}_i \to x_2^*$. By taking subsequences if necessary, we may assume that

$$
\bar{t}_1 < \tilde{t}_1 < \bar{t}_2 < \tilde{t}_2 < \cdots .
$$

Then by Lemma 4.5.2, the sequence $\bar{x}_1, \tilde{x}_1, \bar{x}_2, \tilde{x}_2, \ldots$ is monotone on L and hence has a unique limit. This means that $x_1^* = x_2^*$. □

LEMMA 4.5.4. *Let Γ be an orbit such that $\Omega(\Gamma^+)$ does not contain equilibria.*

(a) *If $\Gamma \cap \Omega(\Gamma^+) \neq \emptyset$, then Γ is a closed orbit.*

(b) *If Γ^+ is bounded and $\Omega(\Gamma^+)$ contains a closed orbit Γ_0, then $\Omega(\Gamma^+) = \Gamma_0$.*

PROOF. (a) Let $x^* \in \Gamma \cap \Omega(\Gamma^+)$. Then x^* is a regular point. Let L be a transversal passing through x^*. By Lemma 4.5.3, Part (a), there exists a sequence of points $\{x_i\}_{i=1}^\infty \subset \Gamma \cap L$ such that $x_i \to x^*$. By Lemma 4.2.3, $\Omega(\Gamma^+)$ is an invariant set. Since $x^* \in \Omega(\Gamma^+)$ and Γ passes through x^*, we see that $\Gamma \subset \Omega(\Gamma^+)$. Hence for $i = 1, 2, \ldots$,

$$
x_i \in \Gamma \cap L \subset \Omega(\Gamma^+) \cap L.
$$

By Lemma 4.5.3, Part (b), $x_i = x^*$ for all $i = 1, 2, \ldots$. Then Γ is a closed orbit.

(b) By Lemmas 4.2.2 and 4.2.3, $\Omega(\Gamma^+)$ is a nonempty, closed, and connected set. Assume $\Gamma_0 \subsetneq \Omega(\Gamma^+)$. Note that Γ_0 is a closed set. It is easy to see that the set $\Gamma_0^c := \Omega(\Gamma^+) \setminus \Gamma_0$ is not closed. Otherwise, both Γ_0 and Γ_0^c are bounded closed sets. By Exercise 4.11, Part (a), we have $d(\Gamma_0, \Gamma_0^c) > 0$. This clearly contradicts the definition of connected sets. Thus, Γ_0 contains an accumulation point

x^* of Γ_0^c. Hence there are points $x_i \in \Gamma_0^c$, $i = 1, 2, \ldots$, such that $x_i \to x^*$. Since x^* is a regular point, there is a transversal L passing through x^* as shown in Fig. 4.15. By Lemma 4.5.1, for $i \in \mathbb{N}$ sufficiently large, the orbit Γ_i passing through x_i will cross L at a point $\bar{x}_i$. Note that $\Gamma_i \subset \Omega(\Gamma^+)$ due to the invariance of $\Omega(\Gamma^+)$. It follows that $\Gamma_i \subset \Gamma_0^c$ and hence $\bar{x}_i \in \Gamma_0^c$ since different orbits do not intersect by Lemma 4.1.1. Then we have that $\bar{x}_i \neq x^*$ and $x^*, \bar{x}_i \in \Omega(\Gamma^+) \cap L$. This contradicts Lemma 4.5.3 and completes the proof. $\qquad \square$

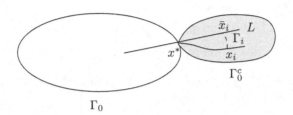

FIGURE 4.15. Set $\Omega(\Gamma^+)$ containing closed orbit Γ_0

Now we are ready to prove the Poincaré–Bendixson theorem.

PROOF OF THEOREM 4.5.1. If Γ is a closed orbit, then we are done.

Assume Γ is not a closed orbit. Since Γ^+ is bounded, by Lemmas 4.2.2 and 4.2.3, $\Omega(\Gamma^+)$ is nonempty, compact, and invariant. Let $x_0 \in \Omega(\Gamma^+)$. Then the orbit Γ_0 passing through x_0 is contained in $\Omega(\Gamma^+)$ due to the invariance of $\Omega(\Gamma^+)$. Clearly, Γ_0 is bounded and hence there exists a point $x^* \in \Omega(\Gamma_0^+)$. Note that $\Omega(\Gamma_0^+) \subset \Omega(\Gamma^+)$ since $\Omega(\Gamma^+)$ is closed. By the assumption, x^* is a regular point. Let L be a transversal passing through x^* as shown in Fig. 4.16. By Lemma 4.5.3, Part (a), there exists a sequence

$$\{x_i\}_{i=1}^{\infty} \subset \Gamma_0 \cap L \subset \Omega(\Gamma^+) \cap L \text{ such that } x_i \to x^*.$$

On the other hand, by Lemma 4.5.3, Part (b), $\Omega(\Gamma^+)$ can only intersect L at one point. It follows that $x_i = x^*$ for all $i \in \mathbb{N}$. As a result, Γ_0 and $\Omega(\Gamma_0^+)$ have the point x^* in common. By Lemma 4.5.4, Part (a), Γ_0 is a closed orbit. Since $\Gamma_0 \subset \Omega(\Gamma_0^+)$, by Lemma 4.5.4, Part (b), $\Omega(\Gamma^+) = \Gamma_0$ and hence is a closed orbit. The proof is complete. $\qquad \square$

The proof presented here is based on that in [11] and its modified versions in [41] and [38].

Considering that $\Omega(\Gamma^+)$ may contain equilibria, we have the following equivalent version of the Poincaré–Bendixson theorem.

THEOREM 4.5.2. *Assume $x(t)$ is a solution of system (A-2) and its positive semi-orbit Γ^+ is bounded. Then one of the following three statements is true:*

(a) $\Omega(\Gamma^+)$ contains an equilibrium of system (A-2);

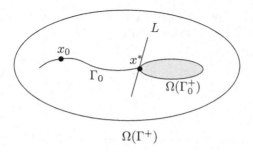

FIGURE 4.16. Set $\Omega(\Gamma^+)$

(b) Γ is a closed orbit;

(c) $\Omega(\Gamma^+)$ is a closed orbit.

The same conclusion holds when Γ^+ and $\Omega(\Gamma^+)$ are replaced by Γ^- and $A(\Gamma^-)$, respectively.

REMARK 4.5.2. By Lemma 4.2.3, we see that if an orbit Γ of system (A-2) is not a closed orbit, but $\Omega(\Gamma)$ $[A(\Gamma)]$ is, then Γ spirals toward $\Omega(\Gamma)$ as $t \to \infty$ $[A(\Gamma)$ as $t \to -\infty]$. Since orbits do not intersect each other, by the continuity of the vector field, we see that all orbits in a neighborhood of $\Omega(\Gamma)$ which are on the same side as Γ spiral toward $\Omega(\Gamma)$ as $t \to \infty$ $[A(\Gamma)$ as $t \to -\infty]$.

Since closed orbits represent periodic solutions, the Poincaré–Bendixson theorem provides a way to determine the existence of periodic solutions of system (A-2).

COROLLARY 4.5.1. *Assume $x(t)$ is a solution of system (A-2) and its positive semi-orbit Γ^+ is bounded. If $\Omega(\Gamma^+)$ does not contain any equilibrium of system (A-2), then system (A-2) has a periodic solution.*

The same conclusion holds when Γ^+ and $\Omega(\Gamma^+)$ are replaced by Γ^- and $A(\Gamma^-)$, respectively.

We observe that when the first statement in Theorem 4.5.2 holds, $\Omega(\Gamma^+)$ may contain both equilibria and nontrivial orbits. To investigate the structure of $\Omega(\Gamma^+)$ for this case, we introduce the following definitions from [41].

DEFINITION 4.5.2. (a) A separatrix cycle of system (A-2) is a continuous image of a circle which consists of a finite number of equilibria x_i, $i = 1, \ldots, k$, and the same number of nontrivial orbits Γ_i, $i = 1, \ldots, k$, such that $A(\Gamma_i) = x_i$ and $\Omega(\Gamma_i) = x_{i+1}$ for $i = 1, \ldots, k$ with $x_{k+1} := x_1$.

(b) A connected geometric structure is called a graphic for System (A-2) if it is composed of one or more separatrix cycles which may mutually share one equilibrium only and are compatibly oriented so that the nearby orbits either from inside or from outside may spiral toward the whole structure as $t \to \infty$ or as $t \to -\infty$.

A graphic may be of one separatrix cycle as shown in Fig. 4.17, or compounded separatrix cycles as shown in Fig. 4.18.

FIGURE 4.17. Graphics with one Separatrix

FIGURE 4.18. Graphics with multiple Separatrices

A nontrivial orbit in a graphic is called a heteroclinic orbit if it is connected with two distinct equilibria, and it is called a homoclinic orbit if the two equilibria coincide.   By definition, a graphic may contain a countably infinite number of homoclinic orbits connecting one equilibrium.

Now we present the generalized Poincaré–Bendixson theorem.

THEOREM 4.5.3. *Let $x(t)$ be a solution of system (A-2) and Γ its orbit. Assume Γ^+ is contained in a compact set $E \subset \mathbb{R}^2$ and system (A-2) has at most a finite number of equilibria in E. Then one of the following four statements is true:*

(a) $\Omega(\Gamma^+)$ contains only one equilibrium of system (A-2);

(b) Γ is a closed orbit;

(c) $\Omega(\Gamma^+)$ is a closed orbit;

(d) $\Omega(\Gamma^+)$ is a graphic for System (A-2).

The same conclusion holds when Γ^+ and $\Omega(\Gamma^+)$ are replaced by Γ^- and $A(\Gamma^-)$, respectively.

PROOF. Clearly, $\Omega(\Gamma^+)$ is a nonempty, invariant, and connected set, and contains at most a finite number of equilibria.

Assume $\Omega(\Gamma^+)$ contains only equilibria. Then it can contain only one equilibrium since $\Omega(\Gamma^+)$ is connected. Hence Case (a) holds.

Assume $\Omega(\Gamma^+)$ contains only regular points. Then by Theorem 4.5.1, either Γ is a closed orbit, or $\Omega(\Gamma^+)$ is a closed orbit. Hence Case (b) or (c) holds.

Assume $\Omega(\Gamma^+)$ contains both equilibria and regular points. Then it contains at least one nontrivial orbit Γ_0 by the invariance of $\Omega(\Gamma^+)$. We note that Γ_0 cannot be a closed orbit, otherwise, $\Omega(\Gamma^+) = \Gamma_0$ by Lemma 4.5.4, Part (b). We claim that $\Omega(\Gamma_0)$ is an equilibrium. If not, then there exists a regular point $x_0 \in \Omega(\Gamma_0)$. Using the same argument as in the proof of Theorem 4.5.1 we derive that Γ_0 is a closed orbit, contradicting the assumption. This shows that every nontrivial orbit in $\Omega(\Gamma^+)$ has one equilibrium in $\Omega(\Gamma^+)$ as its ω-limit set. Similarly, every nontrivial orbit in $\Omega(\Gamma^+)$ has one equilibrium in $\Omega(\Gamma^+)$ as its α-limit set. On the other hand, due to the connectedness of $\Omega(\Gamma^+)$, every equilibrium x^* in $\Omega(\Gamma^+)$ is the ω-limit set [α-limit set] of a nontrivial orbit Γ^* in $\Omega(\Gamma^+)$. We show that x^* is also the α-limit set [ω-limit set] of an orbit in $\Omega(\Gamma^+)$. Otherwise, as $t \to \infty$ [$t \to -\infty$], Γ must spiral down to a structure which contains $\Gamma^* \cup \{x^*\}$ as a "pier". This means that the vector field $f(x)$ has opposite directions to the two sides of Γ^*. By the continuity of the vector field, it follows that $f(x) = 0$ at any point $x \in \Gamma^*$ which contradicts the assumption that Γ^* does not contain equilibria. Therefore, $\Omega(\Gamma^+)$ is a graphic and hence case (d) holds. $\square$

Finally, we point out that the Poincaré–Bendixson theorem introduced in this section cannot be simply extended to autonomous systems defined in $\mathbb{R}^n$ with $n \geq 3$. This is because the unique feature that a Jordan curve dissects $\mathbb{R}^2$ into two disjoint regions, which plays a key role in establishing the Poincaré–Bendixson theorem, fails to hold for simple closed curves in general.

4.6. Periodic Solutions and Orbital Stability

In this section, we study the existence and nonexistence of periodic solutions of autonomous system (A-2). Some special periodic solutions with orbits as circles can be obtained by solving the equations. The "polar-coordinate form" of system (A-2) is often used for this purpose. In fact, let

$$x_1 = r \cos \theta \quad \text{and} \quad x_2 = r \sin \theta.$$

Then

$$r^2 = x_1^2 + x_2^2 \quad \text{and} \quad \tan \theta = \frac{x_2}{x_1}.$$

It follows from system (A-2) that

$$rr' = x_1 x_1' + x_2 x_2' = x_1 f_1(x_1, x_2) + x_2 f_2(x_1, x_2)$$

and

$$r^2 \theta' = x_1 x_2' - x_1' x_2 = x_1 f_2(x_1, x_2) - x_2 f_1(x_1, x_2)$$

which leads to
(4.6.1)
$$r' = \frac{1}{r}(x_1 f_1(x_1, x_2) + x_2 f_2(x_1, x_2)) \quad \text{and} \quad \theta' = \frac{1}{r^2}(x_1 f_2(x_1, x_2) - x_2 f_1(x_1, x_2)).$$

EXAMPLE 4.6.1. Consider the system

(4.6.2)
$$\begin{cases} x_1' = x_2(1 + r^2) \\ x_2' = -x_1(1 + r^2), \end{cases}$$

where $r^2 = x_1^2 + x_2^2$. By way of (4.6.1), Eq. (4.6.2) becomes the system

$$r' = 0 \quad \text{and} \quad \theta' = -(1 + r^2).$$

Then the general solution is $r = r_0$, $\theta = \theta_0 - (1 + r_0)t$ with arbitrary $r_0 \geq 0$ and $\theta_0 \in \mathbb{R}$.

When $r_0 = 0$, then $r \equiv 0$ on $\mathbb{R}$. This shows that $(0,0)$ is an equilibrium. When $r_0 > 0$, the solution is periodic with the period $2\pi/(1 + r_0)$, and hence has a closed orbit given by the circle $r = r_0$. This shows that $(0,0)$ is a center. However, unlike the centers for linear equations, the periods of the periodic solutions are different and are determined by the radii of the closed orbits. See Fig. 4.19.

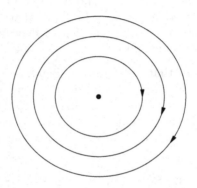

FIGURE 4.19. Periodic solutions for Example 4.6.1

EXAMPLE 4.6.2. Consider the system

(4.6.3)
$$\begin{cases} x_1' = x_2 \\ x_2' = -x_1 - x_2(r^2 - 1), \end{cases}$$

where $r^2 = x_1^2 + x_2^2$. By way of (4.6.1), Eq. (4.6.3) becomes the system

$$r' = -r(r^2 - 1)\sin^2\theta \quad \text{and} \quad \theta' = -1 - \frac{1}{2}(r^2 - 1)\sin 2\theta.$$

The general solution cannot be obtained explicitly. However, we have the following observations:

    (i) $r = 0$ is a solution which corresponds to the equilibrium $(0,0)$.

    (ii) $r = 1$, $\theta = \theta_0 - t$ is a 2π-periodic solution with the orbit given by the circle $r = 1$.

    (iii) When $0 < r(0) < 1$, then $0 < r(t) < 1$ for all $t \in \mathbb{R}$ due to Lemma 4.1.1. Moreover, $r'(t) > 0$ except when $\theta = 0$ [mod π], where the orbit is perpendicular to the x_1-axis. This implies that $r(t)$ is strictly increasing. Thus there is no closed orbit inside the circle $r = 1$. Also, since $\theta' \leq -1 - (r^2 - 1)/2 < -1/2$, we see that $\theta(t)$ is strictly decreasing and $\theta(t) \to -\infty$ as $t \to \infty$.

(iv) When $r(0) > 1$, then as above, $r(t) > 1$ for all $t \in \mathbb{R}$ and $r(t)$ is strictly decreasing. Thus there is no closed orbit outside the circle $r = 1$. Also, since $\theta' = -1 + o(1)$ as $r \to 1$ uniformly for $t \in \mathbb{R}$, we see that $\theta(t) \to -\infty$ as $t \to \infty$ for r close to 1.

From the above we see that the equilibrium $(0,0)$ is an unstable spiral-point and the circle $r = 1$ is the only closed orbit. Furthermore, for any nontrivial solution, the positive semi-orbit Γ^+ is bounded and bounded away from the only equilibrium $(0,0)$. By the Poincaré–Bendixson theorem, $\Omega(\Gamma^+)$ is a closed orbit and hence is the circle $r = 1$. This show that all other orbits from inside and outside of the circle spiral toward this circle as $t \to \infty$. See Fig. 4.20.

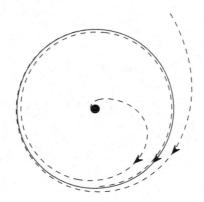

FIGURE 4.20. Periodic solutions for Example 4.6.2

EXAMPLE 4.6.3. Consider the system

$$(4.6.4) \qquad \begin{cases} x_1' = x_2 + x_1 r^2 \sin \frac{\pi}{r} \\ x_2' = -x_1 + x_2 r^2 \sin \frac{\pi}{r} , \end{cases}$$

where $r^2 = x_1^2 + x_2^2$. Although the right-hand functions are not defined at $(0,0)$, we may extend them to $(0,0)$ by continuity. By way of (4.6.1), Eq. (4.6.4) becomes the system

$$r' = r^3 \sin \frac{\pi}{r} \qquad \text{and} \qquad \theta' = -1.$$

Then we have the following observations:

(i) $r = 0$ is a solution which corresponds to the only equilibrium $(0,0)$.

(ii) For each $k \in \mathbb{N}$, $r = 1/k$ is a closed orbit.

(iii) For each $k \in \mathbb{N}$, when $1/(k+1) < r(0) < 1/k$, then $1/(k+1) < r(t) < 1/k$ for all $t \in \mathbb{R}$ due to Lemma 4.1.1. Moreover, $(-1)^k r'(t) > 0$ for $t \in \mathbb{R}$. This shows that $(-1)^k r(t)$ is strictly increasing. Thus there is no closed orbit in the region with $1/(k+1) < r < 1/k$. Note that the orbit Γ of this solution is bounded and bounded away from the only equilibrium $(0,0)$. By the Poincaré–Bendixson theorem, both $\Omega(\Gamma^+)$ and $A(\Gamma^-)$ are closed orbits. It follows that the orbit spirals toward one of its adjacent closed orbits as $t \to \infty$ and as $t \to -\infty$, respectively.

(iv) When $r(0) > 1$, then $r(t) > 1$ and $r'(t) > 0$ for all $t \in \mathbb{R}$. Thus there is no closed orbit outside the circle $r = 1$. Note that its negative semi-orbit Γ^- is bounded and bounded away from $(0,0)$. By the Poincaré–Bendixson theorem, $A(\Gamma^-)$ is a closed orbit and hence is the circle $r = 1$. This shows that all orbits in this region spiral toward this circle as $t \to -\infty$.

By definition we see that $(0,0)$ is a center. However, in any neighborhood of $(0,0)$, there is an infinite number of spirals. See Fig. 4.21.

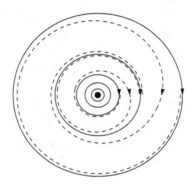

FIGURE 4.21. Periodic solutions for Example 4.6.3

From the above examples, we see that closed orbits may or may not be isolated. If a closed orbit Γ is isolated, i.e., all orbits starting from a neighborhood of Γ spiral toward Γ either as $t \to \infty$ or as $t \to -\infty$, then it is called a limit cycle. Clearly, the closed orbits in Examples 4.6.2 and 4.6.3 are limit cycles, but those in Example 4.6.1 are not.

In general, when the orbits of periodic solutions of system (A-2) are not circles, the above method fails to work. In this case, the Poincaré–Bendixson theorem may be used together with the Lyapunov function method to show the existence, and sometimes the uniqueness, of closed orbits. We explain the ideas with two examples.

EXAMPLE 4.6.4. Consider the system

$$(4.6.5) \qquad \begin{cases} x_1' = x_1(1 - x_1^2 - 2x_2^2) + 2x_2 e^{x_1 x_2} \\ x_2' = x_2(1 - x_1^2 - 2x_2^2) - x_1 e^{x_1 x_2}. \end{cases}$$

It is easy to see that $(0,0)$ is the only equilibrium. Let $V = x_1^2 + 2x_2^2$. Then V is positive definite on $\mathbb{R}^2$. By a simple computation we see that

$$\dot{V} = 2(x_1^2 + 2x_2^2)(1 - x_1^2 - 2x_2^2) = 2V(1 - V).$$

Let $C = \{(x_1, x_2) : x_1^2 + 2x_2^2 = 1\}$. Then we have the following observations:

(i) In the region inside C, $\dot{V} > 0$, hence $V(x_1(t), x_2(t))$ is strictly increasing along any nontrivial solution $(x_1(t), x_2(t))$. This means that there is no closed orbit inside C.

(ii) In the region outside C, $\dot{V} < 0$, hence $V(x_1(t), x_2(t))$ is strictly decreasing along any solution $(x_1(t), x_2(t))$. This means that there is no closed orbit outside C.

(iii) For any orbit Γ not on C, it crosses C at most once. In fact, if Γ crosses C at two points and $x(t) := (x_1(t), x_2(t))$ is a solution represented by Γ, then without loss of generality, we may assume there are $t_1, t_2 \in \mathbb{R}$ with $t_1 < t_2$ such that $V(x(t_1)) = V(x(t_2)) = 1$ and $V(x(t)) < 1$ for $t \in (t_1, t_2)$. It follows that $V(x(t_2)) - V(x(t_1)) = \int_{t_1}^{t_2} [V(x(t))]' \, dt > 0$. We have reached a contradiction.

From the above we see that if a semi-orbit Γ^+ is not on C, it is either eventually inside C and going toward C or eventually outside C and going toward C. In both the cases, Γ^+ is bounded and bounded away from the only equilibrium $(0, 0)$. By the Poincaré–Bendixson theorem, $\Omega(\Gamma^+)$ is a closed orbit. Note from (i)–(iii) that the only possibility for a closed orbit is C. Therefore, system (4.6.5) has a unique closed orbit C.

EXAMPLE 4.6.5 (Lienard equation). We consider the second-order scalar equation

(4.6.6) $$x'' + f(x, x')x' + g(x) = 0$$

under the assumptions that $f \in C(\mathbb{R}^2, \mathbb{R})$ and $g \in C(\mathbb{R}, \mathbb{R})$ satisfying

(i) $f(0, 0) < 0$, and there exists an $a > 0$ such that $f(x_1, x_2) > 0$ when $\sqrt{x_1^2 + x_2^2} > a$; and

(ii) $xg(x) > 0$, $x \neq 0$, and $G(x) := \int_0^x g(s) \, ds \to \infty$ as $x \to \pm\infty$.

Physically, $f(x, x')$ is the damping coefficient which depends on the position x and velocity x'. By assumption (i), it is negative when $x^2 + x'^2$ is small and is positive when $x^2 + x'^2$ is large. Thus, Eq. (4.6.6) is a model of self-excited and self-controlled oscillation.

Let $x_1 = x$, $x_2 = x'$. Then Eq. (4.6.6) becomes the system of equations

(4.6.7) $$x_1' = x_2, \quad x_2' = -f(x_1, x_2)x_2 - g(x_1).$$

It is easy to see that $(0, 0)$ is the only equilibrium. Let $V = G(x_1) + x_2^2/2$. Then V is positive definite on $\mathbb{R}^2$ and

$$\dot{V} = g(x_1)x_2 + x_2[-f(x_1, x_2)x_2 - g(x_1)] = -x_2^2 f(x_1, x_2).$$

By assumption (i), there exists a $b \in (0, a)$ such that

$$\dot{V} \geq 0 \text{ if } x_1^2 + x_2^2 \leq b^2 \quad \text{and} \quad \dot{V} \leq 0 \text{ if } x_1^2 + x_2^2 \geq a^2.$$

Let $x_1 = x_1(t), x_2 = x_2(t)$ be a nontrivial solution of system (4.6.7) and denote $\bar{V}(t) = V(x_1(t), x_2(t))$. We first show that $\bar{V}(t)$ is bounded away from 0 as $t \to \infty$. Otherwise, $\liminf_{t\to\infty} \bar{V}(t) = 0$. In this case, we can always choose a sequence $t_n \to \infty$ such that $\bar{V}(t_n) \to 0$ and $\bar{V}'(t_n) < 0$. Note that

$\bar{V}(t_n) \to 0$ implies that $(x_1(t_n), x_2(t_n)) \to (0,0)$ and hence $x_1^2(t_n) + x_2^2(t_n) \leq b^2$ for large n. It follows that $\bar{V}'(t_n) \geq 0$. We have reached a contradiction. Similarly, with the assumption (ii) we can show that $\bar{V}(t)$ is bounded above as $t \to \infty$. Combining the above, we see that the positive semi-orbit Γ^+ of this solution is bounded and bounded away from the only equilibrium $(0,0)$. By the Poincaré–Bendixson theorem, system (4.6.7) has a closed orbit, and hence Eq. (4.6.6) has a periodic solution.

Next, we derive conditions which guarantee that system (A-2) has no periodic solutions. The theorem below contains ideas that have been applied in Examples 4.6.2–4.6.4.

THEOREM 4.6.1. *Let E be an open subset of $\mathbb{R}^2$. Assume there exists a function $V \in C^1(E, \mathbb{R})$ satisfying $\dot{V}(x_1, x_2) \geq 0 [\leq 0]$ on E and the set $\{(x_1, x_2) \in E : \dot{V}(x_1, x_2) = 0\}$ does not contain closed curves. Then system (A-2) has no closed orbits in E.*

PROOF. Assume system (A-2) has an ω-periodic solution $x_1 = x_1(t)$, $x_2 = x_2(t)$ in E. By the assumptions we have

$$[V(x_1(t), x_2(t))]' \geq 0 [\leq 0] \quad \text{and} \quad [V(x_1(t), x_2(t))]' \not\equiv 0 \text{ for } t \in [0, \omega].$$

Hence

$$V(x_1(\omega), x_2(\omega)) - V(x_1(0), x_2(0)) = \int_0^\omega [V(x_1(t), x_2(t))]' dt \neq 0.$$

This shows that $(x_1(\omega), x_2(\omega)) \neq (x_1(0), x_2(0))$. We have reached a contradiction. $\square$

EXAMPLE 4.6.6. Consider the system

(4.6.8) $$\begin{cases} x_1' = f_1(x_1, x_2) \\ x_2' = x_2 + 3x_2(x_1^2 + 2x_2^2) := f_2(x_1, x_2), \end{cases}$$

where $f_1 \in C(\mathbb{R}^2, \mathbb{R})$ is arbitrary. Let $V(x_1, x_2) := x_2^2/2$. Then $V \in C^1(\mathbb{R}^2, \mathbb{R})$ and

$$\dot{V}(x_1, x_2) = x_2 f_2(x_1, x_2) = x_2^2 + 3x_2^2(x_1^2 + 2x_2^2) = x_2^2(1 + 3(x_1^2 + 2x_2^2)) \geq 0$$

on $\mathbb{R}^2$, and $\dot{V}(x_1, x_2) = 0 \iff x_2 = 0$. Clearly, the set $\{(x_1, x_2) \in \mathbb{R}^2 : \dot{V}(x_1, x_2) = 0\}$ does not contain closed curves. By Theorem 4.6.1, system (4.6.8) has no periodic solutions.

We observe that f_1 is not involved in this process. In other words, f_2 itself determines the nonexistence of closed orbits, no matter what f_1 is. This is easy to understand since the second equation shows that $|x_2(t)|$ increases as t increases which excludes the possibility that a periodic solution exists.

To obtain another result on the nonexistence of periodic solutions for system (A-2), we recall the following definition:

DEFINITION 4.6.1. An open set $E \subset \mathbb{R}^2$ is said to be a simply connected region if

(i) it is path-connected, i.e., any $x_1, x_2 \in E$ can be connected by a path (continuous curve) in E; and
(ii) whenever a Jordan curve lies entirely in E, its interior is contained in E.

Intuitively, a simply connected region is a region of one piece with no holes inside, and hence any simple closed curve can be shrunk to a point continuously in the region. Now, we state the second theorem on the nonexistence of periodic solutions.

THEOREM 4.6.2 (Poicaré–Durac Negative Criterion). *Let E be a simply connected region in $\mathbb{R}^2$. Assume there exists a function $r : E \to \mathbb{R}$, called a Durac function, such that $rf \in C^1(E, \mathbb{R})$ and*
(4.6.9)
$$(rf_1)_{x_1} + (rf_2)_{x_2} \geq 0 \ \text{on} \ E \ \text{and} \ (rf_1)_{x_1} + (rf_2)_{x_2} \not\equiv 0 \ \text{on any subregion of} \ E.$$

Then system (A-2) has no closed orbits in E.

REMARK 4.6.1. With the choices $r(t) \equiv 1 \ [-1]$, condition (4.6.9) reduces to the condition:

$$(f_1)_{x_1} + (f_2)_{x_2} \geq 0 \ [\leq 0] \ \text{on} \ E \ \text{and} \ (f_1)_{x_1} + (f_2)_{x_2} \not\equiv 0 \ \text{on any subregion of} \ E.$$

To prove Theorem 4.6.2, we need to use the following Green's theorem: Suppose that C is a positively oriented and piecewise-smooth Jordan curve with D its interior. Let $M, N \in C^1(C \cup D, \mathbb{R})$. Then

$$\oint_C M \, dx_1 + N \, dx_2 = \iint_D (N_{x_1} - M_{x_2}) \, dx_1 dx_2.$$

PROOF OF THEOREM 4.6.2. Suppose that system (A-2) has a periodic solution $x_1 = x_1(t)$, $x_2 = x_2(t)$ in E with period ω. Then the Jordan curve $C := \{(x_1(t), x_2(t)) : t \in [0, \omega]\}$ is its orbit. Let D be the interior of C. By applying Green's theorem with $M = -rf_2$ and $N = rf_1$ we obtain that

$$0 < \iint_D [(rf_1)_{x_1} + (rf_2)_{x_2}] \, dx_1 dx_2 = \oint_C (-rf_2) \, dx_1 + (rf_1) \, dx_2$$

$$= \int_0^\omega r(x_1(t), x_2(t))[-f_2(x_1(t), x_2(t))x_1'(t) + f_1(x_1(t), x_2(t))x_2'(t)] \, dt$$

$$= \int_0^\omega r(x_1(t), x_2(t))[-f_2(x_1(t), x_2(t))f_1(x_1(t), x_2(t))$$

$$+ f_1(x_1(t), x_2(t))f_2(x_1(t), x_2(t))] \, dt = 0.$$

This contradiction shows that system (A-2) has no periodic solutions.   □

EXAMPLE 4.6.7. Consider the system

(4.6.10)
$$\begin{cases} x_1' = ax_1 + g_1(x_2) := f_1(x_1, x_2) \\ x_2' = bx_2 + g_2(x_1) := f_2(x_1, x_2), \end{cases}$$

where $g_i \in C^1(\mathbb{R}, \mathbb{R})$. Note that $(f_1)_{x_1} + (f_2)_{x_2} = a + b$. By Theorem 4.6.2 with $r \equiv 1$, system (4.6.11) has no periodic solutions if $a + b \neq 0$. We observe that the functions $g_1(x_2)$ and $g_2(x_1)$ are not involved in this process and hence can be arbitrary.

However, when $a + b = 0$, Theorem 4.6.2 fails to apply. In this case, system (4.6.10) may or may not have periodic solutions. To see this, let $a = b = 0$. Then it is easy to verify that system (4.6.10) has an infinite number of periodic solutions if $g_1(x_2) = x_2$ and $g_2(x_1) = -x_1$, and has no periodic solutions if $g_1(x_2) = x_2$ and $g_2(x_1) = x_1$.

EXAMPLE 4.6.8. Consider the system

(4.6.11)
$$\begin{cases} x_1' = 3x_1x_2^2 - x_1 \\ x_2' = -x_1 + 2x_2 + x_1x_2 - 2x_2^3. \end{cases}$$

We note that

$$(f_1)_{x_1} + (f_2)_{x_2} = 1 + x_1 - 3x_2^2$$

which changes sign in $\mathbb{R}^2$. Hence Theorem 4.6.2 fails to apply with $r \equiv 1$. However, if we let $r = x_1$, then

$$(rf_1)_{x_1} + (rf_2)_{x_2} = (3x_1^2x_2^2 - x_1^2)_{x_1} + (-x_1^2 + 2x_1x_2 + x_1^2x_2 - 2x_1x_2^3)_{x_2} = x_1^2$$

which satisfies (4.6.9). By Theorem 4.6.2, system (4.6.11) has no periodic solutions.

REMARK 4.6.2. (i) Both Theorems 4.6.1 and 4.6.2 provide sufficient conditions for system (A-2) not to have periodic solutions. However, they employ different tools. In fact, Theorem 4.6.1 uses the function f and Theorem 4.6.2 uses the divergence of the function rf.

(ii) Neither Theorems 4.6.1 nor 4.6.2 provides necessary conditions for the nonexistence of periodic solutions, and neither can cover the other. For instance, it is impossible for Theorem 4.6.2 to apply to system (4.6.8) with a general f_1 and for Theorems 4.6.1 to apply to system (4.6.10) with general $g_1(x_2)$ and $g_2(x_1)$.

(iii) Both Theorems 4.6.1 and 4.6.2 involve smart choices of functions. There are no regular ways to find appropriate functions $V(x_1, x_2)$ and $r(x_1, x_2)$. Plenty of practice is needed in order to master these methods.

(iv) Another way for proving nonexistence of periodic solutions will be introduced in the next section, see Theorem 4.7.3.

Finally, we discuss the stability of periodic solutions of system (A-2). Recall that the Lyapunov stability of solutions is defined in Definition 3.1.1 which characterizes the closeness of the solution curve of a given solution and those of nearby solutions. We observe that in Example 4.6.1, although the orbits for all solutions except the trivial one are circles centered at $(0,0)$, none of them is stable in the sense of Lyapunov stability. This is because different orbits rotate at different angle speeds and hence the solution curves are getting away from each other as $t \to \infty$. With the same reason, in Example 4.6.2, it is hard to determine the stability of the unique periodic solution in the sense of Lyapunov stability even though all the nearby orbits approach its orbit as $t \to \infty$. However, from the point of view of orbits, we have a reason to treat the orbits of the periodic solutions in Example 4.6.1 as stable and the orbit of the periodic solution in Example 4.6.2 as asymptotically stable. This leads to a new definition for stability.

Let $x \in \mathbb{R}$ and let A be a set in $\mathbb{R}^2$. We recall that $d(x, A)$ stands for the distance between the point x and the set A, i.e., $d(x, A) = \inf\{d(x, a) : a \in A\}$.

DEFINITION 4.6.2. Let $p(t)$ be a periodic solution of system (A-2) and Γ its closed orbit. Then

(a) Γ is said to be orbitally stable if for any $\epsilon > 0$, there exists a $\delta = \delta(\epsilon) > 0$ such that $d(x_0, \Gamma) < \delta$ implies that any solution $x(t)$ of system (A-2) with $x(t_0) = x_0$ for $t_0 \in \mathbb{R}$ satisfies $d(x(t), \Gamma) < \epsilon$ for all $t \geq t_0$.

(b) Γ is said to be orbitally asymptotically stable if it is orbitally stable and there exists a $\delta_1 > 0$ such that $d(x_0, \Gamma) < \delta_1$ implies that any solution $x(t)$ of system (A-2) with $x(t_0) = x_0$ for $t_0 \in \mathbb{R}$ satisfies $d(x(t), \Gamma) \to 0$ as $t \to \infty$.

(c) Γ is said to be orbitally unstable if it is not orbitally stable.

By this definition, any orbitally asymptotically stable closed orbit is a limit cycle.

It is easy to see that all the closed orbits in Example 4.6.1 are orbitally stable, the unique closed orbit in Example 4.6.2 is an orbitally asymptotically stable limit cycle, and the closed orbits in Example 4.6.3 are alternatively orbitally asymptotically stable and orbitally unstable limit cycles.

REMARK 4.6.3. Note that Lyapunov stability denotes stability for solution curves, while orbital stability denotes stability for orbits. Thus, if a periodic solution of system (A-2) is Lyapunov stable, then its orbit is orbitally stable. However, the converse is not true in general.

We also point out that the concept of orbital stabilities for closed orbits can also be extended to one-sided orbital stabilities for closed orbits. In fact, a closed orbit Γ is said to be orbitally stable [orbitally asymptotically stable, orbitally unstable] from inside if x_0 in Definition 4.6.2, Part (a) [(b), (c)] is limited to the interior of Γ. Likewise, Γ is said to be orbitally stable [orbitally asymptotically stable, orbitally unstable] from outside if x_0 in Definition 4.6.2, Part (a) [(b), (c)] is limited to the exterior of Γ.

The following result is obtained from the Poincaré–Bendixson theorem.

THEOREM 4.6.3. *Let C_i, $i = 1, 2$, be Jordan curves with C_1 inside C_2, and E be the region in between. Suppose that there is no equilibrium in E.*

(a) *Assume every positive [negative] semi-orbit starting from a point on $C := C_1 \cup C_2$ will go inside E. Then system (A-2) has closed orbits Γ_i, $i = 1, 2$, in E, with Γ_1 either inside Γ_2 or the same as Γ_2, such that Γ_1 is orbitally asymptotically stable [orbitally unstable] from inside and Γ_2 is orbitally asymptotically stable [orbitally unstable] from outside.*

(b) *If we further assume that $f = (f_1, f_2)^T$ is analytic on E, then system (A-2) has at least one orbitally asymptotically stable [orbitally unstable] limit cycle in E.*

PROOF. (a) Let Γ be an orbit of system (A-2) which enters E from C_1 at $t = t_0$ for some t_0. From the assumptions, Γ will stay in E for all $t > t_0$ and hence Γ^+ is bounded. Clearly, Γ is not a closed orbit. Otherwise, it will cross C_1 at a $t_1 > t_0$, contradicting the assumption. By the Poincaré–Bendixson theorem, $\Gamma_1 := \Omega(\Gamma^+)$ is a closed orbit in E. As shown in Remark 4.5.2, Γ_1 is orbitally asymptotically stable from inside. Similarly, system (A-2) has a closed orbit Γ_2 in E which is orbitally asymptotically stable from outside. It is easy to see that either Γ_1 is in the interior of Γ_2 or Γ_1 and Γ_2 are the same.

(b) It has been proved that if f is analytic on E, then either every closed orbit is a limit cycle or every closed orbit is surrounded by closed orbits only, see [51, p. 236]. Obviously, the latter case must be excluded. Then the conclusion follows from the one-sided orbital stability properties given in Part (a).

$\square$

As a consequence of Theorem 4.6.3, we have the next result.

COROLLARY 4.6.1. *Let D be a simply connected region in $\mathbb{R}^2$ with boundary C. Suppose that there is only one equilibrium x^* in D which is an unstable [stable] spiral-point.*

(a) *Assume every positive [negative] semi-orbit starting from a point on C will go inside D. Then system (A-2) has closed orbits Γ_i, $i = 1, 2$, in D, with Γ_1 either inside Γ_2 or the same as Γ_2, such that Γ_1 is orbitally asymptotically stable [orbitally unstable] from inside and Γ_2 is orbitally asymptotically stable [orbitally unstable] from outside.*

(b) *If we further assume that $f = (f_1, f_2)^T$ is analytic on D, then system (A-2) has at least one orbitally asymptotically stable [orbitally unstable] limit cycle in D.*

PROOF. By the assumptions, there is a Jordan curve C_1 inside C, sufficiently close to x^*, such that every positive semi-orbit starting from a point on C_1 will go outside C_1. Let $C_2 := C$ and E the region between C_1 and C_2.

Then the conditions of Theorem 4.6.3 are satisfied and hence the conclusions follow from Theorem 4.6.3. □

If a periodic solution is given, then we may use the so-called variational equation to determine its orbital stability. To see this, let $p(t)$ be an ω-periodic solution of system (A-2) with Γ its orbit. From system (A-2), $p'(t) = f(p(t))$. Differentiating both sides and using the chain rule we have $[p'(t)]' = f_x(p(t)) p'(t)$, where f_x is the Jacobi matrix of f with respect to x. This shows that $y = p'(t)$ is a solution of the system

$$(4.6.12) \qquad y' = f_x(p(t))y.$$

System (4.6.12) is called the variational equation of system (A-2) with respect to $p(t)$. Since $p(t)$ is ω-periodic, system (4.6.12) is a homogeneous linear equation with periodic coefficients of period ω.

Recall from Sect. 2.3 that if $X(t)$ is a fundamental matrix solution of system (4.6.12), then $V = X^{-1}(0)X(\omega)$ is the transition matrix for $X(t)$ and the eigenvalues μ_1 and μ_2 of V are the characteristic multipliers which are independent of the choice of $X(t)$. By Corollary 2.5.4,

$$(4.6.13) \quad \mu_1\mu_2 = \exp\left\{\int_0^\omega \operatorname{tr}\left(f_x(p(t))\right) dt\right\} = \exp\left\{\int_0^\omega [(f_1)_{x_1} + (f_2)_{x_2}]_{x=p(t)}\, dt\right\}.$$

We see that $p(t)$ is ω-periodic implies that $p'(t)$ is an ω-periodic solution of system (4.6.12). Thus by Exercise 2.18, (b), either $\mu_1 = 1$ or $\mu_2 = 1$. Then we have the following criteria for orbital stability.

THEOREM 4.6.4. *Assume $p(t)$ is an ω-periodic solution of system (A-2) with orbit Γ. Let μ_1 and μ_2 be the characteristic multipliers of the variational equation (4.6.12) with $\mu_1 = 1$. Then Γ is orbitally asymptotically stable limit cycle if $|\mu_2| < 1$, and Γ is orbitally unstable limit cycle if $|\mu_2| > 1$.*

We do not offer the proof here. The interested reader is referred to [22, p. 254–256] or [11, p. 321–327] for the proof. The corollary below is a direct consequence of Theorem 4.6.4 and (4.6.13).

COROLLARY 4.6.2 (Poincaré). *Assume $p(t)$ is an ω-periodic solution of system (A-2) with orbit Γ. Then Γ is an orbitally asymptotically stable limit cycle if*

$$\int_0^\omega [(f_1)_{x_1} + (f_2)_{x_2}]_{x=p(t)}\, dt < 0,$$

and Γ is an orbitally unstable limit cycle if

$$\int_0^\omega [(f_1)_{x_1} + (f_2)_{x_2}]_{x=p(t)}\, dt > 0.$$

4.7. Indices of Equilibria

In this section we study the relations between a closed orbit and the equilibria inside by investigating the index of the closed orbit and the indices of the equilibria. For this purpose, we first discuss the rotation of a vector field.

Let $C = \{x(t) : t \in [a,b]\}$ be a positively oriented Jordan curve. Assume $f \in C([a,b], \mathbb{R}^2)$ such that $f(x(t)) \neq (0,0)$ for $t \in [a,b]$ and $f(a) = f(b)$. Then f is called a vector field on the curve C (Fig. 4.22).

DEFINITION 4.7.1. Let $\phi \in C([a,b], \mathbb{R})$ be the inclination angle of the vector field $f(t)$ on C, i.e., the angle from the positive direction of the x_1-axis to the direction of f. Hence $\tan \phi(t) = f_2(t)/f_1(t)$, $t \in [a,b]$. Then the number $I_f(C) := [\phi(b) - \phi(a)]/2\pi$ along the positive direction of C is called the index of C with respect to f.

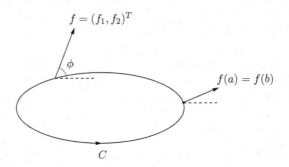

FIGURE 4.22. Inclination of a vector field

REMARK 4.7.1. (i) $I_f(C)$ means the number of rotations of the vector field f along the closed curve C and hence it is an integer.
(ii) If C is a smooth Jordan curve and f is a tangent vector field on C, then $I_f(C) = 1$.

We then introduce a homotopic property of $I_f(C)$ with respect to the closed curve C.

LEMMA 4.7.1. *Assume that*
(a) $C_i = \{x^{[i]}(t) : t \in [a,b]\}$, $i = 1, 2$, *are Jordan curves which are homotopic to each other, i.e., there exists a function* $x(t,s) \in C([a,b] \times [0,1], \mathbb{R}^2)$ *such that*
    (i) *for any* $s \in [0,1]$ $C(s) := \{x(t,s) : t \in [a,b]\}$ *is a Jordan curve;*
    (ii) $C(0) = C_1$ *and* $C(1) = C_2$.
(b) $f \in C\left(\cup_{0 \leq s \leq 1} C(s), \mathbb{R}^2\right)$ *such that* $f(x(t,s)) \neq (0,0)$ *for any* $(t,s) \in [a,b] \times [0,1]$.

Then $I_f(C_1) = I_f(C_2)$.

PROOF. Note that $I_f(C_s)$ is an integer-valued function and is continuous in s for $s \in [0,1]$. Hence $I_f(C_s)$ does not change value for $0 \le s \le 1$. Thus, $I_f(C_1) = I_f(C_2)$. $\qquad\square$

REMARK 4.7.2. The homotopic relation between the closed curves C_1 and C_2 given in Lemma 4.7.1 means that one of them can be deformed continuously to the other. The homotopy function $x(t,s)$ is often chosen to be

$$x(t,s) = (1-s)x^{[1]}(t) + sx^{[2]}(t).$$

With essentially the same proof we get a homotopic property of $I_f(C)$ with respect to the vector field f.

LEMMA 4.7.2. *Assume that*

(a) $C = \{x(t) : t \in [a,b]\}$ *is a Jordan curve;*

(b) $f^{[i]} \in C([a,b], \mathbb{R}^2)$, $i = 1,2$, *are vector fields on C which are homotopic to each other, i.e., there exists a function $f(t,s) \in C([a,b] \times [0,1], \mathbb{R}^2)$ such that*

(i) $f(t,s) \ne (0,0)$ *for any $(t,s) \in [a,b] \times [0,1]$;*

(ii) $f(t,0) = f^{[1]}(t)$ *and $f(t,1) = f^{[2]}(t)$ for $t \in [a,b]$.*

Then $I_{f^{[1]}}(C) = I_{f^{[2]}}(C)$.

The following result is derived from Lemma 4.7.1.

COROLLARY 4.7.1. *Let C be a Jordan curve with D its interior. Assume $f \in C(C \cup D, \mathbb{R}^2)$ such that $f(x) \ne 0$ for $x \in C \cup D$. Then $I_f(C) = 0$.*

PROOF. Let the closed curves C, C_1, and C_2 be shown as in Fig. 4.23. Then C can be deformed continuously to C_1 and to C_2, i.e., C is homotopic to both C_1 and C_2. By Lemma 4.7.1, $I_f(C_1) = I_f(C) = I_f(C_2)$. On the other hand, by looking at the rotations of the vector field over the curves, we see that $I_f(C) = I_f(C_1) + I_f(C_2)$. Combining the above, we obtain that $I_f(C) = 2I_f(C)$ and hence $I_f(C) = 0$. $\qquad\square$

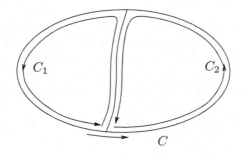

FIGURE 4.23. Index of a closed curve

Based on the indices of closed curves, we may define the indices of equilibria of system (A-2).

DEFINITION 4.7.2. Let x_0 be an isolated equilibrium of system (A-2). Then the index $I(x_0)$ of x_0 is defined to be $I_f(C)$, where f is the right-hand side function of system (A-2) and $C = \{x : |x - x_0| = \epsilon\}$ for an $\epsilon > 0$ such that there are no equilibria other than $(0,0)$ inside and on the curve C.

Definition 4.7.2 makes sense since all Jordan curves surrounding the equilibrium x_0 only are homotopic and hence have the same index with respect to f by Lemma 4.7.1.

Now we present the results on the indices of equilibria. The first one is for the homogeneous linear system we studied in Sect. 4.3, i.e., the system

$$(4.7.1) \qquad \begin{cases} x_1' = ax_1 + bx_2 \\ x_2' = cx_1 + dx_2 \end{cases} \quad \text{or} \quad x' = Ax,$$

where $x = \begin{bmatrix} x_1 \\ x_2 \end{bmatrix}$ and $A = \begin{bmatrix} a & b \\ c & d \end{bmatrix}$. From Sect. 4.3 we know that for the equilibrium $(0,0)$ to be isolated, we must have $q := ad - bc \neq 0$.

LEMMA 4.7.3. *The index $I(O)$ of the equilibrium $O = (0,0)$ satisfies*

$$I(O) = \begin{cases} 1, & ad - bc > 0 \\ -1, & ad - bc < 0. \end{cases}$$

In other words, $I(O) = 1$ when $(0,0)$ is a node, a spiral-point, or a center; and $I(O) = -1$ when $(0,0)$ is a saddle-point.

PROOF. Choose a curve $C = \{x : |x| = r\}$ for an $r > 0$. The C has the parametric equation $x_1 = r \cos t$, $x_2 = r \sin t$ for $t \in [0, 2\pi]$. We treat the numbers a, b, c, d in Eq. (4.7.1) as parameters and denote the right-hand side function f by $f^{[abcd]}$ to emphasize its dependence on the parameters.

Assume $ad - bc > 0$. We claim that the function $f^{[abcd]}$ is homotopic to the function $f^{[1001]}$ on the curve C. In fact, we can define $[\bar{a}\bar{b}\bar{c}\bar{d}] \in C([0,1], \mathbb{R}^4)$ such that

$$[\bar{a}\bar{b}\bar{c}\bar{d}](1) = [abcd] \quad \text{and} \quad [\bar{a}\bar{b}\bar{c}\bar{d}](0) = [1001],$$

and $\bar{q}(s) := (\bar{a}\bar{d} - \bar{b}\bar{c})(s) > 0$ for all $s \in [0, 1]$. We leave this as an exercise, see Exercise 4.35. We see that in this process, $(0,0)$ is the only equilibrium of system (4.7.1) and hence $f^{[abcd]}$ changes homotopically to $f^{[1001]}$ on C. By Lemma 4.7.2, $I_{f^{[abcd]}}(C) = I_{f^{[1001]}}(C)$. Thus, we only need to calculate $I(O)$ for the case when $a, d = 1$ and $b, c = 0$. Note that in this case, $f_1(x_1, x_2) = x_1$ and $f_2(x_1, x_2) = x_2$.

Let $\phi(t)$ be the inclination angle of the function f on C. Then for $t \in [0, 2\pi]$,

$$\tan \phi(t) = \frac{f_2(x_1(t), x_2(t))}{f_1(x_1(t), x_2(t))} = \frac{x_2(t)}{x_1(t)}.$$

and hence

$$\phi'(t) = \frac{x_1(t)x_2'(t) - x_2(t)x_1'(t)}{r^2} = \frac{r^2(\cos^2 t + \sin^2 t)}{r^2} = 1.$$

Integrating both sides from 0 to 2π we have

$$2\pi I_f(C) = \phi(2\pi) - \phi(0) = \int_0^{2\pi} \phi'(t)\,dt = \int_0^{2\pi} dt = 2\pi.$$

This shows that $I_f(C) = 1$.

The proof for the case when $ad - bc < 0$ is similar and is left as an exercise, see Exercise 4.36. □

The second result is for perturbations of linear systems, i.e., for the system

(4.7.2) $$x' = Ax + \rho(x),$$

where $A \in \mathbb{R}^{2\times2}$ is the same as in system (4.7.1) and $\rho \in C(\mathbb{R}^2, \mathbb{R}^2)$ satisfying $|\rho(x)| = o(|x|)$ as $x \to 0$. It follows from the assumption that $\rho(0,0) = 0$ and hence $(0,0)$ is an equilibrium. It is easy to verify that the equilibrium $(0,0)$ is isolated.

LEMMA 4.7.4. *The index $I(O)$ of the equilibrium $O = (0,0)$ satisfies*

$$I(O) = \begin{cases} 1, & ad - bc > 0 \\ -1, & ad - bc < 0. \end{cases}$$

In other words, $I(O) = 1$ when $(0,0)$ is a node, a spiral-point, or a center; and $I(O) = -1$ when $(0,0)$ is a saddle-point.

PROOF. Choose a curve $C = \{x : |x| = r\}$ with a small $r > 0$ such that there are no equilibria other than $(0,0)$ inside and on the curve C. Let D be the interior of the curve C. Denote $f^L(x) = Ax$ and $f(x,s) = f^L(x) + s\rho(x)$ for $0 \le s \le 1$. Then for any $0 \le s \le 1$, $f(x,s) = 0$ for $x \in C \cup D$ if and only if $x = 0$. Hence $f(x,s) \ne 0$ on C for any $0 \le s \le 1$. This shows that $f(x,0)$ and $f(x,1)$ are homotopic on C. Since $f(x,0) = f^L(x)$ and $f(x,1) = f^L(x) + \rho(x)$, we have that $f^L(x) + \rho(x)$ is homotopic to $f^L(x)$ on C. Then the conclusion follows from Lemma 4.7.3. □

REMARK 4.7.3. Although the result in Lemma 4.7.4 is for the equilibrium $(0,0)$ of Eq. (4.7.2), it can be applied to any isolated equilibrium (a,b) of system (A-2). In fact, the equilibrium (a,b) can be moved to $(0,0)$ by shifting which does not change the coefficient matrix in linearization.

Now we are ready to present the relation between the index of a Jordan curve and the indices of equilibria of system (A-2).

THEOREM 4.7.1. *Let C be a Jordan curve in $\mathbb{R}^2$. Assume system (A-2) has a finite number of equilibria x_i, $i = 1,\ldots,k$, in the interior of C and has no equilibria on C. Then $T_f(C) = \sum_{i=1}^k I(x_i)$.*

PROOF. For $i = 1, \ldots, k$, let C_i be small circles inside C centered at x_i such that there are no equilibria other than x_i inside and on the curve C_i. Add $k - 1$ pairs of parallel paths $L_{i,i+1}^{\pm}$, $i = 1, \ldots, k-1$, inside C as shown in Fig. 4.24. By removing the minor parts of the circles C_i between $L_{i,i+1}^{\pm}$ for each $i = 1, \ldots, k-1$, we form a Jordan curve

$$C_* = \left(\cup_{i=1}^{k} \tilde{C}_i \right) \cup \left(\cup_{i=1}^{k-1} L_{i,i+1}^{\pm} \right),$$

where $\tilde{C}_i$ is the remaining major part of the circle C_i. Since $f(x) \neq 0$ between C and C_*, C and C_* are homotopic. By Lemma 4.7.2 we have that $I_f(C) = I_f(C_*)$. By taking limits as $d(L_{i,i+1}^{+}, L_{i,i+1}^{-}) \to 0$ for $i = 1, \ldots, k-1$, we obtain that

$$I_f(C) = \sum_{i=1}^{k-1} I_f(C_i) = \sum_{i=1}^{k-1} I(x_i).$$

$\square$

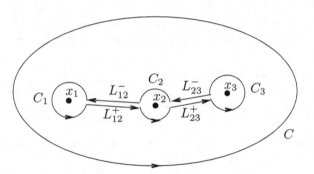

FIGURE 4.24. Indices of closed curve and equilibria

Derived from Theorem 4.7.1, we have the following relations between the closed orbits and the equilibria of system (A-2).

THEOREM 4.7.2. *Assume system* (A-2) *has a closed orbit* Γ. *Then system* (A-2) *must have at least one equilibrium in the interior of* Γ.

PROOF. Since Γ is a closed orbit of system (A-2), f is a tangent vector field on Γ. By Remark 4.7.1, Part (ii), $I_f(\Gamma) = 1$. Then the conclusion follows from Theorem 4.7.1. $\square$

THEOREM 4.7.3. *Suppose that*

 (i) *system* (A-2) *has a closed orbit* Γ *with* D *its interior*,
 (ii) $f \in C^1(\Gamma \cup D, \mathbb{R}^2)$ *satisfying* $\det(\partial f / \partial x) \neq 0$ *whenever* $f = 0$.
Then there are at most finitely many equilibria of system (A-2) *in* D. *More-over, the number of equilibria of system* (A-2) *in* D *must be odd, and among them,*

 the number of saddle-points = *the number of the other equilibria* − 1.

PROOF. Assumption (ii) shows that all equilibria of system (A-2) in D are elementary. Assume there are an infinite number of equilibria of system (A-2) in D. By the Bolzano–Weierstrass theorem, there is a sequence of equilibria $\{x_n\}_{n=0}^{\infty}$ of system (A-2) in D and an $x_0 \in \Gamma \cup D$ satisfying $x_n \neq x_0$ and $x_n \to x_0$. Since $f(x_n) = 0$, we see that $f(x_0) = 0$ which implies that x_0 is an equilibrium of system (A-2). Note that Γ does not contain equilibria, we have $x_0 \in D$. Therefore,

$$0 = f(x_n) - f(x_0) = \left.\frac{\partial f}{\partial x}\right|_{x=x_0} (x_n - x_0) + o(|x_n - x_0|)$$

as $n \to \infty$, and hence $\partial f/\partial x|_{x=x_0} = 0$. This contradicts assumption (ii) and hence confirms that there are at most finitely many equilibria of system (A-2) in D.

As in the proof of Theorem 4.7.2 we have $I_f(\Gamma) = 1$. Since all equilibria x_i, $i = 1, \ldots, k$, of system (A-2) in D are isolated and elementary, by Lemma 4.7.4 and Remark 4.7.2, we see that $I(x_i) = -1$ if x_i is a saddle-point and $I(x_i) = 1$ if x_i is any other equilibrium. Then the conclusion follows from Theorem 4.7.1. □

Theorem 4.7.3 provides us with another way to show the nonexistence of periodic solutions of autonomous systems in $\mathbb{R}^2$.

EXAMPLE 4.7.1. Consider the system

(4.7.3)
$$\begin{cases} x_1' = x_1(1 - x_2) \\ x_2' = x_2(2 - x_1). \end{cases}$$

We observe that the criteria given by Theorems 4.6.1 and 4.6.2 are hard to be applied to this equation.

By letting the right-hand side functions be zero we obtain two equilibria $O = (0,0)$ and $P = (2,1)$. We also note that there are orbits on both sides of the origin on the x_1-axis and on the x_2-axis.

The linearization of Eq. (4.7.3) at $O = (0,0)$ is

$$x_1' = x_1, \quad x_2' = 2x_2$$

with the coefficient matrix $A = \begin{bmatrix} 1 & 0 \\ 0 & 2 \end{bmatrix}$. Thus, $p = 3$, $q = 2$, and $\Delta = 3^2 - 8 = 1 > 0$. As shown in Sect. 4.3, $O = (0,0)$ is an improper node.

Let $y_1 = x_1 - 2$ and $y_2 = x_2 - 1$. Then the equilibrium $P = (2,1)$ of system (A-2) becomes the equilibrium $P_1 = (0,0)$ of the new system

(4.7.4) $$y_1' = -(y_1 + 2)y_2, \quad y_2' = -y_1(y_2 + 1).$$

With similar discussion we find that $P_1 = (0,0)$ is a saddle-point of system (4.7.4) and hence $P = (2,1)$ is a saddle-point of system (4.7.3).

Assume system (4.7.3) has a closed orbit Γ. By Theorem 4.7.3, Γ must surround the equilibrium $O = (0,0)$ but not the equilibrium $P = (2,1)$. Hence it crosses the x_1-axis and x_2-axis. Since both x_1-axis and x_2-axis consist of orbits, this contradicts the uniqueness of orbits given by Lemma 4.1.1.

4.8. Flows for n-Dimensional First-Order Autonomous Equations

In the last section, we further investigate the behavior of the flow for the n-dimensional first-order autonomous equation (A) given in Sect. 4.1. Before discussing the general case, we first consider the autonomous homogeneous linear equation

(H-c) $$x' = Ax,$$

where $A \in \mathbb{R}^{n \times n}$. Assume that

(A) A has m_- eigenvalues with negative real part λ_i^-, $i = 1, \ldots, m_-$;
     m_+ eigenvalues with positive real part λ_j^+, $j = 1, \ldots, m_+$;
   and
     m_0 eigenvalues with zero real part λ_k^0, $k = 1, \ldots, m_0$.

Obviously, $m_- + m_+ + m_0 = n$. From Theorems 2.4.3 and 3.2.2 we know that the asymptotic behavior of all solutions and the stability of the zero solution of Eq. (H-c) are dependent on the above eigenvalues. In particular, if $m_- = n$, then all solutions $x(t)$ satisfy $x(t) \to 0$ as $t \to \infty$ and hence the zero solution is asymptotically stable. However, when $m_+ > 0$, the only thing we know from the above-mentioned theorems is that there exists at least one solution $x(t)$ such that $x(t) \to 0$ as $t \to -\infty$. To depict the complete scenario of the solutions of Eq. (H-c), we need the following lemma and definition from linear algebra, see Coddington and Levinson [11, Exercise 3.40] or Hirsch and Smale [25].

LEMMA 4.8.1. *Let $A \in \mathbb{R}^{n \times n}$ satisfying assumption (A). Then there exists a nonsingular matrix $T \in \mathbb{R}^{n \times n}$ such that*

$$T^{-1}AT = B := \begin{bmatrix} B_- & 0 & 0 \\ 0 & B_+ & 0 \\ 0 & 0 & B_0 \end{bmatrix},$$

where $B_-, B_+,$ and B_0 are $m_-, m_+,$ and m_0 dimensional square matrices whose eigenvalues are λ_i^-, $i = 1, \ldots, m_-$; λ_j^+, $j = 1, \ldots, m_+$; and λ_k^0, $k = 1, \ldots, m_0$; respectively.

We point out that unlike the result in Sect. 2.4, the matrix B may not be the Jordan canonical form of matrix A due to the restriction that the transformation matrix T must be real-valued. Here, B is a real-valued block diagonal matrix. For example, B is allowed to be the matrix $\begin{bmatrix} -3 & 0 & 0 \\ 0 & 0 & 1 \\ 0 & -2 & 0 \end{bmatrix}$

with $B_- = [-3]$, $B_0 = \begin{bmatrix} 0 & 1 \\ -2 & 0 \end{bmatrix}$, and B_+ absent. However, B is not in the Jordan canonical form.

DEFINITION 4.8.1. *Let X be a vector space and X_1, X_2 subspaces of X. Then X is said to be the direct sum of X_1 and X_2, denoted by $X = X_1 \oplus X_2$,* if $X_1 \cap X_2 = \{0\}$ and $X = X_1 + X_2$, i.e., for any $x \in X$, there are unique $x_1 \in X_1$ and $x_2 \in X_2$ such that $x = x_1 + x_2$.

The direct sum of a finite number of subspaces X_i, $i = 1, \ldots, k$, of a vector space X is similarly defined.

Now we have the result on the characterization of solutions of Eq. (H-c).

THEOREM 4.8.1. *Let $A \in \mathbb{R}^{n \times n}$ satisfy assumption (A). Then there exist subspaces E_-, E_+, and E_0 of $\mathbb{R}^n$ satisfying*

(a) $\dim E_- = m_-$, $\dim E_+ = m_+$, $\dim E_0 = m_0$, *and* $\mathbb{R}^n = E_- \oplus E_+ \oplus E_0$;

(b) E_-, E_+, *and E_0 are invariant for the flow* $\Phi := \{\phi_t = e^{At} : t \in \mathbb{R}\}$ *induced by Eq. (A); and*

(c) $\phi_t(x) \to 0$ *exponentially as $t \to \infty$ for any $x \in E_- \setminus \{0\}$, $\phi_t(x) \to 0$ exponentially as $t \to -\infty$ for any $x \in E_+ \setminus \{0\}$, and $\phi_t(x)$ is bounded for $t \in \mathbb{R}$ or grows at a speed of a polynomial as $t \to \pm\infty$ for any $x \in E_0$.*

We call E_-, E_+, and E_0 the stable, unstable, and center subspaces for Eq. (H-c), respectively. When any of E_-, E_+, and E_0 is 0-dimensional, it is regarded as the empty set.

PROOF. Let T and B be defined as in Lemma 4.8.1 and let $y = T^{-1}x$. Then Eq. (H-c) becomes the equation $y' = By$. For any vector $c \in \mathbb{R}^n$, we write it in the form $c = (c_-, c_+, c_0)^T$, where c_-, c_+, and c_0 are respectively m_-, m_+, and m_0 dimensional subvectors of c. Define

$$F_- := \{(c_-, 0, 0)^T : c_- \in \mathbb{R}^{m_-}\}, \quad F_+ := \{(0, c_+, 0)^T : c_+ \in \mathbb{R}^{m_+}\},$$

$$F_0 := \{(0, 0, c_0)^T : c_0 \in \mathbb{R}^{m_0}\}.$$

Clearly, $\mathbb{R}^n = F_- \oplus F_+ \oplus F_0$. Since $e^{Bt} = \begin{bmatrix} e^{B_-t} & 0 & 0 \\ 0 & e^{B_+t} & 0 \\ 0 & 0 & e^{B_0t} \end{bmatrix}$, F_-, F_+, and F_0 are invariant sets for the flow $\Psi := \{\psi_t = e^{Bt} : t \in \mathbb{R}\}$. By the definition of the matrix B, we see that conclusion (c) holds with ϕ_t, E_-, E_+, and E_0 replaced by ψ_t, F_-, F_+, and F_0, respectively. Define

$$E_- := \{Tc : c \in F_-\}, \quad E_+ := \{Tc : c \in F_+\}, \quad E_0 := \{Tc : c \in F_0\}.$$

Clearly, $\mathbb{R}^n = E_- \oplus E_+ \oplus E_0$. Since $e^{At} = Te^{Bt}T^{-1}$, conclusions (b) and (c) follow from the transformation $x = Ty$ immediately. □

EXAMPLE 4.8.1. Consider Eq. (H-c) with $A = \begin{bmatrix} 0 & 1 & 0 \\ -1 & 0 & 0 \\ 0 & 0 & k \end{bmatrix}$ for $k \neq 0$.

It is easy to see that A is already a block diagonal matrix and its eigenvalues are $\lambda_{1,2} = \pm \mathbf{i}$ and $\lambda_3 = k$. It follows that the center subspace for Eq. (H-c) is the x_1x_2-plane and the stable [unstable] subspace for Eq. (H-c) is the x_3-axis if $k < 0$ [$k > 0$].

Now, we turn our attention to the behavior of solutions of the nonlinear autonomous equation

$$\text{(A)} \qquad\qquad x' = f(x),$$

where $f \in C^1$ near an equilibrium. Without loss of generality, we assume the equilibrium is at $x = 0$.

Recall that in Sect. 3.4, we used the linearization method to study the stability of the zero solution of Eq. (A). In the following, we let Eq. (H-c) be the linearization of Eq. (A) at $x = 0$, i.e., $A = \partial f/\partial x|_{x=0}$ in Eq. (H-c). Here we assume the matrix A satisfies assumption $(\mathcal{A})$.

It is reasonable to expect that in a small neighborhood of $x = 0$, the structure of solutions of Eq. (A) has certain similarities to that of Eq. (H-c). However, due to the nonlinearity of the right-hand side function f, we do not expect the existence of the stable, unstable, and center subspaces for Eq. (A). Instead, we tend to believe that there are some surfaces (including curves) in $\mathbb{R}^n$ which are invariant for Eq. (A). To establish the results, we introduce the following definitions:

DEFINITION 4.8.2. Sets E_1, $E_2 \subset \mathbb{R}^n$ are said to be homeomorphic if there exists a one-to-one correspondence $\mathcal{H} : E_1 \to E_2$ such that $\mathcal{H}$ and $\mathcal{H}^{-1}$ are both continuous. The map $\mathcal{H}$ is called a homeomorphism from E_1 onto E_2.

DEFINITION 4.8.3. Let $m \in \{1, \ldots, n\}$. Then an m-dimensional differential manifold M in $\mathbb{R}^n$ is a connected set with an open covering $\cup_\alpha U_\alpha = M$ such that

(a) for each α, U_α is homeomorphic to the open unit ball B_m in $\mathbb{R}^m$, and

(b) for any α_1 and α_2, if $\mathcal{H}_{\alpha_1} : U_{\alpha_1} \to B_m$ and $\mathcal{H}_{\alpha_2} : U_{\alpha_2} \to B_m$ are homeomorphisms, then the map

$$\mathcal{H} := \mathcal{H}_{\alpha_1} \circ \mathcal{H}_{\alpha_2}^{-1} : \mathcal{H}_{\alpha_2}(U_{\alpha_1} \cap U_{\alpha_2}) \to \mathcal{H}_{\alpha_1}(U_{\alpha_1} \cap U_{\alpha_2})$$

is differentiable with $\det \partial f(x)/\partial x \neq 0$ in its domain.

Roughly speaking, an m-dimensional manifold is a connected set which locally has the structure of the m-dimensional Euclidean space. Because of this, we can use coordinates to study the local behavior of manifolds. However, not every manifold has the global structure of a Euclidean space. For instance, A circle in the plane is a 1-dimensional manifold, but it is not homeomorphic to the open interval $(-1, 1)$ in $\mathbb{R}^1$. On the other hand, not every differentiable "curve" or "surface" is a manifold. For instance, the figure eight in the plane is not a 1-dimensional manifold in $\mathbb{R}^2$, and the famous Klein bottle is not a 2-dimensional manifold in $\mathbb{R}^3$, due to the self-intersecting nature of these figures.

Now we extend the concepts for stable, unstable, and center subspaces for linear equation (H-c) to corresponding manifolds for nonlinear equation (A).

DEFINITION 4.8.4. Let Eq. (H-c) be the linearization of Eq. (A) at $x = 0$ satisfying assumption $(\mathcal{A})$, and E_-, E_+, and E_0 are defined as in Theorem 4.8.1. Denote by $\Phi := \{\phi_t : t \in \mathbb{R}\}$ the flow induced by Eq. (A). Then

(a) an m_--dimensional differentiable manifold M_s is said to be a stable manifold for Eq. (A) at $x = 0$ if it is positively invariant for Eq. (A) and tangent to E_- at $x = 0$, and $\phi_t(x) \to 0$ as $t \to \infty$ for any $x \in M_s$;

(b) an m_+-dimensional differentiable manifold M_u is said to be an unstable manifold for Eq. (A) at $x = 0$ if it is negatively invariant for Eq. (A) and tangent to E_+ at $x = 0$, and $\phi_t(x) \to 0$ as $t \to -\infty$ for any $x \in M_u$;

(c) An m_0-dimensional differentiable manifold M_c is said to be a center manifold for Eq. (A) at $x = 0$ if it is invariant for Eq. (A) and tangent to E_0 at $x = 0$.

In the above, if any of M_s, M_u, and M_c is 0-dimensional, it is regarded as the empty set.

Now, we present the results for autonomous nonlinear equations parallel to those in Theorem 4.8.1 for autonomous linear equations. The results below are based on Carr [6], Chicone [9], Chow and Hale [10], Coddington and Levinson [11], Hartman [22], Hirsch, Pugh, and Shub [26], Kelley [30], Perko [41], and Wiggins [46]. Here, we only give a brief presentation of some major results together with a few illustrative examples. Due to their lengths, most proofs are to be omitted in this book.

THEOREM 4.8.2 (Invariant Manifold Theorem). *Let Eq. (H-c) be the linearization of Eq. (A) at $x = 0$ satisfying assumption $(\mathcal{A})$. Then in a neighborhood of $x = 0$, there exist an m_--dimensional stable manifold, an m_+-dimensional unstable manifold, and an m_0-dimensional center manifold for Eq. (A).*

Here, if any of M_s, M_u, and M_c is 0-dimensional, it is regarded as the empty set.

We verify Theorem 4.8.2 with the example below.

EXAMPLE 4.8.2. Consider the system

(4.8.1)
$$\begin{cases} x_1' = -x_1 \\ x_2' = x_2 + x_1^2 \\ x_3' = x_3^2 \ . \end{cases}$$

Clearly, the linearization of Eq. (4.8.1) is the system

(4.8.2)
$$\begin{cases} x_1' = -x_1 \\ x_2' = x_2 \\ x_3' = 0 \end{cases}$$

whose stable, unstable, and center subspaces E_-, E_+, and E_0 are the x_1-axis, x_2-axis, and x_3-axis, respectively. By solving the system, we find that the solution $x(t) := (x_1, x_2, x_3)^T(t)$ of system (4.8.1) satisfying the IC $x(0) = c := (c_1, c_2, c_3)^T$ is

$$\begin{cases} x_1 = c_1 e^{-t} \\ x_2 = (c_1^2/3 + c_2)e^t - (c_1^2/3)e^{-2t} \\ x_3 = c_3/(1 - c_3 t). \end{cases}$$

Define

$$\begin{aligned} M_s &= \{c = (c_1, c_2, c_3)^T : c_2 = -c_1^2/3, c_3 = 0\}, \\ M_u &= \{c = (c_1, c_2, c_3)^T : c_1 = 0, c_3 = 0\}, \quad \text{and} \\ M_c &= \{c = (c_1, c_2, c_3)^T : c_1 = 0, c_2 = 0\}. \end{aligned}$$

It is obvious that M_s, M_u, and M_c are tangent to E_-, E_+, and E_0 at $(0,0,0)$, respectively; M_u is negatively invariant and M_c is invariant for system (4.8.1); and $x(t) \to 0$ as $t \to \infty$ for $c \in M_s$ and $x(t) \to 0$ as $t \to -\infty$ for $c \in M_u$. It is also easy to show that M_s is positively invariant for system (4.8.1). In fact, for any $c \in M_s$, the solution $x(t)$ of system (4.8.1) with the IC $x(0) = c$ satisfies

$$x_2(t) = \left(\frac{c_1^2}{3} + c_2\right)e^t - \left(\frac{c_1^2}{3}\right)e^{-2t} = -\left(\frac{c_1^2}{3}\right)e^{-2t} = -\frac{x_1^2(t)}{3}$$

and $x_3(t) = 0$, and hence $x(t) \in M_s$ for any $t \geq 0$. Therefore, M_s, M_u, and M_c are stable, unstable, and center manifolds of system (4.8.1). See Fig. 4.25, Part (a).

We observe that in this problem, the stable and unstable manifolds are uniquely determined. However, it is not the case for the center manifold. To see this, we note that the equation for orbits in the $x_1 x_3$-plane is $dx_1/dx_3 = -x_1/x_3^2$ whose general solution is $x_1 = ke^{1/x_3}$ for $k \in \mathbb{R}$. Note that the orbits given above for all $k \in \mathbb{R}$ are tangent to the x_3-axis at $(0,0)$. Thus, for any $k \neq 0$, we may patch this curve with the positive x_3-axis to form a differential manifold $\bar{M}_c$ in $\mathbb{R}^2$: $x_1 = \begin{cases} ke^{1/x_3}, & x_3 < 0 \\ 0, & x_3 \geq 0 \end{cases}$, see Fig. 4.25, Part (b). By embedding the manifolds $\bar{M}_c$ with $k \neq 0$ into the $x_1 x_3$-plane in $\mathbb{R}^3$, we find that they are all invariant for system (4.8.1) and tangent to E_0. Therefore, they are all center manifolds for system (4.8.1). However, there is only one analytic center manifold M_c for system (4.8.1).

The result in Theorem 4.8.2 is a summary of several results in the literature. To see how the results were developed, we now prove the important hyperbolic case in Theorem 4.8.2, i.e., the case where the real parts of all eigenvalues of the matrix A are not zero. The proof we adopt here is from Perko [41].

THEOREM 4.8.3 (Stable-Unstable Manifold Theorem). *Let Eq. (H-c) be the linearization of Eq. (A) at $x = 0$ satisfying assumption (A) with m_-, $m_+ > 0$ and $m_0 = 0$. Then in a neighborhood of $x = 0$, there exist an m_--dimensional stable manifold and an m_+-dimensional unstable manifold for Eq. (A).*

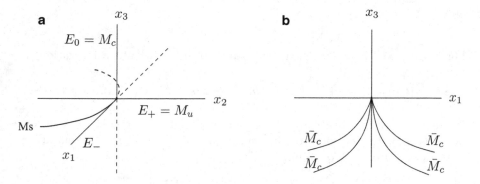

FIGURE 4.25. Manifolds and subspaces

PROOF. From the assumption we may rewrite Eq. (A) as

$$(4.8.3) \qquad x' = Ax + g(x),$$

where the matrix A satisfies assumption $(\mathcal{A})$ with m_-, $m_+ > 0$ and $m_0 = 0$, and $g \in C^1(\mathbb{R}^n, \mathbb{R}^n)$ such that $g(0) = 0$ and $\partial g/\partial x|_{x=0} = 0$. By Lemma 4.8.1, there exists a nonsingular matrix $T \in \mathbb{R}^{n \times n}$ such that

$$T^{-1}AT = B := \begin{bmatrix} B_- & 0 \\ 0 & B_+ \end{bmatrix},$$

where B_- and B_+ are m_- and m_+ dimensional square matrices whose eigenvalues are λ_i^-, $i = 1, \ldots, m_-$, and λ_j^+, $j = 1, \ldots, m_+$; respectively. Let $y = T^{-1}x$. Then Eq. (4.8.3) becomes

$$(4.8.4) \qquad y' = By + h(y),$$

where $h(y) := T^{-1}g(Ty) \in C^1(\mathbb{R}^n, \mathbb{R}^n)$ satisfying $h(0)=0$ and $\partial h/\partial y|_{y=0} = 0$. Let E_- and E_+ be defined as in Theorem 4.8.1. We observe that

$$F_- := \{T^{-1}d : d \in E_-\} \quad \text{and} \quad F_+ := \{T^{-1}d : d \in E_+\}$$

are the stable and unstable subspaces of the linearization $y'=By$ of Eq. (4.8.4), respectively.

We first show that there is m_--dimensional stable manifold for Eq. (4.8.4). For this purpose, we denote

$$U(t) = \begin{bmatrix} e^{B_-t} & 0 \\ 0 & 0 \end{bmatrix} \quad \text{and} \quad V(t) = \begin{bmatrix} 0 & 0 \\ 0 & e^{B_+t} \end{bmatrix}.$$

Then we have

$$U'(t) = BU(t), \ V'(t) = BV(t), \ \text{and} \ U(t) + V(t) = e^{Bt}.$$

Note that the eigenvalues of B_- and B_+ have negative and positive real parts, respectively. As shown in Theorem 3.2.2, Part (b), there exist $k, \alpha > 0$ and $\sigma \in (0, \alpha)$ such that

$$|U(t)| \leq ke^{-(\alpha+\sigma)t} \quad \text{for all } t \geq 0$$

and

$$|V(t)| \le k e^{\sigma t} \quad \text{for all } t \le 0.$$

For any $c \in \mathbb{R}^{m-}$, let $\tilde{c} := (c, 0)^T \in \mathbb{R}^n$ and consider the integral equation

$$(4.8.5) \quad u(t; c) = U(t)\tilde{c} + \int_0^t U(t-s)h(u(s; c))\, ds - \int_t^\infty V(t-s)h(u(s; c))\, ds.$$

It is easy to see that if $u(t; c)$ is a solution of Eq. (4.8.5), then it is a solution of Eq. (4.8.4). In fact, for any solution $u(t; c)$ of Eq. (4.8.5), by differentiating both sides of Eq. (4.8.5) we have

$$
\begin{aligned}
u'(t; c) \;&=\; U'(t)\tilde{c} + \int_0^t U'(t-s)h(u(s; c))\, ds - \int_t^\infty V'(t-s)h(u(s; c))\, ds \\
&\quad + U(0)h(u(t; c)) + V(0)h(u(t; c)) \\
&=\; B\left(U(t)\tilde{c} + \int_0^t U(t-s)h(u(s; c))\, ds - \int_t^\infty V(t-s)h(u(s; c))\, ds \right) \\
&\quad + h(u(t; c)) \\
&=\; B u(t; c) + h(u(t; c)).
\end{aligned}
$$

Hence $u(t; c)$ is a solution of Eq. (4.8.4). We now show that Eq. (4.8.5) has a solution on $[0, \infty)$ by the method of successive approximations. Define a sequence of functions $\{u^{[i]}\}_{i=0}^\infty$ by recurrence as follows:

$$u^{[0]}(t; c) = 0;$$

and for $i = 0, 1, \ldots,$

(4.8.6)

$$u^{[i+1]}(t; c) = U(t)\tilde{c} + \int_0^t U(t-s)h(u^{[i]}(s; c))\, ds - \int_t^\infty V(t-s)h(u^{[i]}(s; c))\, ds.$$

Note that $\partial h/\partial y|_{y=0} = 0$. By the mean value theorem, there exists a $\delta > 0$ such that if $|y_1|, |y_2| \in (0, \delta)$, then

$$|h(y_1) - h(y_2)| < \frac{\sigma}{4k}|y_1 - y_2|.$$

Let $|c| < \delta/(2k)$. By induction we can show that for $i = 1, 2, \ldots$ and $t \ge 0$,

$$(4.8.7) \quad |u^{[i]}(t; c)| < \delta \quad \text{and} \quad |u^{[i]}(t; c) - u^{[i-1]}(t; c)| < \frac{k|c|e^{-\alpha t}}{2^{i-1}} \le \frac{k|c|}{2^{i-1}}.$$

We leave this as an exercise, see Exercise 4.43. Since the constant series $\sum_{i=1}^\infty k|c|/2^{i-1}$ is convergent, by the Weierstrass M-test, we see that the series of functions $\sum_{i=1}^\infty (u^{[i]}(t; c) - u^{[i-1]}(t; c))$ is uniformly convergent for $t \ge 0$. This means that the sequence of functions $\{u^{[i]}(t; c)\}_{i=0}^\infty$ is uniformly convergent for $t \ge 0$. Let $u(t; c)$ be the uniform limit of $\{u^{[i]}(t; c)\}_{i=0}^\infty$. Taking limits on both sides of (4.8.6) as $i \to \infty$ we see that $u(t; c)$ is a solution of

Eq. (4.8.5) and hence a solution of Eq. (4.8.4). By (4.8.7) we see that for $t \geq 0$,

$$(4.8.8) \quad |u(t;c)| \leq \sum_{i=1}^{\infty} |u^{[i]}(t;c) - u^{[i-1]}(t;c)| < \sum_{i=1}^{\infty} \frac{k|c|e^{-\alpha t}}{2^{i-1}} = 2k|c|e^{-\alpha t}.$$

Denote $u(t;c) = (u_-, u_+)^T(t;c)$ with u_- and u_+ the m_- and m_+ dimensional subvectors of u, respectively. From (4.8.5), $u(t;c)$ satisfies the IC

$$u_-(0;c) = c \quad \text{and} \quad u_+(0;c) = -\int_0^{\infty} V(-s)h(u(s;c))\, ds.$$

Let $D := \{c \in \mathbb{R}^{m_-} : \text{solution } u(t;c) \text{ of Eq. (4.8.5) exists}\}$. Define a function $\psi : D \to \mathbb{R}^{m_+}$ by

$$\psi(c) = -\int_0^{\infty} V(-s)h(u(s;c))\, ds$$

and define a differential manifold $N_s := \{(c, \psi(c))^T : c \in D\}$. We show that N_s is a stable manifold for Eq. (4.8.4). For any $c \in D$ let $y(t;c)$ be the solution of Eq. (4.8.4) satisfying the IC $y(0) = (c, \psi(c))$ and $u(t;c)$ the solution of Eq. (4.8.5) with the same c. Then $u(t;c)$ is also a solution of Eq. (4.8.4) with the same IC. By the uniqueness of solutions of IVPs, $y(t;c) \equiv u(t;c)$. Clearly, $y(0;c) \in N_s$. We claim that $y(t;c) \in N_s$ for any $t \geq 0$. In fact, for $t_1 > 0$ let $\bar{u}(t;c) := u(t + t_1;c)$. Then $\bar{u}(t;c)$ is also a solution of Eq. (4.8.5). Hence $u_-(t_1;c) = u_-(0;c) \in D$. From (4.8.5) and using the properties of autonomous equations, we have that for $t \geq 0$

$$\begin{aligned} u_+(t;c) &= -\int_t^{\infty} V(t-s)h(u(s;c))\, ds = -\int_0^{\infty} V(-s)h(u(s+t;c))\, ds \\ &= -\int_0^{\infty} V(-s)h(u(s, u(t;c))\, ds = \psi(u_+(t;c)). \end{aligned}$$

This shows that $u(t;c) \in N_s$ and hence $y(t;c) \in N_s$ for any $t \geq 0$. As a result, N_s is positively invariant. From (4.8.8), $y(t;c) = u(t;c) \to 0$ as $t \to \infty$ for $c \in D$. On the other hand, it can be shown that $y(t;c) \not\to 0$ as $t \to \infty$ for $c \notin D$, see Coddington and Levinson [11, p. 332]. Furthermore, as shown in [11, p. 333], $\partial\psi/\partial c|_{c=0} = \mathbf{0}$. Hence N_s is tangent to F_- at $c = 0$. Therefore, N_s is a stable manifold for Eq. (4.8.4). Consequently, $M_s := \{x = Ty : y \in N_s\}$ is a stable manifold for Eq. (4.8.3).

With t replaced by $-t$, Eq. (4.8.3) becomes the equation $x' = -Ax - g(x)$. By the same argument, we can show that this equation has a m_+-dimensional stable manifold, which is a m_+-dimensional unstable manifold for Eq. (4.8.3). $\qquad \square$

REMARK 4.8.1. (i) It can be further shown that the stable manifold M_s and the unstable manifold M_u given in Theorem 4.8.3 are uniquely determined.

(ii) Although the manifolds M_s and M_u are defined only in a neighborhood of $x = 0$, they can be extended globally. Let $\Phi := \{\phi_t : t \in \mathbb{R}\}$ be the flow induced by Eq. (A). We call the sets

$$W^s(0) := \{\phi_t(x) : x \in M_s, t \leq 0\} \quad \text{and} \quad W^u(0) := \{\phi_t(x) : x \in M_u, t \geq 0\}$$

the global stable and unstable manifolds for Eq. (A) at $x = 0$, respectively. It is easy to see that $W^s(0)$ and $W^u(0)$ are uniquely determined, invariant, and satisfy $\lim_{t \to \infty} \phi(x) = 0$ for $x \in M_s$ and $\lim_{t \to -\infty} \phi(x) = 0$ for $x \in M_u$.

Theorem 4.8.3 depicts the relations between the stable and unstable manifolds for Eq. (A) and the stable and unstable subspaces for its linearized equation (H-c) in a neighborhood of $x = 0$. This result has been strengthened to characterize the relation between the local topological structures of the orbits of the two equations by the following theorem:

THEOREM 4.8.4 (Hartman–Grobman Theorem [19, 23]). *Let Eq. (H-c) be the linearization of Eq. (A) at $x = 0$ satisfying assumption (A) with $m_-, m_+ > 0$ and $m_0 = 0$. Let $\Phi := \{\phi_t : t \in \mathbb{R}\}$ and $\Psi := \{\psi_t : t \in \mathbb{R}\}$ be the flows induced by Eq. (A) and Eq. (H-c), respectively. Then there exists a homeomorphism $\mathcal{H} : U \to V$ with U and V neighborhoods of $x = 0$ in $\mathbb{R}^n$ such that for any $x \in U$, there exists a neighborhood I of $t = 0$ in $\mathbb{R}$ satisfying*

(4.8.9) $$\mathcal{H}(\phi_t(x)) = \psi_t(\mathcal{H}(x)), \quad t \in I;$$

i.e., $\mathcal{H}$ is a one-to-one correspondence between the orbits of Eq. (A) and those of its linearization (H-c) near $x = 0$. Moreover, it preserves the directions of orbits.

The proof of Theorem 4.8.4 given in [22, p. 244–251], also in [41, p. 121–124], provides a constructive way for finding the homeomorphism $\mathcal{H}$ by successive approximations. Here, we only illustrate the result in Theorem 4.8.4 intuitively by an example.

EXAMPLE 4.8.3. Consider the system

(4.8.10) $$\begin{cases} x_1' = x_1 \\ x_2' = -x_2 + x_1^2. \end{cases}$$

Clearly, the linearization of Eq. (4.8.10) is the system

(4.8.11) $$\begin{cases} x_1' = x_1 \\ x_2' = -x_2. \end{cases}$$

By solving the systems with the IC $x_1(0) = c_1$, $x_2(0) = c_2$, we find that the flow induced by Eq. (4.8.10) is given by

$$\phi_t(c_1, c_2) = \left(c_1 e^t, (c_2 - \frac{1}{3}c_1^2)e^{-t} + \frac{1}{3}c_1^2 e^{2t} \right)^T$$

and the flow induced by Eq. (4.8.11) is given by

$$\psi_t(c_1, c_2) = \left(c_1 e^t, c_2 e^{-t} \right)^T.$$

Our purpose is to find a homeomorphism $\mathcal{H} = (\mathcal{H}_1, \mathcal{H}_2)^T : \mathbb{R}^2 \to \mathbb{R}^2$ so that (4.8.9) holds, which implies that $\mathcal{H}(\phi_t(x))$ must be an image of ψ_t for any $t \in \mathbb{R}$. Thus, $\mathcal{H}_1(x_1, x_2) = x_1$ for any $(x_1, x_2) \in \mathbb{R}^2$ and $\mathcal{H}_2(\phi_t(c_1, c_2))$ cannot contain a term of e^{2t}. This motivates us to define the map

$$\mathcal{H}(x_1, x_2) = \left(x_1, x_2 - \frac{1}{3}x_1^2 \right)^T.$$

It is easy to see that $\mathcal{H}$ is a homeomorphism from $\mathbb{R}^2$ to $\mathbb{R}^2$, and

$$\mathcal{H}(\phi_t(c_1, c_2)) = \left(c_1 e^t, (c_2 - \frac{1}{3}c_1^2)e^{-t} \right)^T = \psi_t(\mathcal{H}(c_1, c_2)).$$

Hence (4.8.9) holds.

We observe that the stable and unstable subspaces for system (4.8.11) are respectively

$$E_- = \{(c_1, c_2) : c_1 = 0\} \quad \text{and} \quad E_+ = \{(c_1, c_2) : c_2 = 0\},$$

and the stable and unstable manifolds for system (4.8.10) are respectively

$$M_s = \{(c_1, c_2) : c_1 = 0\} \quad \text{and} \quad M_u = \{(c_1, c_2) : c_2 = \frac{1}{3}c_1^2\}.$$

Therefore, $\mathcal{H}$ maps M_s and M_u onto E_- and E_+, respectively. The phase portraits for systems (4.8.10) and (4.8.11) are given in Fig. 4.26.

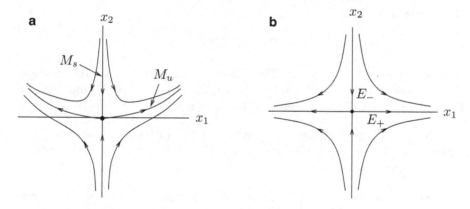

FIGURE 4.26. Homeomorphism between phase portraits. (a) Phase portrait for Eq. (4.8.10). (b) Phase portrait for Eq. (4.8.11)

Another important case of Theorem 4.8.2 is the case with $m_+ = 0$ as shown below, see Carr [6, Theorem 1] for the proof.

THEOREM 4.8.5 (Center Manifold Theorem). *Let Eq. (H-c) be the linearization of Eq. (A) at $x = 0$ satisfying assumption (A) with m_-, $m_0 > 0$ and $m_+ = 0$. Then in a neighborhood of $x = 0$, there exist an m_--dimensional stable manifold and and m_0-dimensional center manifold for Eq. (A).*

Note that $m_+ = 0$ means that all eigenvalues of matrix A have nonpositive real parts. Since $m_0 > 0$, some of them have zero real parts. In this case, the stability of the zero solution of Eq. (A) cannot be determined by linearization. Now, Theorem 4.8.5 can be employed in the further study of this stability problem.

In fact, as shown in the proof of Theorem 4.8.3, by a linear transformation $y = T^{-1}x$ with some nonsingular matrix $T \in \mathbb{R}^{n \times n}$, Eq. (A) can be written as

$$y' = By + h(y)$$

or its component form

$$(4.8.12) \qquad \begin{cases} y_1' = B_- y_1 + h_1(y_1, y_2) \\ y_2' = B_0 y_2 + h_2(y_1, y_2), \end{cases}$$

where $B = \begin{bmatrix} B_- & 0 \\ 0 & B_0 \end{bmatrix}$ with B_- and B_0 being m_- and m_0 dimensional square matrices whose eigenvalues have negative and zero real parts, respectively; $h = (h_1, h_2)^T \in C^1(\mathbb{R}^n, \mathbb{R}^n)$ satisfying $h(0) = 0$ and $\partial h / \partial y |_{y=0} = 0$. We observe that the center manifold given in Theorem 4.8.5 is tangent to the y_2-axis at $(0,0)$. Assume $y_1 = \xi(y_2)$ is the equation for the center manifold in the new coordinate system. Then the orbits on the center manifold satisfy the equation

$$(4.8.13) \qquad u' = B_0 u + h_2(\xi(u), u)$$

in a neighborhood of $u = 0$. The following theorem reveals the relation between the stability of the zero solution of system (4.8.12) and that of Eq. (4.8.13), see Carr [6, Theorem 2] for the proof.

THEOREM 4.8.6. *If the zero solution of Eq. (4.8.13) is stable [asymptotically stable, unstable], then the zero solution of system (4.8.12) is stable [asymptotically stable, unstable].*

Theorem 4.8.6 shows that the stability of the zero solution of system (4.8.12) is determined by that of Eq. (4.8.13). This is natural since the orbits on the stable manifold approach zero as $t \to \infty$. Note that Eq. (4.8.13) is a system of dimension m_0. Therefore, by Theorem 4.8.6, we only need to solve an m_0-dimensional problem for the n-dimensional problem.

We comment that a center manifold may not be unique. However, Theorem 4.8.6 guarantees that the stability of the zero solution of Eq. (4.8.13) remains the same for all center manifolds $\xi(u)$ for system (4.8.12).

To utilize Theorem 4.8.6, we must determine the function $\xi(u)$ for a center manifold and then study the stability of the zero solution of Eq. (4.8.13). However, it is usually difficult or impossible to obtain the exact form for $\xi(u)$, instead, we may approximate $\xi(u)$ by its Taylor polynomials.

Differentiating the equation $y_1 = \xi(y_2)$ with respect to t and using the chain rule we obtain that

$$y_1' = \frac{\partial \xi(y_2)}{\partial y_2} y_2'.$$

Since the center manifold is invariant for Eq. (4.8.12), we have

$$B_- \xi(y_2) + h_1(\xi(y_2), y_2) = \frac{\partial \xi(y_2)}{\partial y_2} [B_0 y_2 + h_2(\xi(y_2), y_2)].$$

Hence $\xi(u)$ is a solution of the equation

(4.8.14) $\qquad \dfrac{\partial \xi(u)}{\partial u} [B_0 u + h_2(\xi(u), u)] - [B_- \xi(u) + h_1(\xi(u), u)] = 0.$

Furthermore, it satisfies the IC

(4.8.15) $\qquad\qquad\qquad \xi(0) = 0, \quad \xi'(0) = 0.$

We then may approximate the solution $\xi(u)$ of IVP (4.8.14), (4.8.15) by its Taylor polynomial to any degree of accuracy, provided the functions h_1 and h_2 are smooth enough. Consequently, we can use Theorem 4.8.6, together with an appropriate approximation of $\xi(u)$, to determine the stability of the zero solution of Eq. (4.8.12). We show this by examples.

EXAMPLE 4.8.4. Consider the system

(4.8.16) $\qquad\qquad\qquad \begin{cases} y_1' = -y_1 \\ y_2' = y_2^2. \end{cases}$

Clearly, from Theorem 4.8.5, system (4.8.16) has a stable manifold and a center manifold. By the argument in Example 4.8.2 we see that there are in fact an infinite number of center manifolds. However, for all of them, the Eq. (4.8.13) for the center manifolds is $u' = u^2$. It is easy to see that its zero solution is unstable. By Theorem 4.8.6, the zero solution of system (4.8.16) is unstable.

EXAMPLE 4.8.5. Consider the system

(4.8.17) $\qquad\qquad\qquad \begin{cases} y_1' = -y_1 + 2y_2^2 - y_1 y_2^4 \\ y_2' = -y_1^2 y_2 + y_1 y_2^3. \end{cases}$

Here

(4.8.18)
$$B_- y_1 + h_1(y_1, y_2) = -y_1 + 2y_2^2 - y_1 y_2^4 \quad \text{and} \quad B_0 y_2 + h_2(y_1, y_2) = -y_1^2 y_2 + y_1 y_2^3.$$

Clearly, from Theorem 4.8.5, system (4.8.17) has a stable manifold and a center manifold. Since h_1 and h_2 are analytic functions, with the IC (4.8.15) we may approximate $\xi(u)$ by finding its Taylor polynomial in the form

$$\xi(u) = \sum_{i=2}^{k} a_i u^i + r_k(u).$$

with any $k \geq 2$ for $u \in \mathbb{R}$ close to 0, where $r_k(u) = O(u^{k+1})$ as $u \to 0$. We now try it with $k = 2$ and let

$$(4.8.19) \qquad\qquad \xi(u) = au^2 + \eta(u).$$

Substituting (4.8.18) and (4.8.19) into (4.8.14) we obtain that

$$(2au + \eta')[-(au^2 + \eta)^2 u + (au^2 + \eta)u^3] - [-(au^2 + \eta) + 2u^2 - (au^2 + \eta)u^4] = 0.$$

By comparing the coefficients of the terms with the lowest degree, we find that $a = 2$ and $\eta(u) = O(u^6)$ as $u \to 0$. Hence $\xi(u) = 2u^2 + O(u^6)$. As a result, the Eq. (4.8.13) for the center manifold becomes

$$u' = -2u^5 + O(u^9).$$

It is easy to see that its zero solution is asymptotically stable. By Theorem 4.8.6, the zero solution of system (4.8.17) is asymptotically stable.

We comment that if the equation for the center manifold with an approximated $\xi(u)$ does not bear enough information for the stability of the zero solution, a better approximation of $\xi(u)$ must be found. To see this, let us look at the next example for a modified equation of Eq. (4.8.17).

EXAMPLE 4.8.6. Consider the system

$$(4.8.20) \qquad \begin{cases} y_1' = -y_1 + y_2^2 - y_1 y_2^4 \\ y_2' = -y_1^2 y_2 + y_1 y_2^3. \end{cases}$$

Here

$(4.8.21)$
$$B_- y_1 + h_1(y_1, y_2) = -y_1 + y_2^2 - y_1 y_2^4 \text{ and } B_0 y_2 + h_2(y_1, y_2) = -y_1^2 y_2 + y_1 y_2^3.$$

By the same way as in Example 4.8.5 we first let $\xi(u) = au^2 + \eta$ and find that $a = 1$ and η satisfies the equation

$$(4.8.22) \qquad (2u + \eta')(-u^3 \eta - u\eta^2) + (\eta + u^6 + u^4 \eta) = 0$$

and hence $\eta(u) = O(u^6)$ as $u \to 0$. As a result, the Eq. (4.8.13) for the center manifold becomes $u' = O(u^9)$ which does not provide enough information for the stability of the zero solution.

To obtain a better approximation for $\xi(u)$, let

$$(4.8.23) \qquad\qquad \eta(u) = bu^6 + \zeta.$$

Substituting (4.8.23) into (4.8.22) we have that as $u \to 0$

$$O(u^{10}) + [bu^6 + \zeta + u^6 + u^4(bu^6 + \zeta)] = 0$$

By comparing the coefficients of the terms with the lowest degree, we find that $b = -1$ and $\zeta = O(u^{10})$ as $u \to 0$. Hence $\xi(u) = u^2 - u^6 + O(u^{10})$. As a result, the Eq. (4.8.13) for the center manifold becomes

$$u' = -[u^2 - u^6 + O(u^{10})]^2 u + [u^2 - u^6 + O(u^{10})]u^3 = u^9 + O(u^{13}).$$

It is easy to see that its zero solution is unstable. By Theorem 4.8.6, the zero solution of system (4.8.20) is unstable.

The last example is for a modified equation of Eq. (4.8.20).

EXAMPLE 4.8.7. Consider the system

(4.8.24)
$$\begin{cases} y_1' = -y_1 + y_2^2 \\ y_2' = -y_1^2 y_2 + y_1 y_2^3. \end{cases}$$

Here

$$B_- y_1 + h_1(y_1, y_2) = -y_1 + y_2^2 \quad \text{and} \quad B_0 y_2 + h_2(y_1, y_2) = y_1 y_2(-y_1 + y_2^2).$$

It is easy to check that $y_1 = y_2^2$ makes both the right-hand side functions zero and hence $\xi(u) = u^2$ satisfies (4.8.14). As a result, the Eq. (4.8.13) for the center manifold becomes $u' = 0$ which is stable but not asymptotically stable. By Theorem 4.8.6, the zero solution of system (4.8.24) is stable but not asymptotically stable. Note that in this problem, the center manifold $y_1 = y_2^2$ is full of equilibria.

Exercises

4.1. Prove that the set of all equilibria of Eq. (A) is a closed set.

4.2. Prove Corollary 4.1.2, Part (b).

4.3. Assume that $x = t + 1$, $y = \sin\left(\dfrac{\pi}{2}t\right)$ is a solution of the system
$\begin{cases} x' = f(x, y) \\ y' = g(x, y) \end{cases}$ where $f, g \in C^1(\mathbb{R}^2, \mathbb{R})$. For each of the following systems, use the properties of the autonomous equations to determine whether or not the set of functions is a solution of the same system. Justify your answers.

(a) $x_1 = t$, $y_1 = -\cos\left(\dfrac{\pi}{2}t\right)$;

(b) $x_2 = t$, $y_2 = \cos\left(\dfrac{\pi}{2}t\right)$.

4.4. Sketch the phase portrait for each of the following systems:

(a) $\begin{cases} x' = -y \\ y' = x, \end{cases}$ (b) $\begin{cases} x' = -y + 2 \\ y' = x - 1, \end{cases}$ (c) $\begin{cases} x' = x^2 - x \\ y' = (2x - 1)y. \end{cases}$

4.5. Assume f_1, f_2, $g \in C(\mathbb{R}^2, \mathbb{R})$ such that $g(x_1, x_2) \neq 0$ on $\mathbb{R}^2$. Show that the two systems

$$x_1' = f_1(x_1, x_2), \quad x_2' = f_2(x_1, x_2)$$

and

$$x_1' = (gf_1)(x_1, x_2), \quad x_2' = (gf_2)(x_1, x_2)$$

have exactly the same orbits in the phase plane. Moreover, the directions of the orbits remain the same when $g(x_1, x_2) > 0$ and become opposite when $g(x_1, x_2) < 0$.

4.6. Use the results in Problems 3 and 4 to sketch the phase portrait for each of the following systems:

(a) $\begin{cases} x' = y \\ y' = -x, \end{cases}$ (b) $\begin{cases} x' = (-y + 2)(x^2 + y^2) \\ y' = (x - 1)(x^2 + y^2), \end{cases}$

(c) $\begin{cases} x' = -y(x^2 + y^2 - 1) \\ y' = x(x^2 + y^2 - 1). \end{cases}$

4.7. (a) Let $A \in \mathbb{R}^{n \times n}$ and $\phi_t = e^{At}$. Show that $\Phi := \{\phi_t : t \in \mathbb{R}\}$ is a dynamical system on $\mathbb{R}^n$.

(b) Sketch the orbit $\phi_t(x_0)$ of Φ in $\mathbb{R}^2$ for each of the following cases:

(i) $A = \begin{bmatrix} 2 & 0 \\ 0 & -1 \end{bmatrix}$, $x_0 = \begin{bmatrix} 3 \\ 1 \end{bmatrix}$;

(ii) $A = \begin{bmatrix} -1 & -2 \\ 0 & 3 \end{bmatrix}$, $x_0 = \begin{bmatrix} -3 \\ 1 \end{bmatrix}$.

4.8. (a) Determine the dynamical system Φ on $\mathbb{R}^2$ induced by the system of differential equations

$$x_1' = -x_1, \quad x_2' = 2x_2 + x_1^2.$$

(b) Show that the set $E := \{(x_1, x_2) : x_2 = -x_1^2/4\}$ is invariant for Φ.

4.9. Prove Remark 4.2.2.

4.10. Show that the set $\{(x, \sin \frac{1}{x}) : x \neq 0\} \cup \{(0,0)\}$ is connected in $\mathbb{R}^2$.

4.11. (a) Assume A and B are bounded closed sets in $\mathbb{R}^n$ such that $A \cap B = \emptyset$. Show that $d(A, B) > 0$, where $d(A, B) := \inf\{d(a, b) : a \in A, b \in B\}$ is the distance between A and B.

(b) Show by example that the conclusion in Part (a) is untrue if the boundedness assumption is removed.

4.12. By sketching the phase portrait for the system (4.2.5), show that any orbit starting from a point $x_0 := (x_{01}, x_{02})$ with $x_{01} \in (-1, 1)$ has the lines $x_1 = \pm 1$ as its ω-limit set, which is obviously not connected.

4.13. Assume Γ is an orbit of Eq. (A) and $\Omega(\Gamma^+) = \{x^*\}$ for some point $x^* \in \mathbb{R}^n$. Show that Γ^+ approaches x^* as $t \to \infty$, i.e., if $x(t)$ is a solution of Eq. (A) with Γ as its orbit, then $x(t) \to x^*$ as $t \to \infty$.

4.14. For each of the following linear systems, find and classify the equilibrium:

(a) $x' = -x + y, \quad y' = x - 4y;$
(b) $x' = 3x - 2y - 7, \quad y' = 2x - 2y - 6;$
(c) $x' = -x + 5y - 2, \quad y' = -2x + 5y + 1;$
(d) $x' = x + y - 1, \quad y' = -2x - y + 2.$

4.15. Discuss the possible phase portraits for the system $x' = ax + by$, $y' = cx + dy$ for the case with $q := ad - bc = 0$.

4.16. Find and classify the equilibrium of the system

$$x' = 1 - xy, \quad y' = x - y^2.$$

4.17. Prove Theorem 4.3 using Remark 4.4.1, Part (b).

4.18. Show that $(0,0)$ is a center for the system

$$x' = y + x^2 y^3 r, \quad y' = -x - x^3 y^2 r,$$

where $r = \sqrt{x^2 + y^2}$.

4.19. Show that $(0,0)$ is a center for the system corresponding to the second-order equation $x'' + kx + h(x) = 0$, where $k > 0$ and $h \in C(\mathbb{R}, \mathbb{R})$ satisfying $|h(x)| = o(|x|)$ as $x \to 0$.

4.20. For $a, b > 0$, consider the system

$$x' = x(y - b), \quad y' = y(x - a).$$

(a) Find all equilibria and determine the type and stability for each of them.
(b) Show that all solutions $(x(t), y(t))$ starting in the region $E = \{(x, y) : x < a, y < b\}$ satisfy $\lim_{t \to \infty} (x(t), y(t)) = (0, 0)$. (In this case, $(0, 0)$ is called an attractor and E is called the domain of attraction of $(0, 0)$.)

4.21. Assume Γ is an orbit of system (A-2) with Γ_1 and Γ_2 its ω-limit set and α-limit set, respectively. Assume Γ_1 is a closed orbit and Γ_2 is not. Show that $\Gamma_1 \cap \Gamma_2 = \emptyset$.

4.22. Assume system (A-2) has only one equilibrium $x^* \in \mathbb{R}^2$ which is stable, and has no closed orbits. Let Γ be an orbit of system (A-2) with $\Omega(\Gamma^+)$ bounded. Show that

(a) Γ^+ approaches x^* as $t \to \infty$,
(b) Γ^- is unbounded.

4.23. For the system $x' = y$, $y' = x - x^2$,

(a) find and classify the equilibria;
(b) find the equation for orbits and solve it for the general solution;
(c) show that the system has a homoclinic orbit and find the equation for it from the general solution.

4.24. The following systems are given by the polar-coordinate form. Based on the Poincaré–Bendixson theorem 4.5.2, sketch the phase portrait for each of the following systems near the unit circle:

(a) $r' = r - r^2$, $\theta' = 1$;
(b) $r' = r - r^2$, $\theta' = 1 - \cos\theta$;
(c) $r' = r - r^2$, $\theta' = 1 - \cos 2\theta$;
(d) $r' = r - r^2$, $\theta' = 1 - \cos n\theta$, $n \in \mathbb{N}$.

4.25. Show that the system

$$x' = x(4 - 2x^2 - y^2) + y, \quad y' = y(4 - 2x^2 - y^2) - 2x$$

has a unique closed orbit.

4.26. Show that the equation

$$x'' - (|x| + |x'| - 1)x' + x|x| = 0$$

has at least one periodic solution.

4.27. Show that the following systems have no periodic solutions:

(a) $x' = y + x^2 y^3, \quad y' = x + x^3 y^2;$
(b) $x' = x - x^3 y^3, \quad y' = y + x^2 y^3;$
(c) $x' = -x^2 y^4, \quad y' = xy - y^2.$

4.28. Show that the system

$$x' = y, \quad y' = x - x^2 + y^2 + y^3$$

has no periodic solutions.

4.29. Let $f, g \in C^1(\mathbb{R}, \mathbb{R})$. Use Theorem 4.6.2 to show that the system

$$x' = ax^3 + f(y), \quad y' = by^5 + g(x)$$

has no periodic solutions if $ab \geq 0$ and $(a, b) \neq (0, 0)$.

4.30. Use Theorem 4.6.2 to show that the system

$$x' = y^2, \quad y' = -2x - ky - 2y^3$$

has no periodic solutions for any $k \in \mathbb{R}$. Can you use Theorem 4.6.1 to solve the same problem?

4.31. Let E be an open subset of $\mathbb{R}^2$ and $f_1, f_2, \in C^1(E, \mathbb{R})$. Consider the system $x_1' = f_1(x_1, x_2), x_2' = f_2(x_1, x_2)$. Assume there exists an $r \in C^1(E, \mathbb{R})$ such that

$$(rf_1)_{x_1} + (rf_2)_{x_2} \geq 0 \text{ on } E \text{ and } (rf_1)_{x_1} + (rf_2)_{x_2} \not\equiv 0 \text{ on any subregion of } E.$$

If E is simply connected, then the Poincaré–Durac negative criterion guarantees that there is no closed orbit of the system lying in E.

(a) If E is doubly connected, i.e., a connected region with one hole, what is the maximum number of closed orbits of the system in E? Give a proof for your answer.
(b) Extend your result to the case when there are n holes in E.

4.32. Consider the system (A-2) $x' = f(x)$, where $f \in C^1(\mathbb{R}^2, \mathbb{R}^2)$. Suppose $\partial f_i / \partial x_j(x) > 0$ for $i \neq j$ and all $x := (x_1, x_2) \in \mathbb{R}^2$. Prove:

(a) If $(\phi_1(t), \phi_2(t)), (\psi_1(t), \psi_2(t))$ are two solutions of Eq. (A-2) on $[a, b]$ with $\phi_i(a) \leq \psi_i(a)$ for $i = 1, 2$, then $\phi_i(t) \leq \psi_i(t)$ on $[a, b]$ for $i = 1, 2$.
(b) Equation (A-2) has no periodic solutions.

4.33. For each of the systems

(a) $x' = y - (r^2 - 1)x^3, \quad y' = -x - (r^2 - 1)x^2 y,$
(b) $x' = y + (r^2 - 1)x^3, \quad y' = -x,$

where $r = \sqrt{x^2 + y^2}$, show that there is a unique limit cycle, and determine the orbital stability of the limit cycle.

4.34. Consider the system

$$x' = -y + x(r^3 - 3r + 1), \quad y' = x + y(r^3 - 3r + 1),$$

where $r = \sqrt{x^2 + y^2}$.

(a) Classify the equilibrium $(0, 0)$.
(b) Show that there are exactly two limit cycles and determine their orbital stability.

4.35. Let $a, b, c, d \in \mathbb{R}$ such that $ad - bc > 0$. Define a function $[\bar{a}\bar{b}\bar{c}\bar{d}] \in C([0, 1], \mathbb{R}^4)$ such that

$$[\bar{a}\bar{b}\bar{c}\bar{d}](1) = [abcd] \quad \text{and} \quad [\bar{a}\bar{b}\bar{c}\bar{d}](0) = [1001],$$

and $\bar{q}(s) := (\bar{a}\bar{d} - \bar{b}\bar{c})(s) > 0$ for all $s \in [0, 1]$.

4.36. Prove Lemma 4.7.3 for the case when $ad - bc < 0$.

4.37. Show that the system

$$x' = x^2 - y^2, \quad y' = 2xy$$

has no closed orbits, and sketch the phase portrait for this system.

4.38. Consider the system

$$x' = 1 - x - 2y, \quad y' = (-3 + x)y.$$

(a) Find all equilibria, and determine their types and indices.
(b) Prove that there are no closed orbits.

4.39. Let $g, h \in C^1(\mathbb{R}^2, \mathbb{R}^2)$ be bounded. Show that the system

$$x' = y + g(x, y), \quad y' = x + h(x, y)$$

has at least one equilibrium.

4.40. Assume $f \in C^1(\mathbb{R}^2, \mathbb{R}^2)$ such that the system $x' = f(x)$ has a closed orbit Γ. Let $g \in C^1(\mathbb{R}^2, \mathbb{R}^2)$ such that g and f are linearly independent on Γ. Show that the system $x' = g(x)$ has at least one equilibrium.

4.41. Find the stable, unstable, and center subspaces for Eq. (H-c) with

$$A = \begin{bmatrix} 2 & 7 & 0 & 0 \\ -1 & -6 & 0 & 0 \\ 0 & 0 & 0 & -2 \\ 0 & 0 & 1 & 0 \end{bmatrix}.$$

4.42. Find the Stable, unstable, and center manifolds for each of the following systems:

(a) $\begin{cases} x_1' = x_1 \\ x_2' = -x_2 + x_1^2 \\ x_3' = x_1^3, \end{cases}$ (b) $\begin{cases} x_1' = x_1 \\ x_2' = -x_2 + x_1^2 \\ x_3' = x_3^3. \end{cases}$

4.43. Prove that (4.8.7) holds.

4.44. For each of the following systems, find a homeomorphism $\mathcal{H}$ between the orbits of the system and those of its linearization:

(a) $\begin{cases} x_1' = -2x_1 \\ x_2' = x_2 + x_1^2, \end{cases}$
(b) $\begin{cases} x_1' = -x_1 \\ x_2' = -x_2 + x_3^2 \\ x_3' = x_3. \end{cases}$

4.45. Use the center manifold method to study the stability of the zero solution of the following systems:

(a) $\begin{cases} x_1' = -x_1 + 2x_1^2 x_2 \\ x_2' = x_1 x_2 - x_1^3, \end{cases}$
(b) $\begin{cases} x_1' = -x_1 - x_2^2 + x_1 x_2^2 \\ x_2' = x_1 x_2 - x_2^2 x_2, \end{cases}$

(c) $\begin{cases} x_1' = -x_1 - x_2^2 - x_1 x_2^2 \\ x_2' = x_1 x_2 + x_2^3 + x_1^2 x_2. \end{cases}$

CHAPTER 5

Introduction to Bifurcation Theory

5.1. Basic Concepts

In this chapter, we study the autonomous equation with a k-dimensional parameter

$$(\mathrm{A}[\mu]) \qquad\qquad x' = f(x, \mu),$$

where $f : D \to \mathbb{R}^n$ with D an open subset of $\mathbb{R}^n \times \mathbb{R}^k$ for some $k \geq 1$. In Sect. 1.5, we studied the dependence of solutions of Eq. $(\mathrm{A}[\mu])$ on the parameter $\mu = (\mu_1, \mu_2, \ldots, \mu_k)^T$. In fact, Theorem 1.5.1 says that if $f \in C(D, \mathbb{R}^n)$ and is locally Lipschitz in x, then the solution of any IVP associated with Eq. $(\mathrm{A}[\mu])$ depends on μ continuously. This result characterizes the continuous behavior of individual solutions on compact intervals of the maximal interval of existence when the parameter μ varies. However, it does not provide us with any information on the structural changes of solutions, i.e., the changes of the phase portrait, when the parameter μ varies.

In application models, we often observe that solutions of differential equations undergo substantial structural changes when the parameters pass through some specific values. A simple example is the second-order differential equation for mechanical vibrations

$$mx'' + cx' + kx = 0,$$

where m, c, and k represent the mass of the vibrator, the damping constant of the medium, and the spring constant, respectively; and the solution $x(t)$ stands for the displacement of the vibrator. From any book on elementary differential equations we can find that the solution structure becomes very different when the values of m, c, and k pass through certain values. In fact, the solution behavior can be classified as harmonic motion when $c = 0$, underdamped motion when $c > 0$ and $c^2 - 4mk < 0$, critically damped motion when $c > 0$ and $c^2 - 4mk = 0$, and overdamped motion when $c > 0$ and $c^2 - 4mk > 0$. If we transfer the second-order equation to a first-order system, we will see that equilibrium $(0, 0)$ is a center when $c = 0$, a stable spiral-point when $c > 0$ and $c^2 - 4mk < 0$, a stable degenerate-node when $c > 0$ and $c^2 - 4mk = 0$, and a stable improper-node when $c > 0$ and $c^2 - 4mk > 0$. Moreover, when $c < 0$, the equation represents a self-excited

© Springer International Publishing Switzerland 2014
Q. Kong, *A Short Course in Ordinary Differential Equations*, Universitext,
DOI 10.1007/978-3-319-11239-8_5

unbounded motion, and hence equilibrium $(0,0)$ is unstable. Therefore, the phase portrait undergoes structural changes at $c = 0$ and at $c^2 - 4mk = 0$.

The bifurcation theory is concerned with the following problem: For a general autonomous equation with parameters, how the topological structure of the phase portrait changes when the parameters vary.

DEFINITION 5.1.1. If the phase portrait of Eq. (A[μ]) undergoes a structural change at $\mu = \mu_0$, we say that a bifurcation occurs at $\mu = \mu_0$, and $\mu = \mu_0$ is called a bifurcation value of Eq. (A[μ]). Such structural changes include

    (i) a change of the number of equilibria,
    (ii) a change of the type or stability of any equilibrium,
    (iii) a change of the number of closed orbits,
    (iv) a change of the orbital stability of any closed orbit;

and more.

To explain the bifurcation phenomena, we look at the examples below.

EXAMPLE 5.1.1. Consider the equation

$$(5.1.1) \qquad x' = f(x, \mu) := \mu - x^k, \quad k \in \mathbb{N}.$$

    (a) k is an odd number.
        In this case, Eq. (5.1.1) has exactly one equilibrium $x_0(\mu) = \sqrt[k]{\mu}$ for each $\mu \in \mathbb{R}$. Moreover, it is easy to see that $f(x, \mu) > 0$ for $x < x_0(\mu)$ and $f(x, \mu) < 0$ for $x > x_0(\mu)$. Therefore, $x_0(\mu)$ is asymptotically stable for every $\mu \in \mathbb{R}$. This means that no bifurcation occurs at any $\mu \in \mathbb{R}$. See Fig. 5.1.

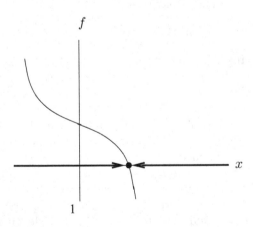

FIGURE 5.1. Example 5.1.1—a (1 equilibrium for all μ)

(b) k is an even number.

In this case, Eq. (5.1.1) has no equilibrium when $\mu < 0$, exactly one equilibrium $x_0(0) = 0$ when $\mu = 0$, and exactly two equilibria $x_\pm(\mu) = \pm\sqrt[k]{\mu}$ when $\mu > 0$. Moreover, it is easy to see that

(i) when $\mu = 0$, $f(x, 0) < 0$ for both $x < x_0(0)$ and $x > x_0(0)$, so $x_0(0)$ is unstable; more specifically, as defined later in Definition 5.2.1, we may say that $x_0(0)$ is unstable from the left and asymptotically stable from the right.

(ii) when $\mu > 0$, $f(t, \mu) < 0$ for $x < x_-(\mu)$ and $x > x_+(\mu)$, and $f(t, \mu) > 0$ for $x_-(\mu) < x < x_+(\mu)$, so $x_-(\mu)$ is unstable and $x_+(\mu)$ is asymptotically stable.

We note that both the number and stability of the equilibria change as μ passes through 0. Therefore, $\mu = 0$ is a bifurcation value. See Fig. 5.2.

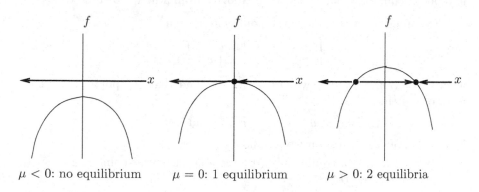

$\mu < 0$: no equilibrium $\mu = 0$: 1 equilibrium $\mu > 0$: 2 equilibria

FIGURE 5.2. Example 5.1.1—b

EXAMPLE 5.1.2. Consider the homogeneous linear system

(5.1.2)
$$\begin{cases} x_1' = ax_1 + bx_2 \\ x_2' = cx_1 + dx_2 \end{cases} \quad \text{or} \quad x' = Ax,$$

where $x = \begin{bmatrix} x_1 \\ x_2 \end{bmatrix}$ and $A = \begin{bmatrix} a & b \\ c & d \end{bmatrix}$. Clearly, $(0, 0)$ is an equilibrium of system (5.1.2). As in Sect. 4.3 we let

$$p = a + d, \quad q = ad - bc, \quad \text{and} \quad \Delta = p^2 - 4q,$$

and treat them as parameters. From the equilibrium classification chart—Fig. 4.12 we have the following observations:

(a) For a fixed $q > 0$, when p changes from negative to zero and then to positive, equilibrium $(0, 0)$ changes from a stable spiral-point to a center and then to a unstable spiral-point. Thus, $p = 0$ is a bifurcation value.

(b) For a fixed $p < 0$, when q changes from positive to negative, equilibrium $(0,0)$ changes from a stable improper-node to a saddle-point which is unstable. Similarly, for a fixed $p > 0$, when q changes from positive to negative, equilibrium $(0,0)$ changes from an unstable improper-node to a saddle-point. In both cases, $q = 0$ is a bifurcation value.

(c) For a fixed $p < 0$, when Δ changes from positive to zero and then to negative, equilibrium $(0,0)$ remains stable, but changes from an improper-node to a proper/degenerate-node and then to a spiral-point. Similarly for the case when $p > 0$ is fixed. In both cases, $\Delta = 0$ is also a bifurcation value.

Therefore, all points on the curves $p = 0$ with $q > 0$, $q = 0$, and $\Delta = 0$ are bifurcation values for system (5.1.2). We note that either the type or the stability of equilibrium $(0,0)$ changes at these points. Furthermore, the number of equilibria changes when q changes from a nonzero value to zero.

EXAMPLE 5.1.3. Consider the nonlinear system

(5.1.3)
$$\begin{cases} x_1' = x_2 + x_1(\mu - r^2) \\ x_2' = -x_1 + x_2(\mu - r^2), \end{cases}$$

where $r = \sqrt{x_1^2 + x_2^2}$. By way of (4.6.1) we find that the polar-coordinate form of system (5.1.3) is

$$r' = f(r, \mu) := r(\mu - r^2) \quad \text{and} \quad \theta' = -1.$$

With a simple computation we see that

(a) when $\mu \le 0$, equilibrium $(0,0)$ is a stable spiral-point and there is no closed orbit;

(b) when $\mu > 0$, equilibrium $(0,0)$ becomes an unstable spiral-point and a closed orbit $r = \sqrt{\mu}$ is created and is orbitally asymptotically stable.

Therefore, $\mu = 0$ is a bifurcation value (Fig. 5.3).

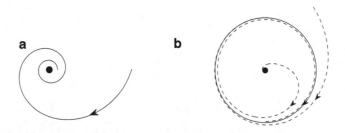

FIGURE 5.3. Example 5.1.3. (a) $\mu \le 0$: 1 equilibrium. (b) $\mu > 0$: 1 equilibrium and 1 closed orbit

Now we consider the structural changes of the phase portrait of the general Eq. (A[μ]) near its equilibria. Let $x_0(\mu)$ be an equilibrium of Eq. (A[μ]), then $f(x_0(\mu), \mu) = 0$. To study the type and stability of $x_0(\mu)$, we may transform $x_0(\mu)$ to the origin by letting $y = x - x_0(\mu)$. Then the transformed equation becomes

$$(5.1.4) \qquad y' = f(y + x_0(\mu), \mu),$$

where $y = 0$ is the new equilibrium. From the linearization method in Sect. 3.3 we see that when $f \in C^1(D, \mathbb{R}^n)$, Eq. (5.1.4) can be written as

$$(5.1.5) \qquad y' = A(\mu)y + r(y, \mu),$$

where $A(\mu) = \partial f/\partial x(x_0(\mu), \mu)$ and $r(y, \mu) = o(|y|)$ as $y \to 0$. From Sects. 4.3 and 4.4, the type and stability of the equilibrium $y = 0$ are largely determined by the eigenvalues $\lambda_i(\mu)$, $i = 1, \ldots, n$, of matrix $A(\mu)$.

To simplify the discussion, throughout this chapter we make the following assumption:

($\mathcal{H}$) $\lambda_i(\mu)$, $i = 1, \ldots, n$, are distinct for all μ in the domain.

We note that assumption ($\mathcal{H}$) implies that bifurcations occur only when $\mathrm{Re}\,\lambda_i(\mu)$ changes sign for some $i \in \{1, \ldots, n\}$, and hence eliminates the possibility of an occurrence of a bifurcation caused by a multiplicity change of eigenvalues such as in Example 5.1.2, Part (c). We comment that in some references, the phase portraits for nodes and spiral-points are regarded as topologically equivalent as long as they have the same stability. Consequently, the structural change in Example 5.1.2, Part (c) is not treated as a bifurcation there.

Since the matrix $A(\mu)$ is real-valued, all nonreal eigenvalues are in complex-conjugate pairs. This means that under assumption ($\mathcal{H}$), a bifurcation may occur at $\mu = \mu_0$ when the two conditions below are satisfied:

(a) Either one of the following holds:
   (i) there exists an $i \in \{1, \ldots, n\}$ such that $\lambda_i(\mu) \in \mathbb{R}$ and $\lambda_i(\mu)$ changes sign at μ_0;
   (ii) there exist j pairs of complex-conjugate eigenvalues $\lambda_{i\pm}(\mu)$, $i = 1, \ldots, j$, such that $\mathrm{Re}\,\lambda_{i\pm}(\mu)$ changes sign at μ_0 and $\mathrm{Im}\,\lambda_{i\pm}(\mu_0) \neq 0$;
   (iii) the combination of (i) and (ii).
(b) $\mathrm{Re}\,\lambda_k(\mu_0) \neq 0$ for all other eigenvalues λ_k.

In particular, a bifurcation where only one real eigenvalue passes through zero is called a one-dimensional bifurcation, and a bifurcation where the real parts of one pair of complex-conjugate eigenvalues pass through zero is called a Poincaré–Andronov–Hopf bifurcation or simply a Hopf bifurcation.

Clearly, the bifurcation in Example 5.1.1, Part (b) is a one-dimensional bifurcation. It is easy to see that in Example 5.1.2, Part (a) and in Example 5.1.3, the eigenvalues of $A(0)$ are $\pm i$. Hence the bifurcations are Hopf bifurcations. If we revisit Example 5.1.2 using the eigenvalues λ_1 and λ_2 as parameters, then the results can be illustrated by Figs. 5.4 and 5.5 below.

The implicit function theorem stated below plays a key role in the development of the bifurcation theory. See [10, Theorem 2.3] for its proof.

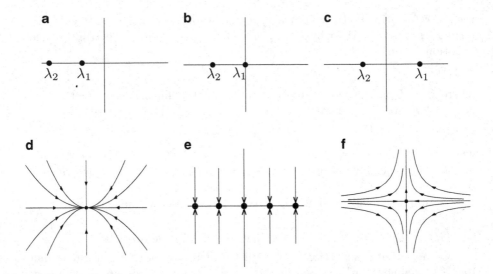

FIGURE 5.4. One-dimensional bifurcation. (a) $\lambda_{1,2} < 0$. (b) $\lambda_2 < 0 = \lambda_1$. (c) $\lambda_2 < 0 < \lambda_1$. (d) Node. (e) Unisolated equilibria. (f) Saddle-point

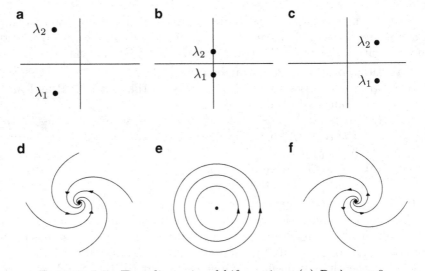

FIGURE 5.5. Two-dimensional bifurcation. (a) $\mathrm{Re}\,\lambda_{1,2} < 0$. (b) $\mathrm{Re}\,\lambda_{1,2} = 0$. (c) $\mathrm{Re}\,\lambda_2 > 0$. (d) Stable spiral-point. (e) Center. (f) Unstable spiral-point

LEMMA 5.1.1 (Implicit Function Theorem). *Let $f \in C^1(D, \mathbb{R}^n)$ with D an open subset of $\mathbb{R}^n \times \mathbb{R}^k$ and $(x_0, \mu_0) \in D$. Assume that*

$$f(x_0, \mu_0) = 0 \quad and \quad \det \frac{\partial f}{\partial x}(x_0, \mu_0) \neq 0.$$

Then there is a neighborhood $D_x \times D_\mu$ of (x_0, μ_0) in D with $D_x \subset \mathbb{R}^n$ and $D_\mu \subset \mathbb{R}^k$ and a unique function $g \in C^1(D_\mu, D_x)$ such that

$$f(g(\mu), \mu) \equiv 0 \quad and \quad g(\mu_0) = x_0.$$

Moreover, if $f \in C^j(D, \mathbb{R}^n)$ for some $j > 1$, then $g \in C^j(D_\mu, D_x)$; and if f is analytic on D, then g is analytic on D_μ.

Based on the implicit function theorem, we have the following result for one-dimensional bifurcations.

THEOREM 5.1.1. *Let $f \in C^1(D, \mathbb{R}^n)$ with D an open subset of $\mathbb{R}^n \times \mathbb{R}^k$, and assume $\mu = \mu_0$ is a one-dimensional bifurcation value for Eq. (A[μ]) with a corresponding equilibrium $x_0(\mu_0)$ such that $(x_0(\mu_0), \mu_0) \in D$. Then*

$$(5.1.6) \qquad f(x_0(\mu_0), \mu_0) = 0 \quad and \quad \det \frac{\partial f}{\partial x}(x_0(\mu_0), \mu_0) = 0.$$

PROOF. Since $x_0(\mu_0)$ is an equilibrium of Eq. (A[μ]) with $\mu = \mu_0$, we see that $f(x_0(\mu_0), \mu_0) = 0$. By contradiction, assume $\det \frac{\partial f}{\partial x}(x_0(\mu_0), \mu_0) \neq 0$. Then from the implicit function theorem, there is a unique C^1-function $x_0(\mu)$ defined in a neighborhood of μ_0 such that $f(x_0(\mu), \mu) \equiv 0$ and $x_0(\mu)$ coincides with $x_0(\mu_0)$ when $\mu = \mu_0$. Hence $x_0(\mu)$ is the only C^1-branch of equilibria of Eq. (A[μ]) with (x, μ) in a neighborhood of $(x_0(\mu_0), \mu_0)$. Consequently, $\det \frac{\partial f}{\partial x}(x_0(\mu), \mu) \neq 0$ for μ near μ_0. As in (5.1.5),

$$\det \frac{\partial f}{\partial x}(x_0(\mu), \mu) = \det A(\mu) = \prod_{i=1}^{n} \lambda_i(\mu),$$

where $\lambda_i(\mu)$, $i = 1, \ldots, n$, are eigenvalues of $A(\mu)$. It follows that $\lambda_i(\mu) \neq 0$ for μ near μ_0 and $i = 1, \ldots, n$, which contradicts the definition of one-dimensional bifurcations. □

REMARK 5.1.1. We point out that condition (5.1.6) is necessary but not sufficient for a one-dimensional bifurcation to occur. In fact, in Example 5.1.1, Part (a), with an odd number $k \geq 3$, (5.1.6) holds for $(x_0, \mu_0) = (0, 0)$. However, $\mu_0 = 0$ is not a bifurcation value for Eq. (5.1.1).

In this chapter we will only introduce the basic bifurcations for Eq. (A[μ]) with $n = 1, 2$, i.e., for scalar equations and planar systems. For further study of the general bifurcation theory for autonomous systems, the reader is referred to Chicone [9], Chow and Hale [10], Guckenheimer and Holmes [20], Hale and Kocak [21], Iooss and Joseph [28], Kuznetsov [34], Ma and Zhou [39], Marsden and McCracken [40], and Zhang [50].

5.2. One-Dimensional Bifurcations for Scalar Equations

In this section, we discuss several types of bifurcations for the scalar equation (A[μ]) by investigating the structure of the equilibrium curves in the μx-plane. Such a figure is called a bifurcation diagram. Let $x_0(\mu)$ be an equilibrium of Eq. (A[μ]). We note that when $n = 1$, the linearization coefficient matrix $A(\mu) = f_x(x_0(\mu), \mu)$ is a scalar and hence is the same as its only eigenvalue. We denote $\lambda(\mu) = f_x(x_0(\mu), \mu)$. Then the stability of the equilibrium $x_0(\mu)$ is determined by the sign of $\lambda(\mu)$. Therefore, if a bifurcation occurs at $\mu = \mu_0$, then $\lambda(\mu_0) = 0$. To see the possible types of bifurcation diagrams for scalar equations, we study more examples. In all the bifurcation diagrams below, (AS) and (NS) stand for asymptotically stable and unstable, respectively.

EXAMPLE 5.2.1. Consider Eq. (5.1.1) with $k = 2m$, i.e., the equation

$$(5.2.1) \qquad x' = f(x, \mu) := \mu - x^{2m}, \quad m \in \mathbb{N}.$$

From example 5.1.1 we see that $\mu = 0$ is a bifurcation value, and Eq. (5.2.1) has no equilibrium when $\mu < 0$, exactly one equilibrium $x_1(0) = 0$ when $\mu = 0$, and exactly two equilibria $x_\pm(\mu) = \pm \mu^{1/(2m)}$ when $\mu > 0$. Moreover, $x_-(\mu)$ is unstable and $x_+(\mu)$ is asymptotically stable. Therefore, the bifurcation diagram and the phase portraits are given in Fig. 5.6.

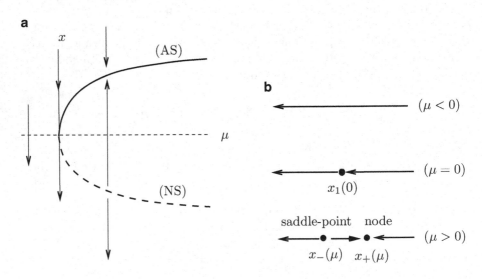

FIGURE 5.6. Saddle-node bifurcation. (a) Bifurcation diagram. (b) Phase portraits

In this diagram, as the parameter μ passes through a bifurcation value in a direction, the number of equilibria changes from zero to two and the two equilibria created at the bifurcation value form a C^1-branch of equilibria whose stability changes from unstable to asymptotically stable at the bifurcation value. Such a bifurcation is called a saddle-node bifurcation.

EXAMPLE 5.2.2. Consider the equation

(5.2.2) $$x' = f(x,\mu) := \mu x - x^{2m}, \quad m \in \mathbb{N}.$$

By letting $f(x,\mu) = 0$ we find two equilibria $x_1(\mu) = 0$ and $x_2(\mu) = \mu^{1/(2m-1)}$. Since $f_x(x,\mu) = \mu - 2mx^{2m-1}$, we have the following:

(a) For $x_1(\mu) = 0$, $\lambda_1(\mu) = \mu$. It follows that $x_1(\mu)$ is asymptotically stable for $\mu < 0$ and unstable for $\mu > 0$.

(b) For $x_2(\mu) = \mu^{1/(2m-1)}$, $\lambda_2(\mu) = -(2m-1)\mu$. Thus, $x_2(\mu)$ is unstable for $\mu < 0$ and asymptotically stable for $\mu > 0$.

As a result, $\mu = 0$ is a bifurcation value, and the bifurcation diagram and the phase portraits are given in Fig. 5.7.

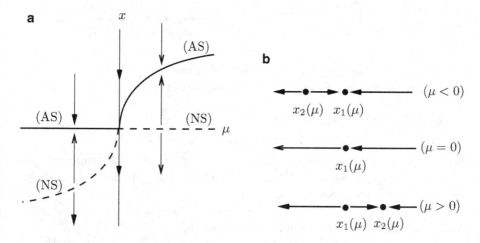

FIGURE 5.7. Transcritical bifurcation. (a) Bifurcation diagram. (b) Phase portraits

In this diagram, two C^1-branches of equilibria with opposite stabilities interchange their stabilities at the bifurcation value. Such a bifurcation is called a transcritical bifurcation.

EXAMPLE 5.2.3. We first consider the equation

(5.2.3) $$x' = f(x,\mu) := \mu x - x^{2m+1}, \quad m \in \mathbb{N}.$$

By letting $f(x,\mu) = 0$ we find that $x_1(\mu) = 0$ is an equilibrium for all $\mu \in \mathbb{R}$ and $x_\pm(\mu) = \pm\mu^{1/(2m)}$ are equilibria for $\mu \geq 0$. Since $f_x(x,\mu) = \mu - (2m+1)x^{2m}$, we have the following:

(a) For $x_1(\mu) = 0$, $\lambda_1(\mu) = \mu$. It follows that $x_1(\mu)$ is asymptotically stable for $\mu < 0$ and unstable for $\mu > 0$;

(b) For $x_\pm(\mu) = \pm\mu^{1/(2m)}$ with $\mu > 0$, $\lambda_\pm(\mu) = -2m\mu < 0$. It follows
that $x_\pm(\mu)$ are asymptotically stable for $\mu > 0$.

Therefore, $\mu = 0$ is a bifurcation value. The bifurcation diagram and phase
portraits are given in Fig. 5.8.

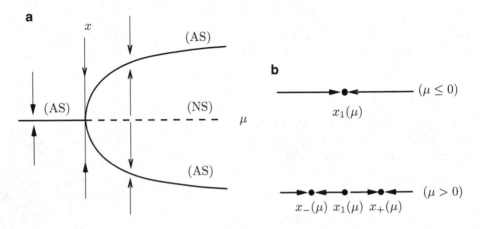

FIGURE 5.8. Supercritical pitchfork bifurcation.
(a) Bifurcation diagram. (b) Phase portraits

We then consider the equation

(5.2.4) $$x' = f(x, \mu) := \mu x + x^{2m+1}, \quad m \in \mathbb{N}.$$

By letting $f(x, \mu) = 0$ we find that $x_1(\mu) = 0$ is an equilibrium for all
$\mu \in \mathbb{R}$ and $x_\pm(\mu) = \pm(-\mu)^{1/(2m)}$ are equilibria for $\mu \le 0$. Since $f_x(x, \mu) =$
$\mu + (2m+1)x^{2m}$, we have the following:

(a) For $x_1(\mu) = 0$, $\lambda_1(\mu) = \mu$. It follows that $x_1(\mu)$ is asymptotically
stable for $\mu < 0$ and unstable for $\mu > 0$;

(b) For $x_\pm(\mu) = \pm(-\mu)^{1/(2m)}$ with $\mu < 0$, $\lambda_\pm(\mu) = -2m\mu > 0$. It
follows that $x_\pm(\mu)$ are unstable for $\mu < 0$.

Therefore, $\mu = 0$ is a bifurcation value. The bifurcation diagram and phase
portraits are given in Fig. 5.9.

The bifurcations for Eqs. (5.2.3) and (5.2.4) are called pitchfork bifurcations because their bifurcation diagrams look like pitchforks. In both of
the two diagrams, one C^1-branch of equilibria $x_1(\mu)$ changes stability at the
bifurcation value, and other two equilibria $x_\pm(\mu)$ are created to one side of
the bifurcation value and form a C^1-branch of equilibria. Moreover, the two
equilibria $x_\pm(\mu)$ have the same stability which is opposite to that of $x_1(\mu)$.
When $x_\pm(\mu)$ are asymptotically stable as in Fig. 5.8, Part (a), the bifurcation
is called a supercritical pitchfork bifurcation; and when $x_\pm(\mu)$ are unstable
as in Fig. 5.9, Part (a), the bifurcation is called a subcritical pitchfork bifurcation. We observe that for the supercritical pitchfork bifurcation case, when
μ passes through the bifurcation value, even though the asymptotic stability

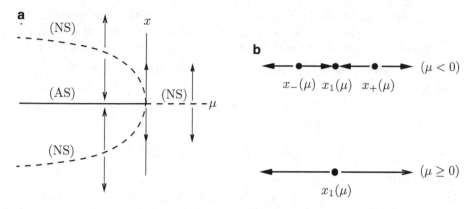

FIGURE 5.9. Subcritical pitchfork bifurcation. (a) Bifurcation diagram. (b) Phase portraits

is transformed from $x_1(\mu)$ to $x_\pm(\mu)$, a stable structure is preserved. However, for the subcritical pitchfork bifurcation case, when μ passes through the bifurcation value, the stability structure for $x_1(\mu)$ is completely lost.

Now we generalize the results on the bifurcation types appeared in Examples 5.1.1–5.1.3 to the scalar equation (A[μ]) given at the beginning of Sect. 5.1. The theorem below is an extension of [18, Theorem 10.1].

THEOREM 5.2.1. *Let $f \in C^2(D, \mathbb{R})$ with D an open subset of $\mathbb{R} \times \mathbb{R}$ and $P_0 := (x_0, \mu_0) \in D$ satisfying*

(5.2.5) $$f(P_0) = 0 \quad and \quad f_x(P_0) = 0.$$

(a) *Assume $f_\mu(P_0) \neq 0$.*
   If there exists a least integer $k \geq 2$ such that $f \in C^k(D, \mathbb{R})$ and $\partial^k f / \partial x^k (P_0) \neq 0$, then Eq. (A[$\mu$]) has a saddle-node bifurcation at μ_0 when k is even and has no bifurcation at μ_0 when k is odd.

(b) *Assume $f_\mu(P_0) = 0$ and denote*

$$\Delta = \det \begin{bmatrix} f_{\mu\mu} & f_{\mu x} \\ f_{x\mu} & f_{xx} \end{bmatrix} (P_0) = (f_{\mu\mu}f_{xx} - f_{x\mu}^2)(P_0).$$

(i) *If $\Delta > 0$, then Eq. (A[μ]) has no equilibrium for μ in a neighborhood of μ_0 except at μ_0.*

(ii) *If $\Delta < 0$ and there exists a least integer $k \geq 2$ such that $f \in C^k(D, \mathbb{R})$ and $\partial^k f / \partial x^k (P_0) \neq 0$, then the bifurcation is transcritical when k is even and pitchfork when k is odd. Moreover, when k is odd, the bifurcation is supercritical pitchfork if $\partial^k f / \partial x^k (P_0) < 0$ and is subcritical pitchfork if $\partial^k f / \partial x^k (P_0) > 0$.*

PROOF. Note that (5.2.5) shows that $\mu = \mu_0$ is a candidate for a bifurcation value with $x = x_0$ the corresponding equilibrium. As shown in Sect. 5.1, we can always transform the point (x_0, μ_0) to $(0,0)$ by the changes of variables $\nu = \mu - \mu_0$ and $y = x - x_0$. Thus, without loss of generality, we only give a proof for the case where $P_0 = (0,0)$. To simplify the notation, we let

$$\alpha = f_\mu(0,0), \quad a = f_{\mu\mu}(0,0), \quad b = f_{x\mu}(0,0), \quad \text{and} \quad c = f_{xx}(0,0).$$

Then $\Delta = ac - b^2$.

Note that $f \in C^2(D, \mathbb{R})$ and $f_x, f_\mu \in C^1(D, \mathbb{R})$. From the Taylor expansions we have that as $(x, \mu) \to (0,0)$

$$(5.2.6) \qquad f(x, \mu) = \alpha\mu + \frac{1}{2}(a\mu^2 + 2bx\mu + cx^2) + o(x^2 + |x\mu| + \mu^2),$$

together with

$$(5.2.7) \qquad\qquad f_x(x, \mu) = b\mu + cx + o(|x| + |\mu|)$$

and

$$f_\mu(x, \mu) = \alpha + a\mu + bx + o(|x| + |\mu|).$$

(a) We first consider the case when $\alpha \neq 0$ and $c \neq 0$. This is the case with $k = 2$. By Lemma 5.1.1, there is a unique C^2-function $\mu = \mu_1(x)$ defined in a neighborhood of $x = 0$ such that

$$f(x, \mu_1(x)) \equiv 0 \quad \text{and} \quad \mu_1(0) = 0.$$

Substituting the Taylor expansion of $\mu_1(x)$ into f given in (5.2.6) we have

$$\mu_1(x) = -\frac{c}{2\alpha}x^2 + o(x^2)$$

as $x \to 0$. This shows that $\mu_1(x)$ is a "quadratic-like" function in a neighborhood of $x = 0$. In view of $c = f_{xx}(0,0) \neq 0$, from (5.2.7) we see that $\lambda(x) := f_x(x, \mu_1(x))$ changes sign at $x = 0$. Hence the C^2-branch of equilibria $\mu = \mu_1(x)$ changes stability at $x = 0$ and hence at $\mu = 0$. Therefore, Eq. (A[μ]) has a saddle-node bifurcation at $\mu = 0$.

The proof for the case with $\alpha \neq 0$, $c = 0$, and $k \geq 3$ is in the same way as above, except that the k-th order Taylor expansion is used. When k is even, the equilibria in a neighborhood of $x = 0$ form a "quadratic-like" curve $\mu = \mu_1(x)$ which changes stability at $\mu = 0$, and hence Eq. (A[μ]) has a saddle-node bifurcation at $\mu = 0$; when k is odd, the equilibria in a neighborhood of $x = 0$ form a "cubic-like" curve $\mu = \mu_1(x)$ which does not change stability at $\mu = 0$, and hence Eq. (A[μ]) has no bifurcation at $\mu = 0$. We omit the details.

(b) We first consider the case when $\alpha = 0$ and $\Delta > 0$. Then $a\mu^2 + 2bx\mu + cx^2$ is a definite quadratic form. From (5.2.6) we see that $(x, \mu) = (0,0)$ is the only solution of $f(x, \mu) = 0$ in a neighborhood of $(0,0)$ and hence Eq. (A[μ]) has no equilibrium for μ in a neighborhood of 0 except at 0.

Then we consider the case when $\alpha = 0$, $\Delta < 0$, and $c \neq 0$. This is the case with $k = 2$. We note that the implicit function theorem cannot be applied to the function f, so we use the zero-factor elimination method first and then apply the implicit function theorem. Let $x = \mu y$ and $g(y, \mu) = f(\mu y, \mu)/\mu^2$. It is easy to see that $g \in C^2(D, \mathbb{R})$. Clearly, when $\mu \neq 0$, $f(x, \mu) = 0 \Leftrightarrow g(y, \mu) = 0$. By (5.2.6),

$$g(y, \mu) = \frac{1}{2}(a + 2by + cy^2) + o(1)$$

as $\mu \to 0$. Since $\Delta < 0$, $g(y, 0) = 0$ has exactly two distinct real roots

$$\gamma_\pm := \frac{-b \pm \sqrt{-\Delta}}{c}$$

and

$$g_y(\gamma_\pm, 0) = b + c\gamma_\pm = \pm\sqrt{-\Delta} \neq 0.$$

By Lemma 5.1.1, there are exactly two distinct C^2-functions $y = y_\pm(\mu)$ defined in a neighborhood of $\mu = 0$ such that

$$g(y_\pm(\mu), \mu) \equiv 0 \quad \text{and} \quad y_\pm(0) = \gamma_\pm.$$

Hence Eq. (A[μ]) has two distinct C^2-branches of equilibria $x = \mu y_\pm(\mu)$ which intersect at $\mu = 0$. To see the stability property of $x_\pm(\mu)$, we let $\lambda_\pm(\mu) := f_x(x_\pm(\mu), \mu)$. Then from (5.2.7),

$$\lambda_\pm(\mu) = b\mu + c\mu y_\pm(\mu) + o(|\mu|) = \pm\mu\sqrt{-\Delta} + o(|\mu|)$$

as $\mu \to 0$. This shows that the two branches of equilibria have opposite stabilities and interchange stability at $\mu = 0$. Therefore, Eq. (A[μ]) has a transcritical bifurcation at $\mu = 0$.

Finally, we consider the case when $\alpha = 0$, $\Delta < 0$, and $c = 0$. It follows that $k \geq 3$. As before, we let $x = \mu y$ and $g(y, \mu) = f(\mu y, \mu)/\mu^2$. From (5.2.6),

$$g(y, \mu) = \frac{1}{2}(a + 2by) + o(1)$$

as $\mu \to 0$. Note that $\Delta < 0$ and $c = 0$ implies that $b \neq 0$. Thus $g(y, 0) = 0$ has exactly one root $\gamma = -a/(2b)$ and $g_y(\gamma, 0) = b \neq 0$. By Lemma 5.1.1, there is a unique C^2-function $y = y_1(\mu)$ defined in a neighborhood of $\mu = 0$ such that

$$g(y_1(\mu), \mu) \equiv 0 \quad \text{and} \quad y_1(0) = \gamma.$$

Hence Eq. (A[μ]) has an C^2-branch of equilibria $x = x_1(\mu) := \mu y_1(\mu)$ satisfying $x_1(0) = 0$. To see the stability property of $x_1(\mu)$, we let $\lambda_1(\mu) := f_x(x_1(\mu), \mu)$. Then from (5.2.7), $\lambda_1(\mu) = b\mu + o(|\mu|)$ as $\mu \to 0$. This shows that this branch of equilibria changes stability at $\mu = 0$.

To obtain a second branch of equilibria, we note that in addition to the assumption that $\alpha = c = 0$ and $\Delta < 0$, we have that there exists a least integer $k \geq 3$ such that $\partial^k f/\partial x^k(0, 0) \neq 0$. Let

$d = \partial^k f/\partial x^k(0,0)$. Then the Taylor expansion for f in (5.2.6) becomes

(5.2.8)
$$f(x,\mu) = \frac{1}{2}(a\mu^2 + 2bx\mu) + \frac{d}{k!}x^k + o(|x|^k + |x\mu| + \mu^2)$$

as $(x,\mu) \to (0,0)$. Let $\mu = x\nu$ and $h(x,\nu) = f(x,x\nu)/x^2$. It is easy to see that $h \in C^{k-2}(D,\mathbb{R})$. Clearly, when $x \neq 0$, $f(x,\mu) = 0 \Leftrightarrow h(x,\nu) = 0$. From the above,

$$h(x,\nu) = \frac{1}{2}(a\nu^2 + 2b\nu) + o(1)$$

as $x \to 0$. Hence $h(0,\nu) = 0$ has exactly one root $\beta_1 = 0$ when $a = 0$ and exactly two roots $\beta_1 = 0$ and $\beta_2 = -2b/a$ when $a \neq 0$.

It can be shown using Lemma 5.1.1 that when $a \neq 0$, there is a function $\nu(x)$ satisfying $h(x,\nu(x)) \equiv 0$ and $\nu(0) = -2b/a$. However, this leads to the same branch of equilibria as $x_1(\mu)$. Now we consider the solution of $h(x,\nu) = 0$ near $(0,0)$.

From (5.2.8) we see that

(5.2.9)
$$h(x,\nu) = b\nu + \frac{d}{k!}x^{k-2} + o(|x|^{k-2} + |\nu|)$$

as $(x,\nu) \to (0,0)$. Since $h(0,0) = 0$ and $h_\nu(0,0) = b \neq 0$, by Lemma 5.1.1, there is a unique C^{k-2}-function $\nu = \nu_2(x)$ defined in a neighborhood of $x = 0$ such that

$$h(x,\nu_2(x)) \equiv 0 \quad \text{and} \quad \nu_2(0) = 0.$$

Substituting the Taylor expansion of $\nu_2(x)$ into h given in (5.2.9) we have that for $x \to 0$

$$\nu_2(x) = -\frac{d}{bk!}x^{k-2} + o(|x|^{k-2}),$$

and hence Eq. (A[μ]) has a C^{k-1}-branch of equilibria

(5.2.10)
$$\mu = \mu_2(x) := x\nu_2(x) = -\frac{d}{bk!}x^{k-1} + o(|x|^{k-1}).$$

Note that $f_x \in C^{k-1}(D,\mathbb{R})$. Then as in (5.2.8), the Taylor expansion of f_x is

(5.2.11)
$$f_x(x,\mu) = b\mu + \frac{d}{(k-1)!}x^{k-1} + o(|x|^{k-1} + |\mu|)$$

as $(x,\mu) \to (0,0)$. Let $\lambda_2(x) := f_x(x,\mu_2(x))$. Then from (5.2.10) and (5.2.11) we have

$$\lambda(x) = (1-k)b\mu_2(x) + o(|\mu_2(x)|) \quad \text{as } \mu_2(x) \to 0.$$

When $k \geq 3$ is even, the C^{k-1}-branch of equilibria $\mu_2(x)$ is a "cubic-like" function in a neighborhood of $x = 0$ and changes stability at $x = 0$. Moreover, the two branches of equilibria $x_1(\mu)$ and $\mu_2(x)$ have opposite stabilities at each value of μ near 0. Therefore, Eq. (A[μ]) has a transcritical bifurcation at $\mu = 0$.

When $k \geq 3$ is odd, the C^{k-1}-branch of equilibria $\mu_2(x)$ is a "quadratic-like" function in a neighborhood of $x = 0$ and have the same stability which is opposite to that of $x_1(\mu)$ at each value of μ. Moreover, $\mu_2(x)$ is asymptotically stable when $d < 0$ and unstable when $d > 0$. Therefore, Eq. (A[μ]) has a supercritical pitchfork bifurcation at $\mu = 0$ when $d < 0$ and has a subcritical pitchfork bifurcation at $\mu = 0$ when $d > 0$.   □

REMARK 5.2.1. We comment that in Theorem 5.2.1, the conditions that $f_\mu(P_0) = 0$ and $\Delta < 0$ are not enough to guarantee that the bifurcation at μ_0 is either transcritical or pitchfork. To see this, let us look at the equation $x' = \mu x$. Clearly, $f_\mu(0,0) = 0$ and $\Delta = -1$. However, neither a transcritical nor a pitchfork bifurcation occurs at $\mu = 0$.

We observe that when all the second partial derivatives of f at P_0 are zero, Theorem 5.2.1 fails to apply directly. However, for certain types of equations, we may use the zero-factor elimination method to reduce the degree of the Taylor expansion of f so that Theorem 5.2.1 can be applied. This is shown in the theorem below.

THEOREM 5.2.2. *Let $f : D \to \mathbb{R}$ with D an open subset of $\mathbb{R} \times \mathbb{R}$ and $P_0 := (x_0, \mu_0) \in D$ satisfying (5.2.5). Consider the equation*

$$(5.2.12) \qquad x' = g(x, \mu) := (x - x_0)^p f(x, \mu), \quad p \in \mathbb{N}.$$

Then μ_0 is a bifurcation value for Eq. (5.2.12) if it is a bifurcation value for Eq. (A[μ]). Furthermore, we have the following:

  (a) *When p is even, then Eq. (5.2.12) keeps all equilibrium branches for Eq. (A[μ]) with the same stability. Moreover, $x = x_0$ is an equilibrium branch whose stability is determined by the sign of $f(x, \mu)$ for x near x_0.*
  (b) *When p is odd, Eq. (5.2.12) keeps all equilibrium branches for Eq. (A[μ]) with the same stability whenever $x > x_0$ and with the opposite stability whenever $x < x_0$. Moreover, $x = x_0$ is an equilibrium branch whose stability is determined by the sign of $f(x, \mu)$ for x near x_0.*

PROOF. Clearly, when $x \neq x_0$, $g(x, \mu) = 0 \Leftrightarrow f(x, \mu) = 0$. This shows that Eq. (5.2.12) keeps all equilibrium branches for Eq. (A[μ]).

  (a) When p is even, then $g(x, \mu)$ has the same sign as $f(x, \mu)$ when $x \neq x_0$. Consequently, the stability of these equilibrium branches remain unchanged.
  (b) When p is odd, then $g(x, \mu)$ has the same sign as $f(x, \mu)$ when $x > x_0$ and the opposite sign as $f(x, \mu)$ when $x < x_0$. Consequently, the stability of these equilibrium branches remain unchanged for $x > x_0$ and become opposite when $x < x_0$.

Obviously, $x = x_0$ is an equilibrium branch for Eq. (5.2.12). Note that the stability of $x = x_0$ cannot be determined by $g_x(x_0, \mu_0)$, but it can be determined by the sign of $f(x, \mu)$ for x near x_0.   □

Before giving examples, we introduce a definition for half-stability for the scalar equation (A$[\mu]$).

DEFINITION 5.2.1. Let $\mu \in \mathbb{R}$ and $x_0(\mu)$ be an equilibrium of Eq. (A$[\mu]$). Then $x_0(\mu)$ is said to be left-asymptotically-stable and right-unstable ((A-N)) provided that the solutions $x(t, \mu)$ of Eq. (A$[\mu]$) in a neighborhood of $x_0(\mu)$ approaches $x_0(\mu)$ as $t \to \infty$ if $x(0, \mu) < x_0(\mu)$ and as $t \to -\infty$ if $x(0, \mu) > x_0(\mu)$, and $x_0(\mu)$ is said to be left-unstable and right-asymptotically-stable ((N-A)) provided that the solutions $x(t, \mu)$ of Eq. (A$[\mu]$) in a neighborhood of $x(t, \mu)$ approaches $x_0(\mu)$ as $t \to -\infty$ if $x(0, \mu) < x_0(\mu)$ and as $t \to \infty$ if $x(0, \mu) > x_0(\mu)$.

Clearly, both the cases in Definition 5.2.1 are special cases of instability.

EXAMPLE 5.2.4. Consider the equation

$$(5.2.13) \qquad x' = x^p(\mu x - x^{2m} + r(x, \mu)), \quad p, m \in \mathbb{N},$$

where $r(x, \mu) = o(x^{2m}, |x\mu|, \mu^2) \in C^{2m}(\mathbb{R} \times \mathbb{R}, \mathbb{R})$. We note that when the zero-factor x^p is removed, Eq. (5.2.13) becomes

$$(5.2.14) \qquad x' = \mu x - x^{2m} + r(x, \mu).$$

It is easy to see from Theorem 5.2.1 that Eq. (5.2.14) has a transcritical bifurcation at $\mu = 0$ with two equilibrium branches

$$x = x_1(\mu) = 0 \quad \text{and} \quad x = x_2(\mu) = \mu^{1/(2m-1)}(1 + o(1)) \quad \text{as } \mu \to 0.$$

As shown in Example 5.2.2, $x_1(\mu)$ is asymptotically stable for $\mu < 0$ and unstable for $\mu > 0$; and $x_2(\mu)$ is unstable for $\mu < 0$ and asymptotically stable for $\mu > 0$. By Theorem 5.2.2, we have the following:

(a) When p is even, Eq. (5.2.13) has exactly the same bifurcation diagram as Eq. (5.2.14).

(b) When p is odd, the stability of $x_1(\mu)$ becomes (N-A) for $\mu < 0$ and (A-N) for $\mu > 0$, and the stability of $x_2(\mu)$ becomes asymptotically stable for both $\mu < 0$ and $\mu > 0$.

See Fig. 5.10 for the bifurcation diagrams.

EXAMPLE 5.2.5. Consider the equation

$$(5.2.15) \qquad x' = x^p(\mu x - x^{2m+1} + r(x, \mu)), \quad p, m \in \mathbb{N}.$$

where $r(x, \mu) = o(|x^{2m+1}|, |x\mu|, \mu^2) \in C^{2m+1}(\mathbb{R} \times \mathbb{R}, \mathbb{R})$ as $(x, \mu) \to (0, 0)$. We note that when the zero-factor x^p is removed, Eq. (5.2.15) becomes

$$(5.2.16) \qquad x' = \mu x - x^{2m+1} + r(x, \mu).$$

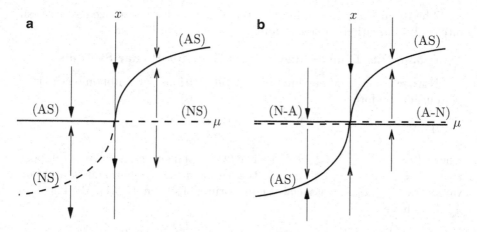

FIGURE 5.10. Generalized transcritical bifurcation. (a) p is even. (b) p is odd

It is easy to see from Theorem 5.2.1 that Eq. (5.2.14) has a supercritical pitchfork bifurcation at $\mu = 0$ with two equilibrium branches

$$x = x_1(\mu) = 0 \quad \text{and} \quad \mu = \mu_2(x) = x^{2m}(1 + o(1)) \quad \text{as } x \to 0.$$

As shown in Example 5.2.2, $x_1(\mu)$ is asymptotically stable for $\mu < 0$ and unstable for $\mu > 0$; and $x_2(\mu)$ is asymptotically stable for $x < 0$ and $x > 0$. By Theorem 5.2.2, we have the following:

(a) When p is even, then Eq. (5.2.16) has exactly the same bifurcation diagram as Eq. (5.2.15).

(b) When p is odd, then the stability of $x_1(\mu)$ becomes (N-A) for $\mu < 0$ and (A-N) for $\mu > 0$, and the stability of $\mu_2(x)$ becomes unstable for $x < 0$ and asymptotically stable for $x > 0$.

See Fig. 5.11 for the bifurcation diagrams.

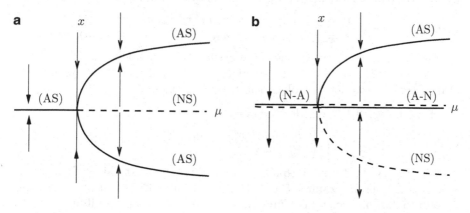

FIGURE 5.11. Generalized supercritical pitchfork bifurcation. (a) p is even. (b) p is odd

The discussions on the bifurcation diagrams for generalized subcritical pitchfork bifurcation are left as an exercise, see Exercise 5.3.

5.3. One-Dimensional Bifurcations for Planar Systems

Now, we consider one-dimensional bifurcations for autonomous planar systems of the form

$$(5.3.1) \qquad \begin{cases} x_1' = f_1(x_1, x_2, \mu) \\ x_2' = f_2(x_1, x_2, \mu), \end{cases}$$

where $f := (f_1, f_2)^T \in C^1(D, \mathbb{R}^2)$ with D an open subset of $\mathbb{R}^2 \times \mathbb{R}$. Suppose that $P_0 = (x_{01}, x_{02}, \mu_0) \in D$ such that μ_0 is a one-dimensional bifurcation value with (x_{01}, x_{02}) an associated equilibrium of system (5.3.1). By Theorem 5.1.1 we have

$$f(P_0) = 0 \quad \text{and} \quad \det \frac{\partial(f_1, f_2)}{\partial(x_1, x_2)}(P_0) = 0.$$

We will study the topological structure of the orbits near the point P_0.

As shown in Sect. 5.2, without loss of generality we may assume $P_0 = (0, 0, 0)$. From the linearization method introduced in Sect. 3.3, system (5.3.1) can be written as

$$\begin{bmatrix} x_1 \\ x_2 \end{bmatrix}' = \begin{bmatrix} a_1 \mu \\ a_2 \mu \end{bmatrix} + A \begin{bmatrix} x_1 \\ x_2 \end{bmatrix} + \begin{bmatrix} r_1(x_1, x_2, \mu) \\ r_2(x_1, x_2, \mu) \end{bmatrix},$$

where $A = \partial(f_1, f_2)/\partial(x_1, x_2)|_{(0,0,0)}$, and for $i = 1, 2$, $a_i = (f_i)_\mu(0, 0, 0)$ and $r_i = o(|x_1| + |x_2| + |\mu|)$ as $(x_1, x_2, \mu) \to (0, 0, 0)$. Let λ_1 and λ_2 be the eigenvalues of matrix A. Since $\mu = 0$ is a one-dimensional bifurcation value, we see that one of λ_1 and λ_2 is zero and the other is not. So we let $\lambda_1 = 0$ and $\lambda_2 \neq 0$. By linear algebra, we can always make a linear transformation in (x_1, x_2) so that the matrix A in the resulting equation becomes diagonal. Thus, we may assume $A = \begin{bmatrix} \lambda_1 & \\ & \lambda_2 \end{bmatrix} = \begin{bmatrix} 0 & \\ & \lambda_2 \end{bmatrix}$. With the above manipulations, system (5.3.1) takes the form

$$(5.3.2) \qquad \begin{cases} x_1' = g_1(x_1, x_2, \mu) := a_1 \mu + r_1(x_1, x_2, \mu) \\ x_2' = g_2(x_1, x_2, \mu) := a_2 \mu + \lambda_2 x_2 + r_2(x_1, x_2, \mu), \end{cases}$$

where $a_i, r_i, i = 1, 2$, satisfy the same assumptions as above.

Before we investigate the topological structure of the orbits for the general system (5.3.2), let us look at the following example.

EXAMPLE 5.3.1. Consider the system

$$(5.3.3) \qquad \begin{cases} x_1' = \mu - x_1^{2m}, \quad m \in \mathbb{N}, \\ x_2' = -x_2. \end{cases}$$

We observe that system (5.3.3) is a simple system-extension of the scalar equation (5.2.1) in Example 5.2.1. Hence $\mu = 0$ is a bifurcation value, and system (5.3.3) has no equilibrium when $\mu < 0$, exactly one equilibrium $(0, 0)$ when $\mu = 0$, and exactly two equilibria $(\pm\mu^{1/(2m)}, 0)$ when $\mu > 0$. Note that for any solution $(x_1(t), x_2(t))$ of system (5.3.3) with any μ, we have $x_2(t) \to 0$ as $t \to \infty$. Thus the stability of the equilibrium branch is determined by the behavior of $x_1(t)$. In light of Example 5.2.1, the phase portraits for system (5.3.3) are given in Fig. 5.12.

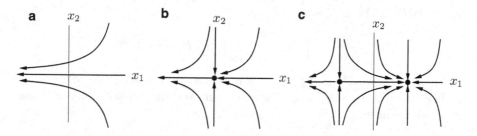

FIGURE 5.12. Planar saddle-node bifurcation. (a) $\mu < 0$. (b) $\mu = 0$. (c) $\mu > 0$

Such a bifurcation is called a planar saddle-node bifurcation. Actually, Fig. 5.12 reflects the original sense of the saddle and node as defined in Sect. 4.4.

We can also extend Examples 5.2.2 and 5.2.3 to create planar transcritical and pitchfork bifurcations. However, this method does not work for the general system (5.3.1). This is because the unknown functions $x_1(t)$ and $x_2(t)$ are generally coupled rather than separated in the two equations of the system.

To deal with the bifurcation at the point $(0, 0, 0)$, we note that $(g_2)_{x_2}$ $(0, 0, 0) = \lambda_2 \neq 0$ which allows us to solve the equation $g_2(x_1, x_2, \mu) = 0$ for x_2 near $(0, 0, 0)$ using the implicit function theorem. In fact, by Lemma 5.1.1, there is a unique C^1-function $x_2 = x_2(x_1, \mu)$ defined in a neighborhood of $(0, 0)$ satisfying

$$(5.3.4) \qquad g_2(x_1, x_2(x_1, \mu), \mu) \equiv 0, \quad x_2(0, 0) = 0.$$

Equation (5.3.4) can be further used to estimate the function $x_2(x_1, \mu)$. When substituting $x_2 = x_2(x_1, \mu)$ into the equation $g_1(x_1, x_2, \mu) = 0$, we obtain that

$$(5.3.5) \quad h(x_1, \mu) := g_1(x_1, x_2(x_1, \mu), \mu) = a_1\mu + r_1(x_1, x_2(x_1, \mu), \mu) = 0.$$

Clearly, the equilibrium branches for system (5.3.2) near $(0, 0, 0)$ is determined by Eq. (5.3.5). Therefore, Eq. (5.3.5) is called the bifurcation equation for system (5.3.2). Since the equation

$$(5.3.6) \qquad x_1' = h(x_1, \mu)$$

is a scalar equation, we may use the results and methods in Sect. 5.2 for scalar equations to study the bifurcation for Eq. (5.3.6) and hence obtain the bifurcation result for system (5.3.2). The method for investigating the qualitative properties of a system of equations by investigating a system with a reduced dimension is called the Lyapunov–Schmidt method. For a general discussion on the applications of the Lyapunov–Schmidt method to the bifurcation problems, see Chow and Hale [10, Section 2.4].

Finally, we comment on the stability of any point $P = (x_1, x_2(x_1, \mu), \mu)$ on an equilibrium branch for system (5.3.2) near $(0, 0, 0)$. Let $\eta_1(P)$ and $\eta_2(P)$ be the eigenvalues of the matrix $\partial(g_1, g_2)/\partial(x_1, x_2)|_P$. Then by continuity, $\eta_2(P) \to \lambda_2$ as $P \to (0, 0, 0)$, and

$$\left. \frac{\partial(g_1, g_2)}{\partial(x_1, x_2)} \right|_P \to \begin{bmatrix} \lambda_1 & \\ & \lambda_2 \end{bmatrix} \quad \text{as } P \to (0, 0, 0).$$

Since $\lambda_2 \neq 0$, $\eta_2(P)$ and $(g_2)_{x_2}(P)$ have the same sign as λ_2 when P is sufficiently close to $(0, 0, 0)$. By the chain rule and the implicit differentiation method, we have

$$
\begin{aligned}
h_{x_1}(P) &= [(g_1)_{x_1} + (g_1)_{x_2}(x_2)_{x_1}](P) = \left[(g_1)_{x_1} + (g_1)_{x_2} \left(-\frac{(g_2)_{x_1}}{(g_2)_{x_2}} \right) \right](P) \\
&= \left[\frac{1}{(g_2)_{x_2}} \det \frac{\partial(g_1, g_2)}{\partial(x_1, x_2)} \right](P) = \left[\frac{\eta_1 \eta_2}{(g_2)_{x_2}} \right](P).
\end{aligned}
$$

It shows that $h_{x_1}(P)$ has the same sign as $\eta_1(P)$. Consequently, we have the following relations between the stability of system (5.3.2) and that for Eq. (5.3.6):

(a) When $\lambda_2 < 0$, the stability of the point P for system (5.3.2) is exactly the same as that of the point (x_1, μ) on the corresponding equilibrium branch for Eq. (5.3.6).

(b) When $\lambda_2 > 0$, then $\eta_2(P) > 0$ for all equilibria P in a neighborhood of $(0, 0, 0)$. Therefore, all equilibria in this neighborhood are unstable. Moreover, from Sect. 4.4 we have that if (x_1, μ) is asymptotically stable for Eq. (5.3.6), then $P = (x_1, x_2(x_1, \mu), \mu)$ is a saddle-point for system (5.3.2); and if (x_1, μ) is unstable for Eq. (5.3.6), then $P = (x_1, x_2(x_1, \mu), \mu)$ is an unstable node for system (5.3.2).

The examples below are used to explain how this method is implemented.

EXAMPLE 5.3.2. Consider the system

(5.3.7)
$$
\begin{cases}
x_1' = f_1(x_1, x_2, \mu) := x_1(\mu - x_1 - x_2^2), \\
x_2' = f_2(x_1, x_2, \mu) := x_2(x_1 - 1).
\end{cases}
$$

This system is used as a predator-prey model and is called a Lotka–Volterra equation. Here x_1 and x_2 stand for the populations of the prey and predator, respectively. When $x_2 \equiv 0$ and $\mu > 0$, the system reduces to the logistic equation for the prey $x_1' = x_1(\mu - x_1)$. By Example 3.1.1 we have that $x_1(t) \to \mu$ as $t \to \infty$, and hence μ is the carrying capacity of the environment. When $x_1 \equiv 0$, the system reduces to the decay equation for the predator $x_2' = -x_2$, and hence $x_2(t) \to 0$ as $t \to \infty$. Now we study the behavior of $x_1(t)$ and $x_2(t)$ for the general case.

It is easy to see that system (5.3.7) has three sets of equilibria:

$$x_1 = 0, x_2 = 0; \quad x_1 = \mu, x_2 = 0; \quad \text{and } x_1 = 1, x_2^2 = \mu - 1.$$

Since these curves in $\mathbb{R}^3$ mutually intersect at the point $(0,0,0)$ and $(1,0,1)$, $\mu = 0$ and $\mu = 1$ are the only candidates for the bifurcation values with corresponding equilibria $(0,0)$ and $(1,0)$, respectively.

(a) We first consider the bifurcation at $\mu = 0$.

Clearly, $f_2(x_1, x_2, \mu) = 0$ determines that $x_2(t) = 0$ in a neighborhood of $(0,0,0)$. Substituting it into $f_1(x_1, x_2, \mu) = 0$ we obtain the bifurcation equation $h(x_1, \mu) := x_1(\mu - x_1) = 0$. We see that the equation

$$x_1' = h(x_1, \mu) = x_1(\mu - x_1)$$

is of the form (5.2.2) in Example 5.2.2 with $m = 1$. It follows that the bifurcation for system (5.3.7) at $(0,0,0)$ is a transcritical bifurcation in the $x_1 x_2 \mu$-space. Note that $\lambda_2 = (f_2)_{x_2}(0,0,0) = -1 < 0$. The stabilities for equilibrium branches for system (5.3.7) near $(0,0,0)$ remain the same as those for the equation $x_1' = h(x_1, \mu)$ near $(0,0)$.

(b) Then we consider the bifurcation at $\mu = 1$.

Let $y_1 = x_1 - 1$, $y_2 = x_2$, and $\nu = \mu - 1$. Then system (5.3.7) is transformed to the system

(5.3.8)
$$\begin{cases} y_1' = g_1(y_1, y_2, \mu) := (y_1 + 1)(\nu - y_1 - y_2^2), \\ y_2' = g_2(y_1, y_2, \nu) := y_1 y_2, \end{cases}$$

and hence the point $(x_1, x_2, \mu) = (1,0,1)$ becomes the point $(y_1, y_2, \nu) = (0,0,0)$. Since

$$(g_1)_{y_1}(0,0,0) = [\nu - 1 - 2y_1 - y_2^2](0,0,0) = -1 \neq 0,$$

by Lemma 5.1.1, there is a unique analytic function $y_1 = y_1(y_2, \nu)$ defined in a neighborhood of $(0,0)$ satisfying

(5.3.9)
$$g_1(y_1(y_2, \nu), y_2, \nu) = (y_1(y_2, \nu) + 1)(\nu - y_1(y_2, \nu) - y_2^2) \equiv 0, \quad y_1(0,0) = 0.$$

Clearly, $y_1(y_2, \nu) = \nu - y_2^2$. From the second equation in (5.3.8), the bifurcation equation for system (5.3.8) is

$$h(y_2, \nu) := \nu y_2 - y_2^3.$$

It is easy to see that $h_\nu(0,0) = 0$, $h_{\nu\nu}(0,0) = 0$, $h_{y_2\nu}(0,0) = 1$, $h_{y_2 y_2}(0,0) = 0$, and $h_{y_2 y_2 y_2}(0,0) = -6 < 0$. By Theorem 5.2.1, the bifurcation for the equation $y_2' = h(y_2, \nu)$ at $(0,0)$ is a supercritical pitchfork bifurcation in the $y_2\nu$-plane. It follows that the bifurcation for system (5.3.8) at $(0,0,0)$ is a pitchfork bifurcation in the $y_1 y_2 \nu$-space. Note that $\lambda_1 = (g_1)_{y_1}(0,0,0) = -1 < 0$. The stabilities for equilibrium branches for system (5.3.8) near $(0,0,0)$ remain the same as those for the equation $y_2' = h(y_2, \nu)$ near $(0,0)$. Therefore, system (5.3.8) has a supercritical pitchfork bifurcation at $\nu = 0$, i.e., system (5.3.7) has a supercritical pitchfork bifurcation at $\mu = 1$.

For the bifurcation diagram for system (5.3.7), see Fig. 5.13, where (AS) and (NS) stand for asymptotically stable and unstable, respectively.

EXAMPLE 5.3.3. Consider the system

$$(5.3.10) \qquad \begin{cases} x_1' = f_1(x_1, x_2, \mu) := \mu x_1 - x_1^3 x_2 + x_1^5 + x_1^6 - 2x_1^2 x_2^2 \\ x_2' = f_2(x_1, x_2, \mu) := x_2 + x_1^2 - \mu x_1^2 + 3x_2^3. \end{cases}$$

By Theorem 5.1.1, $\mu = 0$ is a candidate for a bifurcation value with the corresponding equilibrium $(x_1, x_2) = (0, 0)$. Since $(f_2)_{x_2}(0, 0, 0) = 1 \neq 0$, by Lemma 5.1.1, there is a unique analytic function $x_2 = x_2(x_1, \mu)$ defined in a neighborhood of $(0, 0)$ satisfying

$$f_2(x_1, x_2(x_1, \mu), \mu) = x_2(x_1, \mu) + x_1^2 - \mu x_1^2 + 3x_2^3(x_1, \mu) \equiv 0, \quad x_2(0, 0) = 0.$$

It follows that as $(x, \mu) \to (0, 0)$

$$x_2(x_1, \mu) = -x_1^2 + o(x_1^2 + |\mu x_1| + \mu^2),$$

and hence by (5.3.5), the bifurcation equation for system (5.3.10) is

$$h(x_1, \mu) := \mu x_1 + 2x_1^5 + o(|x_1|^5 + |\mu x_1| + \mu^2).$$

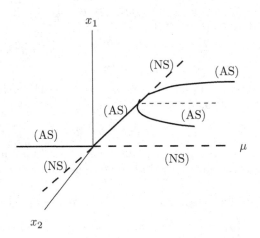

FIGURE 5.13. Example 5.3.2

It is easy to see that $h_\mu(0, 0) = 0$, $h_{\mu\mu}(0, 0) = 0$, $h_{x_1\mu}(0, 0) = 1$, $\partial^i h/\partial x_1^i(0, 0) = 0$ for $i = 2, 3, 4$, hence $\Delta = -1 < 0$. Note that $\partial^5 h/\partial x_1^5(0, 0) = 2 \cdot 5! > 0$. By Theorem 5.2.1, the bifurcation for the equation $x_1' = h(x_1, \mu)$ at $(0, 0)$ is a subcritical pitchfork bifurcation in the $x_1\mu$-plane. Therefore, the bifurcation for system (5.3.10) at $(0, 0, 0)$ is a pitchfork bifurcation in the $x_1 x_2 \mu$-space. Note that $\lambda_2 = (f_2)_{x_2}(0, 0, 0) = 1 > 0$. The equilibrium branches for system (5.3.10) near $(0, 0, 0)$ are all unstable. Moreover, the asymptotically stable equilibria and the unstable equilibria for the equation $x_1' = h(x_1, \mu)$ become saddle-points and unstable nodes, respectively, for system (5.3.10).

5.4. Hopf Bifurcations for Planar Systems

In the last section, we study the Poincaré–Andronov–Hopf bifurcations or Hopf bifurcations for the system

$$(5.4.1\text{-}\mu) \qquad \begin{cases} x_1' = f_1(x_1, x_2, \mu) \\ x_2' = f_2(x_1, x_2, \mu). \end{cases}$$

For the convenience of proofs, we assume that $f := (f_1, f_2)^T : D \to \mathbb{R}^2$ is analytic with D an open subset of $\mathbb{R}^2 \times \mathbb{R}$ and $(0, 0, 0) \in D$. Without loss of generality, we suppose that

$$(5.4.2) \qquad f_1(0, 0, \mu) \equiv 0 \quad \text{and} \quad f_2(0, 0, \mu) \equiv 0,$$

i.e., $(0, 0)$ is an equilibrium for all $\mu \in \mathbb{R}$. Let

$$(5.4.3) \qquad A(\mu) = \left. \frac{\partial(f_1, f_2)}{\partial(x_1, x_2)} \right|_{(0,0,\mu)}$$

and $\alpha(\mu) \pm \mathbf{i}\beta(\mu)$ be the eigenvalues of matrix $A(\mu)$. For $\mu = 0$ to be a Hopf bifurcation value for system (5.4.1-μ), we must have

$$(5.4.4) \qquad \alpha(0) = 0 \quad \text{and} \quad \beta(0) := \beta \neq 0.$$

We first use the Lyapunov function method to derive a Hopf bifurcation result for system (5.4.1-μ).

THEOREM 5.4.1. *Assume* (5.4.2) *and* (5.4.4) *hold.*

(a) *Suppose that the equilibrium* $(0, 0)$ *is a stable spiral-point for system (5.4.1-0) and* $(0, 0)$ *is an unstable spiral-point for system (5.4.1-μ) for* $\mu > 0$ *[$\mu < 0$]. Then for* $\mu > 0$ *[$\mu < 0$] sufficiently close to zero, system (5.4.1-μ) has an orbitally asymptotically stable limit cycle in a neighborhood of* $(0, 0)$.

(b) *Suppose that the equilibrium* $(0, 0)$ *is an unstable spiral-point for system (5.4.1-0) and* $(0, 0)$ *is a stable spiral-point for system (5.4.1-μ) for* $\mu > 0$ *[$\mu < 0$]. Then for* $\mu > 0$ *[$\mu < 0$] sufficiently close to zero, system (5.4.1-μ) has an orbitally unstable limit cycle in a neighborhood of* $(0, 0)$.

To prove Theorem 5.4.1, we need the following lemma for the system

$$(5.4.5) \qquad \begin{cases} x_1' = -\beta x_2 + g_1(x_1, x_2) \\ x_2' = \beta x_1 + g_2(x_1, x_2), \end{cases}$$

where $\beta \neq 0$ and $g := (g_1, g_2)^T : E \to \mathbb{R}^2$ is analytic with E an open subset of $\mathbb{R}^2$ and $(0, 0) \in E$ such that for $i = 1, 2$

$$g_i(0, 0) = 0 \quad \text{and} \quad \left. \frac{\partial(g_1, g_2)}{\partial(x_1, x_2)} \right|_{(0,0)} = \mathbf{0}.$$

LEMMA 5.4.1. *Suppose that the equilibrium $(0,0)$ is a stable spiral-point for system (5.4.5). Then there exists a positive definite Lyapunov function in the form*

$$V(x_1, x_2) = \frac{1}{2}(x_1^2 + x_2^2) + \sum_{n=3}^{2k} V_n(x_1, x_2)$$

for some $k \geq 2$, where $V_n(x_1, x_2)$ is the n-th degree homogeneous polynomial of (x_1, x_2), such that the derivative of $V(x_1, x_2)$ along system (5.4.5)

$$\dot{V}(x_1, x_2) = (c + o(1))(x_1^2 + x_2^2)^k \quad as \ (x_1, x_2) \to (0, 0)$$

for some $c < 0$.

PROOF. Since $g = (g_1, g_2)^T$ is analytic, system (5.4.5) can be written as

$$(5.4.6) \qquad \begin{cases} x_1' = -\beta x_2 + \sum_{j=2}^{\infty} g_{1j}(x_1, x_2) \\ x_2' = \beta x_1 + \sum_{j=2}^{\infty} g_{2j}(x_1, x_2), \end{cases}$$

where for $i = 1, 2$, $g_{ij}(x_1, x_2)$ is a j-th degree homogeneous polynomial of (x_1, x_2). We look for a Lyapunov function in the form

$$(5.4.7) \qquad F(x_1, x_2) = \frac{1}{2}(x_1^2 + x_2^2) + \sum_{i=3}^{\infty} F_i(x_1, x_2),$$

where $F_i(x_1, x_2)$ is an i-th degree homogeneous polynomial of (x_1, x_2). If such a function exists and is convergent in a neighborhood of $(0,0)$, then along system (5.4.5)

$$\dot{F}(x_1, x_2) = \left(x_1 + \sum_{i=3}^{\infty}(F_i)_{x_1}(x_1, x_2) \right) \left(-\beta x_2 + \sum_{j=2}^{\infty} g_{1j}(x_1, x_2) \right)$$
$$+ \left(x_2 + \sum_{i=3}^{\infty}(F_i)_{x_2}(x_1, x_2) \right) \left(\beta x_1 + \sum_{j=2}^{\infty} g_{2j}(x_1, x_2) \right).$$

By multiplying out the right-hand side and rearranging the terms in accordance with the degrees we obtain that

$$(5.4.8) \ \dot{F}(x_1, x_2) = \sum_{n=3}^{\infty} \left[-\beta(x_2(F_n)_{x_1} - x_1(F_n)_{x_2}) + H_n(x_1, x_2) \right],$$

where $H_n(x_1, x_2)$ is an n-th degree homogeneous polynomial of (x_1, x_2) which involves $(F_i)_{x_j}(x_1, x_2)$, $j = 1, 2$, only for $i \leq n - 1$. Now we use the polar-coordinate transformation

$$x_1 = r \cos \theta \quad \text{and} \quad x_2 = r \sin \theta$$

to investigate the existence of F_n for $n \geq 3$. If F_i exists for $i \leq n - 1$, then

$$x_1(F_n)_{x_2} - x_2(F_n)_{x_1} = r \cos \theta (F_n)_{x_2} - r \sin \theta (F_n)_{x_1} = \frac{\partial F_n}{\partial \theta}(r \cos \theta, r \sin \theta)$$
$$= r^n \frac{\partial F_n}{\partial \theta}(\cos \theta, \sin \theta)$$

and
$$H_n(x_1, x_2) = r^n H_n(\cos\theta, \sin\theta).$$
Thus, F_n satisfies
$$-\beta(x_2(F_n)_{x_1} - x_1(F_n)_{x_2}) + H_n(x_1, x_2) = 0$$
if and only if

(5.4.9) $$\beta\frac{\partial F_n}{\partial\theta}(\cos\theta, \sin\theta) + H_n(\cos\theta, \sin\theta) = 0.$$

Note that when $n = 2l - 1$ is odd, $H_n(-x_1, -x_2) = -H_n(x_1, x_2)$. Hence $H_n(\cos\theta, \sin\theta)$ has the Fourier expansion

$$H_n(\cos\theta, \sin\theta) = \sum_{i=1}^{l}[a_i\cos(2i-1)\theta + b_i\sin(2i-1)\theta]$$

which guarantees that

$$\int_0^{2\pi} H_n(\cos\theta, \sin\theta)\, d\theta = 0.$$

By integrating both sides of (5.4.9) with respect to θ, we obtain an n-th degree polynomial $F_n(\cos\theta, \sin\theta)$ of $\cos\theta$ and $\sin\theta$ which contains only odd power terms. Note that we can always make it homogeneous by multiplying the lower power terms by an appropriate degree of $\cos^2\theta + \sin^2\theta$. Then an n-th degree homogeneous polynomial $F_n(x_1, x_2)$ is determined by replacing $\cos\theta$ and $\sin\theta$ in the above function by x_1 and x_2, respectively.

Also, when $n = 2l$ is even, $H_n(-x_1, -x_2) = H_n(x_1, x_2)$. Hence $H_n(\cos\theta, \sin\theta)$ has the Fourier expansion

$$H_n(\cos\theta, \sin\theta) = a_0 + \sum_{i=1}^{l}[a_i\cos 2i\theta + b_i\sin 2i\theta].$$

It follows that

$$\int_0^{2\pi} H_n(\cos\theta, \sin\theta)\, d\theta = 0 \iff a_0 = 0.$$

Consequently, Eq. (5.4.9) determines an n-th degree homogeneous polynomial $F_n(x_1, x_2)$ if and only if $a_0 = 0$.

From (5.4.8) we find that

$$H_3(x_1, x_2) = x_1 g_{12}(x_1, x_2) + x_2 g_{22}(x_1, x_2)$$

is known. Then a third degree homogeneous polynomial $F_3(x_1, x_2)$ is determined by (5.4.9) with $n = 3$. By recursion, for $n \geq 3$, an n-th degree homogeneous polynomial $F_n(x_1, x_2)$ is determined by (5.4.9) as long as $\int_0^{2\pi} H_i(\cos\theta, \sin\theta)\, d\theta = 0$ for $i = 3, \ldots, n$. We claim that there exists an even number $n = 2k$ such that $\int_0^{2\pi} H_{2k}(\cos\theta, \sin\theta)\, d\theta \neq 0$ and hence no function $F_{2k}(x_1, x_2)$ is determined by (5.4.9). Otherwise, there exists an $F_n(x_1, x_2)$ for every $n = 3, 4, \ldots$ satisfying (5.4.9). In this case, it has been shown by Lyapunov (but we omit the details here) that the function $F(x_1, x_2)$ defined by (5.4.7) is convergent and hence is positive definite in a neighborhood of $(0,0)$. By (5.4.8), $\dot{F}(x_1, x_2) = 0$ along system (5.4.5). This means that for any small $l > 0$, $F(x_1, x_2) \equiv l$ as a level curve for system (5.4.5) is closed and hence $(0,0)$ is a center for system (5.4.5). This contradicts the assumption that $(0,0)$ is a spiral-point for system (5.4.5).

Let $V_n(x_1, x_2) := F_n(x_1, x_2)$, $n = 3, \ldots, 2k - 1$, as defined above and let $V_{2k}(\cos\theta, \sin\theta)$ satisfy

$$\beta \frac{\partial V_{2k}}{\partial \theta}(\cos\theta, \sin\theta) = -H_{2k}(\cos\theta, \sin\theta) + c,$$

where $c = \frac{1}{2\pi} \int_0^{2\pi} H_{2k}(\cos\theta, \sin\theta)d\theta$. Define a function

$$V(x_1, x_2) = \frac{1}{2}(x_1^2 + x_2^2) + \sum_{n=3}^{2k} V_n(x_1, x_2).$$

Then $V(x_1, x_2)$ is positive definite in a neighborhood of $(0,0)$ and by (5.4.8),

$$\dot{V}(x_1, x_2) = c(x_1^2 + x_2^2)^k + o((x_1^2 + x_2^2)^k) = (c + o(1))(x_1^2 + x_2^2)^k$$

as $(x_1, x_2) \to (0,0)$. Since $(0,0)$ is asymptotically stable, we have $c < 0$. This completes the proof. $\qquad\square$

PROOF OF THEOREM 5.4.1. We only prove Part (a). Part (b) will become Part (a) after the change of the independent variable $t = -s$.

Under the assumptions (5.4.2) and (5.4.4), without loss of generality we may assume that system (5.4.1-0) is in the form of (5.4.5). Then by Lemma 5.1.1, there is a positive definite Lyapunov function $V(x_1, x_2)$ in a neighborhood $\mathcal{N}$ of $(0,0)$ such that its derivative along system (5.4.1-0)

$$\dot{V}|_{(5.4.1\text{-}0)}(x_1, x_2) = (c + o(1))(x_1^2 + x_2^2)^k \quad \text{as } (x_1, x_2) \to (0,0)$$

for some $c < 0$ and an integer $k \geq 2$. Hence we can make $\mathcal{N}$ so small that

$$\dot{V}|_{(5.4.1\text{-}0)}(x_1, x_2) \leq \frac{c}{2}(x_1^2 + x_2^2)^k \quad \text{in } \mathcal{N}.$$

Choose an $a > 0$ such that the simple closed level curve

$$C_1 := \{(x_1, x_2) : V(x_1, x_2) = a\}$$

is contained in $\mathcal{N}$ with $(0,0)$ the only equilibrium inside C_1, and then choose a $b > 0$ such that the circle

$$C_2 := \{(x_1, x_2) : x_1^2 + x_2^2 = b\}$$

is inside C_1. Let G be the region bounded by C_1 and C_2. Note that $c < 0$ implies that

$$\dot{V}|_{(5.4.1\text{-}0)}(x_1, x_2) \leq \frac{c}{2}b^k \quad \text{in } G.$$

Let $\dot{V}|_{(5.4.1\text{-}\mu)}$ be the derivative of $V(x_1, x_2)$ along system $(5.4.1\text{-}\mu)$. Then

$$\dot{V}|_{(5.4.1\text{-}\mu)}(x_1, x_2) - \dot{V}|_{(5.4.1\text{-}0)}(x_1, x_2)$$
$$= V_{x_1}(x_1, x_2)[f_1(x_1, x_2, \mu) - f_1(x_1, x_2, 0)]$$
$$+ V_{x_2}(x_1, x_2)[f_2(x_1, x_2, \mu) - f_2(x_1, x_2, 0)].$$

Since V_{x_1} and V_{x_2} are bounded on G and $f_i(x_1, x_2, \mu)$, $i = 1, 2$, are continuous in μ uniformly for (x_1, x_2) in G, we have that there exists a $\mu^* > 0$ $[\mu_* < 0]$ such that for $\mu \in (0, \mu^*)$ $[\mu \in (\mu_*, 0)]$

$$\left| \dot{V}|_{(5.4.1\text{-}\mu)}(x_1, x_2) - \dot{V}|_{(5.4.1\text{-}0)}(x_1, x_2) \right| < -\frac{c}{2} b^k \quad \text{in } G.$$

It follows that in the region G

$$\dot{V}|_{(5.4.1\text{-}\mu)}(x_1, x_2)$$
$$= \dot{V}|_{(5.4.1\text{-}0)}(x_1, x_2) + \left(\dot{V}|_{(5.4.1\text{-}\mu)}(x_1, x_2) - \dot{V}|_{(5.4.1\text{-}0)}(x_1, x_2) \right)$$
$$< \frac{c}{2} b^k - \frac{c}{2} b^k = 0.$$

Thus, the positive semi-orbit starting from any point on C_1 will stay inside C_1. Note that $(0, 0)$ is the only equilibrium of system $(5.4.1\text{-}\mu)$ inside C_1 which is an unstable spiral-point. By Corollary 4.6.1, system $(5.4.1\text{-}\mu)$ has an orbitally asymptotically stable limit cycle inside C_1. $\square$

We observe that in addition to (5.4.4), if we assume $\alpha'(0) > 0$ $[\alpha'(0) < 0]$, then $(0, 0)$ is an unstable spiral-point for system $(5.4.1\text{-}\mu)$ for $\mu > 0$ $[\mu < 0]$. Therefore, the lemma below is an immediate consequence of Theorem 5.4.1.

COROLLARY 5.4.1. *Assume (5.4.2) and (5.4.4) hold with $\alpha'(0) > 0$ $[\alpha'(0) < 0]$.*

   (a) *Suppose that the equilibrium $(0, 0)$ is a stable spiral-point for system (5.4.1-0). Then for $\mu > 0$ $[\mu < 0]$ sufficiently close to zero, system $(5.4.1\text{-}\mu)$ has an orbitally asymptotically stable limit cycle in a neighborhood of $(0, 0)$.*

   (b) *Suppose that the equilibrium $(0, 0)$ is an unstable spiral-point for system (5.4.1-0). Then for $\mu < 0$ $[\mu > 0]$ sufficiently close to zero, system $(5.4.1\text{-}\mu)$ has an orbitally unstable limit cycle in a neighborhood of $(0, 0)$.*

To apply Theorem 5.4.1 and Corollary 5.4.1, we need to determine the stability of the equilibrium $(0, 0)$ for system (5.4.1-0). This can usually be done by constructing a Lyapunov function. However, these results fail to apply when the equilibrium $(0, 0)$ is a center for system (5.4.1-0). Moreover, they do not address on the number of closed orbits that system $(5.4.1\text{-}\mu)$ may have.

The next result is derived using the Friedrich method for Hopf-bifurcations.

THEOREM 5.4.2. *Assume* (5.4.2) *and* (5.4.4) *hold with* $\alpha'(0) \neq 0$. *Then either*

> (a) *all orbits of system* (5.4.1-0) *in a neighborhood of* $(0,0)$ *are closed orbits and system* (5.4.1-μ) *does not have closed orbits for* $\mu \neq 0$ *in a neighborhood of* $\mu = 0$, *or*
>
> (b) *for* $\mu > 0$ *sufficiently close to zero only or for* $\mu < 0$ *sufficiently close to zero only, system* (5.4.1-μ) *has a unique limit cycle* $\Gamma(\mu)$ *satisfying* $\Gamma(\mu) \to (0,0)$ *with its period* $T(\mu) \to 2\pi/\beta$ *as* $\mu \to 0$.

PROOF. Let $x = (x_1, x_2)^T$. Then by (5.4.2), system (5.4.1-μ) can be written into the form

$$(5.4.10) \qquad x' = A(\mu)x + g(x, \mu),$$

where $A(\mu)$ is defined by (5.4.3) and $g : D \to \mathbb{R}^2$ is analytic such that

$$g_i(0, 0, \mu) = 0 \quad \text{and} \quad \frac{\partial(g_1, g_2)}{\partial(x_1, x_2)}\bigg|_{(0,0,\mu)} = \mathbf{0},$$

i.e., the Maclaurin series of g with respect to x contains only powers of x with degrees two or higher. In light of (5.4.4) with $\alpha'(0) \neq 0$, without loss of generality we may assume

$$(5.4.11) \qquad A(\mu) = A(0) + \mu B(\mu),$$

where $A(0) = \begin{bmatrix} 0 & -\beta \\ \beta & 0 \end{bmatrix}$ and $\operatorname{tr} B(0) \neq 0$.

For any $r > 0$, denote by $x(t, r, \mu) = (x_1, x_2)^T(t, r, \mu)$ the solution of system (5.4.1-μ) passing through the point $(r, 0) \in \mathbb{R}^2$. Then for any $\mu \neq 0$, system (5.4.1-μ) has a periodic solution $x(t, r(\mu), \mu)$ if and only if there exist $r(\mu), T(\mu) > 0$ such that

$$\begin{cases} x_1(T(\mu), r(\mu), \mu) = r(\mu) \\ x_2(T(\mu), r(\mu), \mu) = 0. \end{cases}$$

Since we expect to find a periodic solution of system (5.4.1-μ) satisfying $r(\mu) \to 0$ and $T(\mu) \to T_0 := 2\pi/\beta$ as $\mu \to 0$, we may let

$$(5.4.12) \qquad \mu = rp(r) \quad \text{and} \quad T(\mu) = T_0(1 + rq(r))$$

for some functions $p(r)$ and $q(r)$ to be determined. If we make the transformation

$$(5.4.13) \qquad t = (1 + rq(r))s \quad \text{and} \quad (x_1, x_2) = r(y_1, y_2),$$

then $y := (y_1, y_2)^T$ becomes a function of (s, r), i.e.,

$$y(s, r) = r^{-1}x((1 + rq(r))s, r, rp(r)).$$

With a simple computation we see that system (5.4.10) for x is transformed to a system for y

$$r\frac{dy}{ds} = (1 + rq(r))[A(rp(r))ry + g(ry, rp(r))]$$

or

$$\frac{dy}{ds} = (1 + rq(r))[A(rp(r))y + rg^*(y, rp(r))],$$

where $g^*(y, rp(r)) = g(ry, rp(r))/r^2$. Note that g^* is analytic by the assumptions for g. It is easy to see that $g^*(y, 0)$ is a second degree homogeneous polynomial of y. By rearranging the right-hand side of the above equation using (5.4.11), we obtain that

(5.4.14-r)
$$\frac{dy}{ds} = A(0)y + rh(y, r),$$

where

$$h(y, r) = [q(r)A(rp(r)) + p(r)B(rp(r))]y + (1 + rq(r))g^*(y, rp(r)).$$

By (5.4.12) and (5.4.13) we see that system (5.4.1-μ) has a periodic solution with period $T(\mu)$ if and only if the solution of system (5.4.14-r) passing through $(1, 0) \in \mathbb{R}^2$ is a periodic solution with period T_0. We will now use the implicit function theorem to show that for sufficiently small $r > 0$, there exist unique $p(r)$ and $q(r)$ such that the solution of system (5.4.14-r) passing through $(1, 0) \in \mathbb{R}^2$ is a periodic solution with period T_0.

We first note that when $r = 0$, (5.4.14-r) becomes the homogeneous linear system

(5.4.14-0)
$$\frac{dy}{ds} = A(0)y.$$

By Exercise 2.14, the fundamental matrix solution of system (5.4.14-0) is

$$Y(s) = e^{A(0)s} = \exp\left\{ \begin{bmatrix} 0 & -\beta \\ \beta & 0 \end{bmatrix} s \right\} = (\cos \beta s)I + \frac{1}{\beta}(\sin \beta s)A(0).$$

It follows that the components of all solutions of system (5.4.14-0) are linear combinations of $\cos \beta s$ and $\sin \beta s$.

By the variation of parameters formula (2.3.2), the solution of system (5.4.14-r) passing through $(1, 0) \in \mathbb{R}^2$ at $s = 0$ is given by

$$y(s, r) = e^{A(0)s} \begin{bmatrix} 1 \\ 0 \end{bmatrix} + r \int_0^s e^{A(0)(s-\tau)} h(y(\tau, r), r) \, d\tau.$$

It is easy to see that $y(s, r)$ is periodic with period T_0 if and only if

$$\int_0^{T_0} e^{-A(0)\tau} h(y(\tau, r), r) \, d\tau = 0,$$

i.e.,

$$(5.4.15) \quad H(p,q,r) := \int_0^{T_0} e^{-A(0)\tau} [q(r)A(rp(r))y(\tau,r) + p(r)B(rp(r))y(\tau,r)$$
$$+ (1 + rq(r))g^*(y(\tau,r), rp(r))]\, d\tau = 0.$$

We claim that system (5.4.15) determines a unique solution for $p(r)$ and $q(r)$ in a neighborhood of $r = 0$. In fact,

$$H(0,0,0) = \int_0^{T_0} e^{-A(0)\tau} g^*(y(\tau,0),0)\, d\tau = 0.$$

This is due to the facts that $y(\tau,0) = e^{A(0)\tau} \begin{bmatrix} 1 \\ 0 \end{bmatrix}$, $g^*(y,0)$ is a second degree homogeneous polynomial of y, and hence the integrand is a third degree homogeneous polynomial of $\cos \beta s$ and $\sin \beta s$. Then we show that with $H = (H_1, H_2)^T$, $\det \left. \dfrac{\partial(H_1,H_2)}{\partial(p,q)} \right|_{(0,0,0)} \neq 0$. In fact, from (5.4.15)

$$\left. \frac{\partial(H_1,H_2)}{\partial(p,q)} \right|_{(0,0,0)} = \left[\int_0^{T_0} e^{-A(0)\tau} B(0) e^{A(0)\tau} \begin{bmatrix} 1 \\ 0 \end{bmatrix} d\tau, \; \int_0^{T_0} e^{-A(0)\tau} A(0) e^{A(0)\tau} \begin{bmatrix} 1 \\ 0 \end{bmatrix} d\tau \right].$$

Note that $T_0 = 2\pi/\beta$. Then

$$\int_0^{T_0} e^{-A(0)\tau} A(0) e^{A(0)\tau} \begin{bmatrix} 1 \\ 0 \end{bmatrix} d\tau = \frac{2\pi}{\beta} A(0) \begin{bmatrix} 1 \\ 0 \end{bmatrix} = 2\pi \begin{bmatrix} 0 \\ 1 \end{bmatrix}.$$

With $B(0) = \begin{bmatrix} b_{11} & b_{12} \\ b_{21} & b_{22} \end{bmatrix}$, by multiplying the integrand out and doing component-wise integration we find that

$$\int_0^{T_0} e^{-A(0)\tau} B(0) e^{A(0)\tau} \begin{bmatrix} 1 \\ 0 \end{bmatrix} d\tau = \frac{\pi}{\beta} \begin{bmatrix} b_{11} + b_{22} \\ b_{21} - b_{12} \end{bmatrix}.$$

Since $\alpha(\mu) \pm i\beta(\mu)$ are the eigenvalues of $A(\mu)$, from (5.4.11) we see that

$$\alpha(\mu) = \frac{1}{2} \operatorname{tr} A(\mu) = \mu \operatorname{tr} B(\mu)$$

and hence

$$\alpha'(0) = \operatorname{tr} B(0) = (b_{11} + b_{22})/2.$$

Thus

$$\det \left. \frac{\partial(H_1,H_2)}{\partial(p,q)} \right|_{(0,0,0)} = \frac{2\pi^2}{\beta} \det \begin{bmatrix} b_{11} + b_{22} & 0 \\ b_{21} - b_{12} & 1 \end{bmatrix} = \frac{2\pi^2}{\beta}(b_{11} + b_{22})$$

$$= \frac{4\pi^2}{\beta} \alpha'(0) \neq 0.$$

By Lemma 5.1.1, there exist an $r^* > 0$ and unique analytic functions $p(r)$ and $q(r)$ such that

$$H(p(r),q(r),r) \equiv 0 \text{ for } 0 \le r < r^* \quad \text{and} \quad p(0) = 0, \; q(0) = 0.$$

Note that for any $r \in (0, r^*)$, system (5.4.1-μ) for $\mu = rp(r)$ has a closed orbit passing through the point $(r,0)$ with period $T(\mu) = T_0(1 + rq(r))$ if and only if $(p(r),q(r),r)$ is a solution of Eq. (5.4.15). Then from the above discussion, we have the following conclusion: If $p(r) \equiv 0$ on $[0, r^*)$, then $\mu = rp(r) \equiv 0$ on $[0, r^*)$. It follows that all orbits of system (5.4.1-0) near the equilibrium $(0,0)$ are closed orbits. Since $p(r)$ is uniquely determined, system (5.4.1-μ) does not have closed orbits for $\mu \neq 0$ in a neighborhood of $\mu = 0$. If $p(r) \not\equiv 0$ on $[0, r^*)$, then by the Taylor expansion of $p(r)$, $\mu = rp(r)$ is monotone on $[0, r^*)$ for r^* small enough. Then $\mu(r) > 0$ or $\mu(r) < 0$ for $r \in (0, r^*)$. If $\mu(r) > 0$ for $r \in (0, r^*)$, then the solution of system (5.4.1-μ) passing through the point $(r, 0) \in \mathbb{R}^2$ is periodic with period $T(\mu) \to 2\pi/\beta$ as $\mu \to 0$. However, system (5.4.1-μ) has no periodic solutions for $\mu < 0$ sufficiently close to $(0,0)$. Similarly for $\mu(r) < 0$ for $r \in (0, r^*)$. This completes the proof. $\qquad\square$

With the help of Theorem 5.4.2, Corollary 5.4.1 can be strengthened to the following result.

THEOREM 5.4.3. *Assume* (5.4.2) *and* (5.4.4) *hold with* $\alpha'(0) > 0$ *[*$\alpha'(0) < 0$*]*.

(a) *Suppose that the equilibrium* $(0,0)$ *is a stable spiral-point for system* (5.4.1-0). *Then for* $\mu > 0$ *[*$\mu < 0$*] sufficiently close to zero, system* (5.4.1-μ) *has a unique orbitally asymptotically stable limit cycle* $\Gamma(\mu)$ *satisfying* $\Gamma(\mu) \to (0,0)$ *with its period* $T(\mu) \to 2\pi/\beta$ *as* $\mu \to 0$.

(b) *Suppose that the equilibrium* $(0,0)$ *is an unstable spiral-point for system* (5.4.1-0). *Then for* $\mu < 0$ *[*$\mu > 0$*] sufficiently close to zero, system* (5.4.1-μ) *has a unique orbitally unstable limit cycle* $\Gamma(\mu)$ *satisfying* $\Gamma(\mu) \to (0,0)$ *with its period* $T(\mu) \to 2\pi/\beta$ *as* $\mu \to 0$.

PROOF. (a) By the assumptions, system (5.4.1-0) does not have closed orbits in a neighborhood of $(0,0)$. By Theorem 5.4.2, for $\mu > 0$ sufficiently close to zero or for $\mu < 0$ sufficiently close to zero, but not for both, system (5.4.1-μ) has a unique limit cycle $\Gamma(\mu)$ satisfying $\Gamma(\mu) \to (0,0)$ with its period $T(\mu) \to 2\pi/\beta$ as $\mu \to 0$. By Corollary 5.4.1, the limit cycle exists only for $\mu > 0$ [$\mu < 0$] sufficiently close to zero and is orbitally asymptotically stable.

(b) The proof is similar and hence omitted.

$\qquad\square$

To demonstrate the applications of the above results, we give a few examples. The first one is an application of Theorem 5.4.3.

EXAMPLE 5.4.1. Consider the system

(5.4.16-μ)
$$\begin{cases} x_1' = f_1(x_1, x_2, \mu) := x_2 - x_1^3 x_2^2 \\ x_2' = f_2(x_1, x_2, \mu) := -2x_1 + \mu x_2 - x_1^2 x_2^5. \end{cases}$$

Clearly, $f_i(0, 0, \mu) \equiv 0$ for $i = 1, 2$. Let $A(\mu)$ be defined by (5.4.3). Then

$$A(\mu) = \begin{bmatrix} 0 & 1 \\ -2 & \mu \end{bmatrix}$$

and hence its eigenvalues are

$$\alpha(\mu) \pm i\beta(\mu) = \frac{\mu}{2} \pm \frac{\sqrt{8 - \mu^2}}{2} i \quad \text{for } |\mu| < 2\sqrt{2}.$$

Hence

$$\alpha(0) = 0, \quad \beta(0) = \sqrt{2} \neq 0, \quad \text{and} \quad \alpha'(0) = 1/2.$$

It follows that (5.4.2) and (5.4.4) hold with $\beta = \sqrt{2}$ and $\alpha'(0) > 0$.

When $\mu = 0$, system (5.4.16-μ) becomes the system

$$(5.4.16\text{-}0) \qquad \begin{cases} x_1' = x_2 - x_1^3 x_2^2 \\ x_2' = -2x_1 - x_1^2 x_2^5. \end{cases}$$

To see if $\mu = 0$ is a Hopf-bifurcation value, we need to determine the type and stability of the equilibrium $(0, 0)$ for system (5.4.16-0). For this purpose, we utilize the Lyapunov function method. Let $V(x_1, x_2) = 2x_1^2 + x_2^2$. Then V is positive definite on $D = \mathbb{R}^2$, and the derivative of V along system (5.4.16-0)

$$\dot{V}(x_1, x_2) = 4x_1(x_2 - x_1^3 x_2^2) + 2x_2(-2x_1 - x_1^2 x_2^5) = -2x_1^2 x_2^2 (2x_1^2 + x_2^4)$$

which is negative semi-definite on D. Denote

$$D_0 := \{(x_1, x_2) : V(x_1, x_2) = 0\} = \{(x_1, x_2) : x_1 = 0 \text{ or } x_2 = 0\}.$$

For any solution $(x_1(t), x_2(t))$ of system (5.4.16-0) contained in D_0, we have $x_1(t) \equiv 0$ or $x_2(t) \equiv 0$. Then by substituting each of them into system (5.4.16-0), we have $x_1(t) \equiv 0$ and $x_2(t) \equiv 0$. This implies that D_0 does not contain nontrivial orbit of system (5.4.16-0). Therefore by Theorem 3.5.2, the zero solution $(0, 0)$ of system (5.4.16-0) is asymptotically stable. Considering that $(0, 0)$ is a center for the linearized system of (5.4.16-0), by Remark 4.4.1 we know that $(0, 0)$ is a stable spiral-point for system of (5.4.16-0). Now, by applying Theorem 5.4.3 we find that for $\mu > 0$ sufficiently close to zero, system (5.4.16-μ) has a unique orbitally asymptotically stable limit cycle $\Gamma(\mu)$ satisfying $\Gamma(\mu) \to (0, 0)$ with its period $T(\mu) \to 2\pi/\sqrt{2}$ as $\mu \to 0$.

The next example is for an application of Theorem 5.4.1.

EXAMPLE 5.4.2. Consider the system

$$(5.4.17\text{-}\mu) \qquad \begin{cases} x_1' = x_2 + \mu x_1^3 - x_1^3 x_2^2 \\ x_2' = -2x_1 + \mu x_2^3 - x_1^2 x_2^5. \end{cases}$$

Similar to Example 5.4.1, we see that (5.4.2) and (5.4.4) hold with $\beta = \sqrt{2}$. However, since $\alpha'(0) = 0$, Theorem 5.4.3 fails to apply to this equation.

We note that system (5.4.17-0) is the same as system (5.4.16-0). Thus, as shown in Example 5.4.1, the equilibrium $(0, 0)$ is a stable spiral-point for system (5.4.17-0). Here, we use the same method to show that $(0, 0)$ is an unstable spiral-point for system (5.4.17-μ) with $\mu > 0$. In fact, if we let

$V(x_1, x_2) = 2x_1^2 + x_2^2$, then V is positive definite on $D = \mathbb{R}^2$, and for $\mu > 0$, the derivative of V along system (5.4.17-μ)

$$
\begin{aligned}
\dot{V}(x_1, x_2) &= 4x_1(x_2 + \mu x_1^3 - x_1^3 x_2^2) + 2x_2(-2x_1 + \mu x_2^3 - x_1^2 x_2^5) \\
&= 2\mu(2x_1^4 + x_2^4) - 2x_1^2 x_2^2(2x_1^2 + x_2^4) \\
&= (\mu + o(1))(4x_1^4 + 2x_2^4)
\end{aligned}
$$

as $(x_1, x_2) \to (0, 0)$. Thus, $\dot{V}$ is positive definite in a neighborhood of $(0, 0)$. By Theorem 3.5.1, Part (c), $(0, 0)$ is unstable for system (5.4.17-μ). Considering that $(0, 0)$ is a center for the linearized system of (5.4.17-0), by Remark 4.4.1 we know that for $\mu > 0$, $(0, 0)$ is an unstable spiral-point for system (5.4.17-μ). Now, by applying Theorem 5.4.1 we find that for $\mu > 0$ sufficiently close to zero, system (5.4.16-μ) has an orbitally asymptotically stable limit cycle $\Gamma(\mu)$ satisfying $\Gamma(\mu) \to (0, 0)$.

However, we are not able to determine the number of limit cycles and their periods using this method.

The last example shows when the case in Theorem 5.4.2, Part (a) may occurs.

EXAMPLE 5.4.3. Consider the system

(5.4.18-μ)
$$
\begin{cases}
x_1' = x_2 + \mu x_1^2 \\
x_2' = -2x_1 + \mu x_2 - 3x_1^2.
\end{cases}
$$

Similar to Example 5.4.1, we see that (5.4.2) and (5.4.4) hold with $\beta = \sqrt{2}$ and $\alpha'(0) = 1/2 > 0$.

When $\mu = 0$, system (5.4.18-μ) becomes the system

(5.4.18-0)
$$
\begin{cases}
x_1' = x_2 \\
x_2' = -2x_1 - 3x_1^2.
\end{cases}
$$

From Example 3.5.6 we see that all orbits of system (5.4.18-0) in a neighborhood of $(0, 0)$ are closed orbits. Therefore, system (5.4.18-μ) satisfies Theorem 5.4.2, Part (a). As a result, system (5.4.18-μ) does not have closed orbits for $\mu \neq 0$ in a neighborhood of $\mu = 0$.

Finally, we comment that in addition to the approaches for the Hopf-bifurcations by the Lyapunov function method and the Friedrich method as used in this section, there are some other approaches such as the normal form method. See for example, [18, Section 10.5].

Exercises

5.1. Use Theorem 5.2.1 to determine the bifurcations for the following equations:

(a) $x'=(3-\mu)x-x^{\alpha}$ with $\alpha=3,4$;

(b) $x'=-\mu^{\alpha}+x^2+r(x)$ with $\alpha=1,2$, where $r(x)=\begin{cases} x^5\sin(1/x), & x\neq 0, \\ 0, & x=0. \end{cases}$

5.2. For each of the following equations, by sketching the bifurcation diagram, find the bifurcation values and determine their types:

(a) $x' = x(\mu - 2x + x^2)$,

(b) $x' = (x - \mu)(\mu - x^2)$,

(c) $x' = x(x^2 - \mu)(x + \mu - 2)$,

(d) $x' = x(\mu + x - x^2)(1 - \mu + x^2)$.

5.3. Use Theorems 5.2.1 and 5.2.2 to determine the bifurcation at $\mu = 0$ and sketch the bifurcation diagram for each of the following equations:

(a) $x' = x^p(\mu x + x^{2m+1} + r(x,\mu))$,

where $p, m \in \mathbb{N}$ and $r(x,\mu) = o(|x^{2m+1}| + |x\mu| + \mu^2) \in C^{2m+1}(\mathbb{R} \times \mathbb{R}, \mathbb{R})$ as $(x,\mu) \to (0,0)$;

(b) $x' = x^p(\mu \pm x^{2m} + r(x,\mu))$,

where $p, m \in \mathbb{N}$ and $r(x,\mu) = o(x^{2m} + |\mu|) \in C^{2m}(\mathbb{R} \times \mathbb{R}, \mathbb{R})$ as $(x,\mu) \to (0,0)$.

5.4. For the system of equations
$$\begin{cases} x_1' = x_1(\mu - 2x_1) - x_1 x_2 \\ x_2' = x_2(x_1 - 1) + x_2^2, \end{cases}$$
determine the bifurcation points and their types by sketching the bifurcation diagram and determining the stability of the equilibria.

5.5. Determine the one-dimensional bifurcation for the system
$$\begin{cases} x_1' = \mu - x_1^{2m}, \ m \in \mathbb{N}, \\ x_2' = x_2. \end{cases}$$

5.6. Discuss the one-dimensional bifurcation for each of the systems

(a) $\begin{cases} x_1' = \mu x_1 - x_1^{2m+1}, \ m \in \mathbb{N}, \\ x_2' = -x_2; \end{cases}$

(b) $\begin{cases} x_1' = \mu^2 x_1 - x_1^{2m+1}, \ m \in \mathbb{N}, \\ x_2' = x_2. \end{cases}$

5.7. Discuss the one-dimensional bifurcation for each of the systems

(a) $\begin{cases} x_1' = \mu x_1 + x_2 - x_1^6 + x_2^7 \\ x_2' = x_2 - x_1^2 - \sin^4 x_1 + x_2^5; \end{cases}$

(b) $\begin{cases} x_1' = \mu x_1 + x_2 - x_1^6 + x_2^7 \\ x_2' = -x_2 + x_1^3 + \mu \sin x_1 + x_2^5. \end{cases}$

5.8. Use the bifurcation diagram and phase portrait to study the Hopf bifurcation for each of the following systems of polar equations:

(a) $r' = r^3(\mu - r^2)$, $\quad \theta' = -1 + r$;

(b) $r' = \mu r + r^3 - r^5$, $\quad \theta' = 1$.

at $\mu = 0$. Determine the stability of the closed orbits for $\mu > 0$.

5.9. Determine the Hopf bifurcation for each of the following systems:

(a) $\begin{cases} x_1' = -x_2 - x_1 x_2^2 \\ x_2' = x_1 + \mu x_2; \end{cases}$
$\qquad$ (b) $\begin{cases} x_1' = \mu x_1 + 4x_2 - x_1 x_2^2 \\ x_2' = -x_1 + \mu x_2 - x_1^4 x_2. \end{cases}$

5.10. Determine the Hopf bifurcation for each of the following systems:

(a) $\begin{cases} x_1' = x_2 + \mu x_1 \\ x_2' = -\sin x_1; \end{cases}$
$\qquad$ (b) $\begin{cases} x_1' = x_2 + \mu x_1 \\ x_2' = -\sin x_1 - x_2^3; \end{cases}$

(c) $\begin{cases} x_1' = x_2 + \mu x_1 \\ x_2' = -\sin x_1 - \mu x_2^3; \end{cases}$
$\qquad$ (d) $\begin{cases} x_1' = x_2 + \mu x_1 \\ x_2' = -\sin x_1 - \mu(x_2 + x_2^3). \end{cases}$

CHAPTER 6

Second-Order Linear Equations

6.1. Introduction

Second-order linear differential equations have broad applications in science, engineering, and many other fields. Therefore, we will further study the behavior of solutions of second-order linear differential equations and several related problems in this chapter.

The standard form of a second-order nonhomogeneous linear differential equation is

(nh) $$x'' + a_1(t)x' + a_2(t)x = f(t)$$

with corresponding homogeneous linear differential equation

(h) $$x'' + a_1(t)x' + a_2(t)x = 0,$$

where a_1, a_2, $f \in C((a,b), \mathbb{R})$ with $\infty \leq a < b \leq \infty$. However, the most commonly used form of second-order linear differential equations is the so-called self-adjoint form

(nh-s) $$(p(t)x')' + q(t)x = h(t)$$

with corresponding homogeneous linear differential equation

(h-s) $$(p(t)x')' + q(t)x = 0,$$

where

(6.1.1) p, q, $h \in C((a,b), \mathbb{R})$ with $-\infty \leq a < b \leq \infty$ such that $p(t) > 0$.

It is easy to see that if, in addition, $p \in C^1((a,b), \mathbb{R})$, then Eq. (nh-s) can be written in the form of Eq. (nh). However, the condition $p \in C^1((a,b), \mathbb{R})$ is not required here for the general setting of Eq. (h-s). To see the reason, we change Eq. (nh-s) into a system of first-order equations in the following way: Let $x_1 = x$ and $x_2 = px'$. Then Eq. (nh-s) becomes the system

$$\begin{cases} x_1' = \dfrac{1}{p(t)}x_2 \\ x_2' = -q(t)x_1 + h(t), \end{cases}$$

i.e.,

(6.1.2) $$\begin{bmatrix} x_1 \\ x_2 \end{bmatrix}' = \begin{bmatrix} 0 & 1/p(t) \\ -q(t) & 0 \end{bmatrix} \begin{bmatrix} x_1 \\ x_2 \end{bmatrix} + \begin{bmatrix} 0 \\ h(t) \end{bmatrix}.$$

© Springer International Publishing Switzerland 2014
Q. Kong, *A Short Course in Ordinary Differential Equations*, Universitext,
DOI 10.1007/978-3-319-11239-8_6

Note that the ICs associated with Eq. (6.1.2) are of the form

(6.1.3) $$x_1(t_0) = z_1 \quad \text{and} \quad x_2(t_0) = z_2$$

for $t_0 \in (a, b)$ and $z_1, z_2 \in \mathbb{R}$. Naturally, we assign ICs associated with Eq. (nh-s) as follows:

(6.1.4) $$x(t_0) = z_1 \quad \text{and} \quad (px')(t_0) = z_2,$$

where $t_0 \in (a, b)$ and $z_1, z_2 \in \mathbb{R}$. From the relation between Eqs. (nh-s) and (6.1.2) we have the definition below:

DEFINITION 6.1.1. A function $\phi(t)$ is said to be a solution of Eq. (nh-s) on (a, b) if $\phi(t)$ and $(p\phi')(t)$ are differentiable and $\phi(t)$ satisfies Eq. (nh-s) on (a, b). $\phi(t)$ is said to be a solution of IVP (nh-s), (6.1.4) if, in addition, it satisfies IC (6.1.4).

We comment that for a solution $\phi(t)$ of Eq. (nh-s) or of IVP (nh-s), (6.1.4), $\phi'(t)$ need not be differentiable on (a, b), but $(p\phi')(t)$ must be.

We observe that under the assumption (6.1.1), IVP (6.1.2), (6.1.3) has a unique solution on (a, b). As a result, we have the following existence-uniqueness result for IVP (nh-s), (6.1.4).

LEMMA 6.1.1. *Under the assumption (6.1.1), IVP (nh-s), (6.1.4) has a unique solution which exists on the whole interval (a, b).*

If we let $p(t) = 1$ in Eq. (nh-s), we obtain

(6.1.5) $$x'' + q(t)x = h(t).$$

REMARK 6.1.1. (i) Equation (nh) can be transformed to the form of Eq. (nh-s). In fact, multiplying both sides of Eq. (nh) by $e^{\int_{t_0}^{t} a_1(s)\,ds}$ with $t_0 \in (a, b)$, we have

$$\left(e^{\int_{t_0}^{t} a_1(s)\,ds} x'\right)' + e^{\int_{t_0}^{t} a_1(s)\,ds} a_2(t)x = e^{\int_{t_0}^{t} a_1(s)\,ds} f(t)$$

which is in the form of Eq. (nh-s). Therefore, Eq. (nh-s) is a more general form than Eq. (nh).

(ii) Although Eq. (6.1.5) is a special case of Eq. (nh-s), by the independent variable substitution $s = s(t) := \int_{t_0}^{t} 1/p(\tau)\,d\tau$ with $t_0 \in (a, b)$, Eq. (nh-s) can be transformed to the form of Eq. (6.1.5). In fact, $s = s(t)$ is strictly increasing on (a, b) and hence has an inverse $t = t(s)$ which satisfies $dt/ds = (ds/dt)^{-1} = p(t)$. Let $y(s) = x(t(s))$. Then

$$\frac{dy}{ds} = \frac{dx}{dt}\frac{dt}{ds} = p(t)x'.$$

It follows that

$$\frac{d^2 y}{ds^2} = \frac{d}{ds}(p(t)x') = \frac{d}{dt}(p(t)x')\frac{dt}{ds} = p(t)(p(t)x')'.$$

Multiplying both sides of Eq. (nh-s) by $p(t)$ and using the above with $t = t(s)$ we have

$$\frac{d^2y}{ds^2} + p(t(s))q(t(s))y = p(t(s))h(t(s))$$

which is in the form of Eq. (6.1.5). Therefore, in some circumstances, Eq. (6.1.5) can be used to represent the general case of second-order linear equations.

Now, we introduce several properties of the homogeneous linear equation (h-s).

LEMMA 6.1.2. *Let* $\phi(t)$ *be a nontrivial solution of Eq.* (h-s). *Then the zeros of* $\phi(t)$ *do not have accumulation points in its domain* (a, b). *In other words, if* t^* *is an accumulation point of the zeros of* $\phi(t)$, *then either* $t^* = a$ *or* $t^* = b$.

PROOF. Assume the contrary and let $t^* \in (a, b)$ be an accumulation point of the zeros of $\phi(t)$. Then without loss of generality, we may assume there is a sequence of zeros $\{t_n\}_{n=1}^{\infty}$ of $\phi(t)$ in (a, b) with $t_1 < t_2 < \cdots$ such that $t_n \to t^*$. Since $\phi(t_n) = 0$ for all $n \in \mathbb{N}$, we have $\phi(t^*) = 0$. By Rolle's theorem, for each $n \in \mathbb{N}$, there exists a $\tau_n \in (t_n, t_{n+1})$ such that $\phi'(\tau_n) = 0$. Clearly, $\tau_n \to t^*$. It follows that $\phi'(t^*) = 0$ and hence $(p\phi')(t^*) = 0$. Therefore, $\phi(t)$ is the solution of Eq. (h-s) satisfying the IC

$$\phi(t^*) = 0 \quad \text{and} \quad (p\phi')(t^*) = 0.$$

By the uniqueness of solutions of IVPs, $\phi(t) \equiv 0$ on (a, b) which contradicts the assumption that $\phi(t)$ is nontrivial. □

LEMMA 6.1.3. *Let* $\phi_1(t)$ *and* $\phi_2(t)$ *be solutions of Eq.* (h-s). *Then the Wronskian* $W(t)$ *of* $\phi_1(t)$ *and* $\phi_2(t)$ *is given by*

$$(6.1.6) \qquad W(t) = \phi_1(t)\big[p(t)\phi_2'(t)\big] - \big[p(t)\phi_1'(t)\big]\phi_2(t) \equiv c \quad \text{on } (a, b)$$

for some $c \in \mathbb{R}$.

PROOF. We see that $X(t) = \begin{bmatrix} \phi_1 & \phi_2 \\ p\phi_1' & p\phi_2' \end{bmatrix}(t)$ is a matrix solution of Eq. (6.1.2). Let $W(t) = \det X(t)$. Then

$$W(t) = \phi_1(t)(p\phi_2')(t) - (p\phi_1')(t)\phi_2(t).$$

Since $\operatorname{tr}\begin{bmatrix} 0 & 1/p(t) \\ -q(t) & 0 \end{bmatrix} \equiv 0$ on (a, b), by Theorem 2.2.3 we have $W(t) \equiv W(t_0)$ on (a, b). Thus, (6.1.6) holds with $c = W(t_0)$. □

We comment that the Wronskian of solutions of Eq. (h) is not constant on (a, b) in general.

LEMMA 6.1.4. *The set of solutions of Eq.* (h-s) *form a* 2-*dimensional vector space.*

PROOF. Based on the relation between Eq. (h-s) and its corresponding first-order system, this is an immediate consequence of Theorem 2.2.2. □

By Lemma 6.1.4, the solution space of Eq. (h-s) is determined by two linearly independent solutions. However, under certain circumstances, with the help of the "reduction of order method," one nontrivial solution of Eq. (h-s) may determine all solutions of this equation. This is shown by the following lemma.

LEMMA 6.1.5. *Let $\phi(t)$ be a solution of Eq. (h-s) such that $\phi(t) \neq 0$ on (a, b). Let $t_0 \in (a, b)$. Then the general solution of Eq. (h-s) is given by*

$$(6.1.7) \qquad x(t) = c_1 \phi(t) + c_2 \phi(t) \int_{t_0}^{t} \frac{ds}{p(s)\phi^2(s)}, \qquad t \in (a, b).$$

PROOF. Let $x(t)$ be any solution of Eq. (h-s). By Lemma 6.1.3,

$$p(t)[\phi(t)x'(t) - \phi'(t)x(t)] = c_2, \qquad t \in (a, b),$$

for some $c_2 \in \mathbb{R}$. It follows that

$$\frac{\phi(t)x'(t) - \phi'(t)x(t)}{\phi^2(t)} = \frac{c_2}{p(t)\phi^2(t)}, \qquad t \in (a, b).$$

Noting that

$$\frac{\phi(t)x'(t) - \phi'(t)x(t)}{\phi^2(t)} = \left(\frac{x}{\phi}\right)'(t),$$

we have

$$\left(\frac{x}{\phi}\right)'(t) = \frac{c_2}{p(t)\phi^2(t)}, \qquad t \in (a, b).$$

Integrating both sides from t_0 to t we obtain that

$$\frac{x(t)}{\phi(t)} = c_2 \int_{t_0}^{t} \frac{ds}{p(s)\phi^2(s)} + c_1, \qquad t \in (a, b),$$

for some $c_1 \in \mathbb{R}$. This shows that (6.1.7) holds. $\square$

• Finally, we recall that if $\phi_1(t)$ and $\phi_2(t)$ are linearly independent solutions of the homogeneous linear equation (h-s), then the general solution $x(t)$ of the nonhomogeneous linear equation (nh-s) is given by the variation of parameters formula in Corollary 2.3.1. More specifically, by Lemma 6.1.3 we see that

$$(6.1.8) \quad x(t) = c_1 \phi_1(t) + c_2 \phi_2(t) + \int_{t_0}^{t} \frac{h(s)}{c} \left[-\phi_1(t)\phi_2(s) + \phi_2(t)\phi_1(s) \right] ds,$$

where c is a constant given by $c = \phi_1(t)(p\phi_2')(t) - \phi_2(t)(p\phi_1')(t)$ for any $t \in (a, b)$.

6.2. Prüfer Transformation

The Prüfer transformation is an important tool in the study of second-order linear differential equations, especially in the development of the Sturmian oscillation theory and the Sturm–Liouville spectral theory. Simply speaking, the Prüfer transformation transforms a second-order linear equation in the Cartesian coordinates to a system of first-order equations in the polar coordinates.

Consider the homogeneous linear equation

(h-s)
$$(p(t)x')' + q(t)x = 0,$$

where

$$p, q \in C([a, b], \mathbb{R}) \text{ with } -\infty < a < b \leq \infty \text{ such that } p(t) > 0.$$

For a nontrivial solution $x(t)$ of Eq. (h-s), define functions $r, \theta \in C([a, b), \mathbb{R})$ satisfying

(6.2.1)
$$r^2(t) = x^2(t) + (px')^2(t) \quad \text{with } r(t) > 0$$

and

(6.2.2)
$$\tan \theta(t) = \frac{x(t)}{(px')(t)} \quad \text{with } \theta(a) \in [0, \pi).$$

Clearly, $r(t)$ and $\theta(t)$ are uniquely determined on $[a, b)$. We call $r(t)$ the Prüfer distance and $\theta(t)$ the Prüfer angle, of the solution $x(t)$.

From (6.2.1) and (6.2.2),

(6.2.3)
$$\cos \theta(t) = \frac{(px')(t)}{r(t)} \quad \text{and} \quad \sin \theta(t) = \frac{x(t)}{r(t)}.$$

By differentiating (6.2.1) with respect to t and applying Eq. (h-s) we see that for $t \in [a, b)$

$$rr' = xx' + (px')(px')' = xx' - qx(px').$$

Then by (6.2.3)

$$r' = \frac{1}{r}(xx' - qx(px')) = \frac{r}{p} \sin \theta \cos \theta - qr \sin \theta \cos \theta = \left(\frac{1}{p} - q\right) r \sin \theta \cos \theta.$$

By differentiating (6.2.2) with respect to t and applying Eqs. (h-s) and (6.2.3) we see that for $t \in [a, b)$

$$(\sec^2 \theta) \, \theta' = \frac{x'(px') - x(px')'}{(px')^2} = \frac{1}{p} + q\frac{x^2}{(px')^2} = \frac{1}{p} + q \tan^2 \theta.$$

Thus

$$\theta' = \cos^2 \theta \left(\frac{1}{p} + q \tan^2 \theta\right) = \frac{1}{p} \cos^2 \theta + q \sin^2 \theta.$$

Therefore, $r(t)$ and $\theta(t)$ satisfy the following differential equations on $[a, b)$:

$$r' = \left(\frac{1}{p(t)} - q(t)\right) r \sin \theta \cos \theta$$

and

(6.2.4)
$$\theta' = \frac{1}{p(t)} \cos^2 \theta + q(t) \sin^2 \theta.$$

From the definition of $\theta(t)$ we see that $x(\tilde{t}) = 0$ for some $\tilde{t} \Leftrightarrow \theta(\tilde{t}) = 0$ [mod π]. More precisely, we have the following result.

THEOREM 6.2.1. *Let $x(t)$ be a nontrivial solution of Eq. (h-s) on $[a, b]$ with $a < b \leq \infty$ and $\theta(t)$ its Prüfer angle. Assume there exist t_k, $k = 1, \ldots, n$, with $a < t_1 < \cdots < t_n < b$ such that*

$$x(t) = 0 \ \ for \ t = t_k \quad and \quad x(t) \neq 0 \ \ for \ t \neq t_k, \quad k = 1, \ldots, n.$$

Then for $k = 1, \ldots, n$, $\theta(t_k) = k\pi$ and $\theta(t) > k\pi$ for $t > t_k$.

PROOF. We have seen that for $k = 1, \ldots, n$, $x(t_k) = 0 \Leftrightarrow \theta(t_k) = 0$ [mod π]. Note from (6.2.4) that $\theta'(t_k) = 1/p(t_k) > 0$. Since $\theta'(t)$ is continuous on $[a, b)$, $\theta(t)$ is increasing in a neighborhood of t_k. In view of the fact that $\theta(t) \neq 0$ [mod π] for $t \in (t_k, t_{k+1})$ and $\theta'(t_{k+1}) > 0$, by the continuity of $\theta(t)$ on $[a, b)$ we see that

$$\theta(t_{k+1}) = \theta(t_k) + \pi, \quad and \quad \theta(t) > \theta(t_k) \ \ for \ t > t_k.$$

Considering $\theta(a) \in [0, \pi)$, we have $\theta(t_k) = k\pi$. See Fig. 6.1. □

6.3. Sturm Comparison Theorems

The results in this section are on the comparisons of zeros of solutions of two equations in the form of Eq. (h-s).

Consider the two equations

(h-s-1)
$$(p_1(t)x')' + q_1(t)x = 0$$

and

(h-s-2)
$$(p_2(t)x')' + q_2(t)x = 0,$$

where $p_i, q_i \in C([a, b), \mathbb{R})$ with $-\infty < a < b \leq \infty$ such that $p_i(t) > 0$ for $i = 1, 2$.

The following definition is to be used in the comparison theorems.

DEFINITION 6.3.1. *If $p_1(t) \geq p_2(t)$ and $q_1(t) \leq q_2(t)$ on $[a, b)$, then Eq. (h-s-2) is said to be a Sturm-majorant of Eq. (h-s-1);*

If in addition, there exists a $c \in [a, b)$ such that either $\{q_1(c) < q_2(c)\}$ or $\{p_1(c) > p_2(c)$ and $q_2(c) \neq 0\}$, then Eq. (h-s-2) is said to be a strict-Sturm-majorant of Eq. (h-s-1).

The first comparison theorem is on the Prüfer angles of solutions of Eqs. (h-s-1) and (h-s-2).

THEOREM 6.3.1. *Let $x_1(t)$ and $x_2(t)$ be nontrivial solutions of Eqs. (h-s-1) and (h-s-2), respectively, and let $\theta_i(t)$ be the Prüfer angle of $x_i(t)$, $i = 1, 2$.*

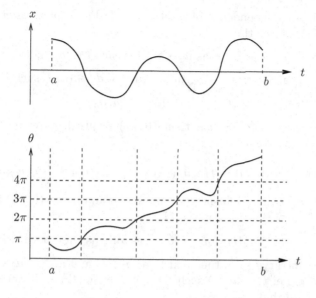

FIGURE 6.1. Prüfer angle

(a) *Assume Eq. (h-s-2) is a Sturm-majorant of Eq. (h-s-1). Then $\theta_1(a)$ $\leq \theta_2(a)$ implies that $\theta_1(t) \leq \theta_2(t)$ for $t \in [a, b)$; and $\theta_1(a) < \theta_2(a)$ implies that $\theta_1(t) < \theta_2(t)$ for $t \in [a, b)$.*

(b) *Assume Eq. (h-s-2) is a strict-Sturm-majorant of Eq. (h-s-1) with $c \in [a, b)$ as defined in Definition 6.3.1. Then $\theta_1(a) \leq \theta_2(a)$ implies that $\theta_1(t) < \theta_2(t)$ for $t \in (c, b)$.*

REMARK 6.3.1. From the proof of Theorem 6.3.1 below, we will see that if $p_i,\ q_i \in C([a, b], \mathbb{R})$ with $-\infty < a < b < \infty$, then all conclusions in Theorem 6.3.1 also hold at b.

To prove this theorem, we need the lemma below for differential inequalities.

LEMMA 6.3.1 (Differential Inequality). *Assume $f(t, x) \in C([a, b) \times \mathbb{R}, \mathbb{R})$ for $-\infty < a < b \leq \infty$ and is locally Lipschitz in x. Suppose that*

(i) *$u(t)$ is the unique solution of the IVP $u' = f(t, u)$, $u(a) = u_0$; and*
(ii) *$v(t)$ is a solution of the IVP $v' \leq f(t, v)$, $v(a) \leq u_0$.*

Then $v(t) \leq u(t)$ for $t \geq a$ as long as both exist.

PROOF. Assume the maximal interval of existence of $u(t)$ is $[a, d)$ for some $d \in (a, b]$. For any $n \in \mathbb{N}$, let $u_n(t)$ be the solution of the IVP

$$u' = f(t, u) + \frac{1}{n}, \quad u(a) = u_0.$$

By the continuous dependence of solutions of IVPs on parameters given by Theorem 1.5.2, we see that

$$u_n(t) \to u(t) \quad \text{as } n \to \infty \quad \text{for each } t \in [a, d).$$

We first show that for any $n \in \mathbb{N}$, there exists a $d_n \in (a, d]$ such that

(6.3.1) $v(t) \le u_n(t) \quad \text{for } t \in [a, d_n).$

In fact, if $v(a) < u_n(a) = u_0$, then (6.3.1) is justified by continuity. If $v(a) = u_n(a) = u_0$, then

$$v'(a) \le f(a, v(a)) = f(a, u_n(a)) < f(a, u_n(a)) + \frac{1}{n} = u_n'(a).$$

Thus, $v'(t) < u_n'(t)$ in a right-neighborhood of a and hence (6.3.1) is justified. Next, we show that for any $n \in \mathbb{N}$,

(6.3.2) $v(t) \le u_n(t) \quad \text{for } t \in [a, d)$ as long as they both exist.

Assume the contrary. Then there exists an $n \in \mathbb{N}$ and a $t^* \in [a, d)$ such that $v(t^*) > u_n(t^*)$. Let $t_* \in [a, t^*)$ such that $v(t_*) = u_n(t_*)$ and $v(t) < u_n(t)$ for $t \in (t_*, t^*)$. Note that

$$v'(t_*) \le f(t_*, v(t_*)) = f(t_*, u_n(t_*)) < f(t_*, u_n(t_*)) + \frac{1}{n} = u_n'(t_*).$$

This implies that $v(t) < u_n(t)$ in a right-neighborhood of t_*, which contradicts the definition of t_* and hence confirms (6.3.2).

By taking the limit as $n \to \infty$ in (6.3.2) we complete the proof.   □

PROOF OF THEOREM 6.3.1. (a) For $i = 1, 2$, let

$$f_i(t, \theta) = \frac{1}{p_i(t)} \cos^2 \theta + q_i(t) \sin^2 \theta.$$

Then for $i = 1, 2$, $\theta_i(t)$ is the unique solution of IVP consisting of the equation

$$\theta_i'(t) = f_i(t, \theta), \quad t \in [a, b),$$

and the given IC at a. By the assumptions, $\theta_1(t)$ satisfies

$$\theta_1' = f_1(t, \theta_1) \le f_2(t, \theta_1).$$

Assume $\theta_1(a) \le \theta_2(a)$. Then using Lemma 6.3.1 with $u = \theta_2$ and $v = \theta_1$ we obtain that $\theta_1(t) \le \theta_2(t)$, $t \in [a, b)$.

Assume $\theta_1(a) < \theta_2(a)$. Let $x_3(t)$ be a nontrivial solution of Eq. (h-s-1) such that its Prüfer angle $\theta_3(t)$ satisfies $\theta_3(a) = \theta_2(a)$. Then $\theta_3(t) \le \theta_2(t)$ for $t \in [a, b)$ as above, and $\theta_1(a) < \theta_3(a)$. We claim that $\theta_1(t) < \theta_3(t)$ for $t \in [a, b)$. Otherwise, there exists a $\tilde{t} \in [a, b)$ such that $\theta_1(\tilde{t}) = \theta_3(\tilde{t})$. Then

$$\frac{u_1(\tilde{t})}{(p_1 u_1')(\tilde{t})} = \frac{u_3(\tilde{t})}{(p_1 u_3')(\tilde{t})} \quad \text{or} \quad (p_1 u_1')(\tilde{t}) = (p_1 u_3')(\tilde{t}).$$

which means that the Wronskian of u_1 and u_3 satisfies $W[u_1, u_3](\tilde{t})$ $= 0$, and hence u_1 and u_3 are linearly dependent on $[a, b)$. As a result, $\theta_1(t) \equiv \theta_3(t)$ on $[a, b)$. This contradicts the assumption that $\theta_1(a) < \theta_3(a)$.

(b) Since strict-Sturm-majorance guarantees Sturm-majorance, by Part (a), we have that $\theta_1(t) \leq \theta_2(t)$, $t \in [a, b)$. Assume the conclusion of Part (b) does not hold. Then there exists a $d \in (c, b)$ such that $\theta_1(d) = \theta_2(d)$.

We claim that $\theta_1(t) \equiv \theta_2(t)$ on $[a, d]$. Otherwise, there exists a $\bar{t} \in (a, d]$ such that $\theta_1(\bar{t}) < \theta_2(\bar{t})$. Applying Part (a) we see that $\theta_1(t) < \theta_2(t)$ for $t \in [\bar{t}, b)$ which contradicts the assumption that $\theta_1(d) = \theta_2(d)$.

Let $\theta(t) := \theta_i(t)$ for $t \in [a, d]$ and $i = 1, 2$. Then for $t \in [a, d]$,

(6.3.3)
$$0 = (\theta_1(t) - \theta_2(t))' = \left(\frac{1}{p_1(t)} - \frac{1}{p_2(t)} \right) \cos^2 \theta(t) + (q_1(t) - q_2(t)) \sin^2 \theta(t).$$

Note that Eq. (h-s-2) is a strict-Sturm-majorant of Eq. (h-s-1) with $c \in [a, b)$ implies that on a right-neighborhood $\mathcal{N}$ of c in $[a, d]$, either $\{q_1(t) < q_2(t)\}$ or $\{p_1(t) > p_2(t)$ and $q_2(t) \neq 0\}$.

If $q_1(t) < q_2(t)$ on $\mathcal{N}$, then by (6.3.3), we must have $\sin \theta(t) \equiv 0$ on $\mathcal{N}$ and hence $\theta_i(t) \equiv 0$ [mod π] on $\mathcal{N}$. Thus $x_i(t) \equiv 0$ on $\mathcal{N}$, and hence $x_i(t) \equiv 0$ on $[a, b)$ by the uniqueness argument. If $p_1(t) > p_2(t)$ and $q_2(t) \neq 0$, then by (6.3.3), we must have $\cos \theta(t) \equiv 0$ on $\mathcal{N}$. It follows that $\theta_i(t) \equiv \pi/2$ [mod π] on $\mathcal{N}$. Thus $(px_2')(t) \equiv 0$ on $\mathcal{N}$. Applying Eq. (h-s) we obtain that $q(t)x_2(t) \equiv 0$ on $\mathcal{N}$. Note that $q_2(t) \neq 0$ on $\mathcal{N}$, we have that $x_2(t) \equiv 0$ on $\mathcal{N}$, and hence $x_2(t) \equiv 0$ on $[a, b)$ by the uniqueness argument. In both cases, we have reached a contradiction.

$\square$

REMARK 6.3.2. In Definition 6.3.1, for Eq. (h-s-2) to be a Sturm-majorant of Eq. (h-s-1), the condition that $p_1(c) > p_2(c)$ and $q_2(c) \neq 0$ cannot simply be replaced by $p_1(c) > p_2(c)$. In fact, if $p_1(c) > p_2(c)$ and $q_1(c) = q_2(c)$, then Theorem 6.3.1, Part (b) may fail to apply. For example, let $p_1(t) = 2$ and $p_2(t) = 1$ on $[0, \infty)$. For $i = 1, 2$, let $x_i(t)$ be solutions of the equations $(p_i u')' = 0$ with Prüfer angle $\theta_i(t)$ satisfying $\theta_i(0) = \pi/2$. Thus $(p_i u')(t)$ is constant on $[0, \infty)$ which implies that $\theta_i(t) = \pi/2$ on $[0, \infty)$. Therefore, $\theta_1(t) \equiv \theta_2(t)$ on $[0, \infty)$.

Based on theorem 6.3.1, we obtain a comparison theorem on the numbers of zeros of solutions of Eqs. (h-s-1) and (h-s-2).

THEOREM 6.3.2 (Sturm Comparison Theorem). *Let* $x_1(t)$ *be a nontrivial solution of Eq.* (h-s-1) *with two consecutive zeros* t_1 *and* t_2 *in* $[a, b)$ *such that* $t_1 < t_2$.

(a) *Assume Eq. (h-s-2) is a Sturm-majorant of Eq. (h-s-1). Then every solution of Eq. (h-s-2) has at least one zero in $(t_1, t_2]$.*

(b) *Assume Eq. (h-s-2) is a strict-Sturm-majorant of Eq. (h-s-1). Then every solution of Eq. (h-s-2) has at least one zero in (t_1, t_2).*

PROOF. (a) Let $x_2(t)$ be any nontrivial solution of Eq. (h-s-2). For $i = 1, 2$, let $\theta_i(t)$ be the Prüfer angle of $x_i(t)$. By Theorem 6.2.1, without loss of generality, we may assume that $\theta_1(t_1) = 0$, $\theta_1(t_2) = \pi$, and $\theta_2(t_1) = \alpha \in [0, \pi)$. By Theorem 6.3.1, Part (a), $\theta_2(t_2) \geq \theta_1(t_2) = \pi$. Since $\theta_2(t)$ is continuous on $[t_1, t_2]$, there exists a $\tau \in (t_1, t_2]$ such that $\theta(\tau) = \pi$. Therefore, τ is a zero of $x_2(t)$.

(b) This can be proved similarly using Theorem 6.3.1, Part (b).

$\square$

Applying Theorem 6.3.2 to the case when Eqs. (h-s-1) and (h-s-2) are the same, we obtain a separation theorem for solutions of the equation

(h-s) $$(p(t)x')' + q(t)x = 0,$$

where

$$p, q \in C([a, b), \mathbb{R}) \text{ with } -\infty < a < b \leq \infty \text{ such that } p(t) > 0.$$

THEOREM 6.3.3 (Sturm Separation Theorem). *Let $x_1(t)$ and $x_2(t)$ be two linearly independent solutions of Eq. (h-s). Then between each pair of consecutive zeros of $x_1(t)$ in $[a, b)$, there is a zero of $x_2(t)$, and vice versa.*

PROOF. Note that Eq. (h-s) is a Sturm-majorant of itself. Let t_1, t_2 are consecutive zeros of $x_1(t)$ in $[a, b)$ such that $t_1 < t_2$. By Theorem 6.3.2, $x_2(t)$ must have at least one zero $\tilde{t}$ in $(t_1, t_2]$. We claim that $\tilde{t} \neq t_2$. Otherwise, $x_1(t)$ and $x_2(t)$ are linearly dependent, contradicting the assumption. Therefore, $\tilde{t} \in (t_1, t_2)$. The vice versa part can be proved in the same way. $\square$

For further discussions on Sturmian theory, see Reid [42].

Now we apply the Sturm comparison theorem to study the distributions of zeros of solutions of the equation

(6.3.4) $$x'' + q(t)x = 0,$$

where $q \in C([0, \infty), \mathbb{R})$.

COROLLARY 6.3.1. *(a) Assume $q(t) \leq 0$ for $t \in [0, \infty)$. Then every nontrivial solution of Eq. (6.3.4) has at most one zero in $[0, \infty)$.*

(b) *Assume $0 < m_1 \leq q(t) \leq m_2 < \infty$ for $t \in [0, \infty)$. Then every nontrivial solution of Eq. (6.3.4) has a countably infinite number of zeros $\{t_n\}_{n=0}^{\infty}$ in $[0, \infty)$ such that*

(6.3.5) $$\frac{\pi}{\sqrt{m_2}} \leq t_{n+1} - t_n \leq \frac{\pi}{\sqrt{m_1}}.$$

(c) *Assume* $\lim_{t\to\infty} q(t) = m$ *for some* $m > 0$. *Then every nontrivial solution of Eq.* (6.3.4) *has a countably infinite number of zeros* $\{t_n\}_{n=0}^{\infty}$ *in* $[0, \infty)$ *such that*

$$\lim_{n\to\infty} (t_{n+1} - t_n) = \frac{\pi}{\sqrt{m}}.$$

PROOF. (a) We compare Eq. (6.3.4) with the equation

(6.3.6) $$x'' = 0.$$

From the condition we see that Eq. (6.3.6) is a Sturm-majorant of Eq. (6.3.4). Assume Eq. (6.3.4) has a nontrivial solution $x_1(t)$ with two zeros $t_1, t_2 \in [0, \infty)$ such that $t_1 < t_2$. By Theorem 6.3.2, every solution of Eq. (6.3.6) has at least one zero in $(t_1, t_2]$. However, $x_2(t) \equiv 1$ is a solution of Eq. (6.3.6) which has no zero in $(t_1, t_2]$. We have reached a contradiction.

(b) We first compare Eq. (6.3.4) with the equation

(6.3.7) $$x'' + m_1 x = 0.$$

The general solution of Eq. (6.3.7) is

$$x_1(t) = A \sin(\sqrt{m_1}\, t - \phi),$$

where $A \geq 0$ and $\phi \in [0, 2\pi)$. It is easy to see that x_1 has zeros $T_k = (k\pi + \phi)/\sqrt{m_1}$ for all $k \in \mathbb{N}_0$. Thus, $T_{k+1} - T_k = \pi/\sqrt{m_1}$.

From the condition we see that Eq. (6.3.4) is a Sturm-majorant of Eq. (6.3.7). Let $x(t)$ be any nontrivial solution of Eq. (6.3.4). By Theorem 6.3.2, $x(t)$ has at least one zero in $(T_k, T_{k+1}]$ for each $k \in \mathbb{N}_0$. Thus, $x(t)$ has an infinite number of zeros on $[0, \infty)$. Moreover, from Lemma 6.1.2 we see that the zeros of $x(t)$ do not have accumulation points in $[0, \infty)$. Thus, the zeros of $x(t)$ in $[0, \infty)$ form a countably infinite set, denoted by $\{t_n\}_{n=0}^{\infty}$, which satisfies $t_0 < t_1 < t_2 < \cdots$.

Then we show that $t_{n+1} - t_n \leq \pi/\sqrt{m_1}$ for all $n \in \mathbb{N}_0$. Assume the contrary, i.e., $t_{n+1} - t_n > \pi/\sqrt{m_1}$ for some $n \in \mathbb{N}_0$. Then we may choose $\phi \in [0, 2\pi)$ such that $x_1(t)$ has a zero $T_k = t_n$ for some $k \in \{0, 1, \dots\}$. Hence $T_{k+1} < t_{n+1}$. Therefore, $x(t) \neq 0$ on $(T_k, T_{k+1}]$. This contradicts the fact that Eq. (6.3.4) is a Sturm majorant of Eq. (6.3.7).

The other side of (6.3.5) can be obtained by comparing Eq. (6.3.5) with the equation

$$x'' + m_2 x = 0.$$

We omit the details.

(c) For any $\epsilon \in (0, m)$, there exits a $T \geq 0$ such that $m - \epsilon \leq q(t) \leq m + \epsilon$ for $t \geq T$. By Part (b) we see that

$$\frac{\pi}{\sqrt{m + \epsilon}} \leq t_{n+1} - t_n \leq \frac{\pi}{\sqrt{m - \epsilon}}, \quad t_n \geq T.$$

Taking the limits as $n \to \infty$ and using the Squeeze theorem we obtain that $t_{n+1} - t_n \to \pi/\sqrt{m}$ as $n \to \infty$.

$\square$

REMARK 6.3.3. In view of Remark 6.1.1, Part (b), when $\int_0^\infty 1/p(t)\,dt = \infty$, Eq. (h-s) can be transformed to the form of Eq. (6.3.4) with a new q still defined on $[0,\infty)$. In this case, the results in Corollary 6.3.1 can be transformed back to Eq. (h-s). We leave the work to the reader.

6.4. Nonoscillation and Oscillation

In this section we study the nonoscillation and oscillation problems for the homogeneous linear differential equation

(h-s) $$(p(t)x')' + q(t)x = 0,$$

where

$$p, q \in C([a,b), \mathbb{R}) \text{ with } -\infty \le a < b \le \infty \text{ such that } p(t) > 0.$$

From the Sturm separation theorem we see that if one nontrivial solution of Eq. (h-s) has infinitely many zeros in $[a,b)$, then every nontrivial solution of Eq. (h-s) has infinitely many zeros in $[a,b)$.

DEFINITION 6.4.1. Equation (h-s) is said to be oscillatory on $[a,b)$ if one nontrivial solution, and hence every nontrivial solution, has infinitely many zeros in $[a,b)$.

Equation (h-s) is said to be nonoscillatory on $[a,b)$ otherwise, i.e., if one nontrivial solution, and hence every nontrivial solution, has finitely many zeros in $[a,b)$.

REMARK 6.4.1. From the definition and Lemma 6.1.2 we see that

(i) if Eq. (h-s) is oscillatory on $[a,b)$ and $x(t)$ is a nontrivial solution of Eq. (h-s) with zeros $\{t_n\}_{n=1}^\infty \subset [a,b)$, then $t_n \to b$ as $n \to \infty$;

(ii) if Eq. (h-s) is nonoscillatory on $[a,b)$ and $x(t)$ is a nontrivial solution of Eq. (h-s), then there exists a $T \in [a,b)$ such that $x(t) \ne 0$ on $[T,b)$.

The following theorem and definition provide a classification of the nontrivial solutions of Eq. (h-s) when it is nonoscillatory.

THEOREM 6.4.1. *Assume Eq. (h-s) is nonoscillatory on $[a,b)$. Then there exists a nontrivial solution $x_0(t)$ of Eq. (h-s), which is unique up to a constant multiple, such that for any solution $x_1(t)$ linearly independent of $x_0(t)$ we have*

(a) $\lim_{t \to b-} \dfrac{x_0(t)}{x_1(t)} = 0,$ *and*

(b) $\int_T^b \dfrac{1}{p(t)x_0^2(t)}\,dt = \infty$ *and* $\int_T^b \dfrac{1}{p(t)x_1^2(t)}\,dt < \infty$ *for some* $T \in [a,b)$.

It is easy to see that the nontrivial solution $x_0(t)$ in Theorem 6.4.1 is determined uniquely, up to a constant multiple, by each of conditions (a) and (b).

DEFINITION 6.4.2. A nontrivial solution $x_0(t)$ of Eq. (h-s) satisfying condition (a) or (b) in Theorem 6.4.1 is called a principal solution of Eq. (h-s) at b, and any solution which is linearly independent of $x_0(t)$ is called a nonprincipal solution Eq. (h-s) at b.

REMARK 6.4.2. From the definition, the principal solution of Eq. (h-s) at b means the "smallest solution" of Eq. (h-s) as t is sufficiently close to b in certain sense.

PROOF OF THEOREM 6.4.1. Let $u(t)$ and $v(t)$ be linearly independent solutions of Eq. (h-s) on $[a, b)$. Then by Remark 6.3.3, there exists a $T \in [a, b)$ such that $u(t) \neq 0$ and $v(t) \neq 0$ on $[T, b)$. From Lemma 6.1.3, we have that for $t \in [T, b)$,

$$p(t)[u'(t)v(t) - u(t)v'(t)] = c$$

for some $c \neq 0$. Thus

(6.4.1)
$$\left(\frac{u(t)}{v(t)}\right)' = \frac{c}{p(t)v^2(t)} \neq 0,$$

which means that $u(t)/v(t)$ is monotone on $[T, b)$. Therefore,

$$\lim_{t \to b-} \frac{u(t)}{v(t)} = k \in \bar{\mathbb{R}} := [-\infty, \infty] \text{ (the extended reals)}.$$

We claim that we can make $k = 0$ by choosing appropriate u and v. In fact, if $k = \pm\infty$, then by interchanging u and v we have that $\lim_{t \to b-} u(t)/v(t) = 0$; if $k \in (-\infty, \infty)$, then by letting $\tilde{u} = u - kv$ we have that $\tilde{u}(t)$ is also a solution of Eq. (h-s) and $\lim_{t \to b-} \tilde{u}(t)/v(t) = 0$. Now, we assume that $u(t)$ and $v(t)$ are linearly independent solutions of Eq. (h-s) satisfying

(6.4.2)
$$\lim_{t \to b-} u(t)/v(t) = 0.$$

Denote $x_0(t) := u(t)$ for the solution $u(t)$ satisfying (6.4.2).

(a) For any solution $x(t)$ of Eq. (h-s) linearly independent of $x_0(t)$,

$$x(t) = c_1 u(t) + c_2 v(t)$$

for some $c_1, c_2 \in \mathbb{R}$ with $c_2 \neq 0$. This shows that

$$\frac{x_0(t)}{x(t)} = \frac{u(t)}{c_1 u(t) + c_2 v(t)} \to 0 \text{ as } t \to b-.$$

(b) Let $x(t)$ be any solution of Eq. (h-s) linearly independent of $x_0(t)$. Then by applying (6.4.1) to x_0 and x we have

$$\frac{x_0(t)}{x(t)} = \frac{x_0(T)}{x(T)} + c \int_T^t \frac{1}{p(s)x^2(s)} ds.$$

In view of the fact that $x_0(t)/x(t) \to 0$ as $t \to b-$, we have $\int_T^b 1/(p(s)x^2(s))\,ds < \infty$.

On the other hand, by applying (6.4.1) to x and x_0, we have

$$\frac{x(t)}{x_0(t)} = \frac{x(T)}{x_0(T)} + c\int_T^t \frac{1}{p(s)x_0^2(s)}\,ds.$$

In view of the fact that $x(t)/x_0(t) \to \infty$ as $t \to b-$, we have $\int_T^b 1/(p(s)x_0^2(s))\,ds = \infty$.

$\square$

The next theorem shows how to construct nonprincipal and principal solutions at b when Eq. (h-s) is nonoscillatory on $[a, b)$.

THEOREM 6.4.2. *Assume Eq. (h-s) is nonoscillatory on $[a, b)$.*

(a) *Let $x(t)$ be any solution of Eq. (h-s) such that $x(t) \neq 0$ on $[T, b)$ for some $T \in (a, b)$. Then*

(6.4.3) $$x_1(t) := x(t)\int_T^t \frac{1}{p(s)x^2(s)}\,ds, \quad t \in [T, b),$$

is a nonprincipal solution of Eq. (h-s) at b.

(b) *Let $x(t)$ be any nonprincipal solution of Eq. (h-s) such that $x(t) \neq 0$ on $[T, b)$ for some $T \in (a, b)$. Then*

(6.4.4) $$x_0(t) := x(t)\int_t^b \frac{1}{p(s)x^2(s)}\,ds, \quad t \in [T, b),$$

is a principal solution of Eq. (h-s) at b.

PROOF. For any solution $x(t)$ of Eq. (h-s) such that $x(t) \neq 0$ on $[T, b)$, by Lemma 6.1.5 with $t_0 = T$, $c_1 = 0$, and $c_2 = 1$, the function $x_1(t)$ as defined in (6.4.3) is a solution of Eq. (h-s). Obviously, $x_1(t)$ and $x(t)$ are linearly independent on $[T, b)$.

(a) Assume $x(t)$ is a principal solution. Since $x_1(t)$ and $x(t)$ are linearly independent on $[T, b)$, by Theorem 6.3.3, $x_1(t)$ is a nonprincipal solution of Eq. (h-s). Assume $x(t)$ is a nonprincipal solution. By Theorem 6.3.3, $\int_T^b 1/(p(t)x^2(t))\,dt < \infty$. Let $x_0(t) = x(t) - kx_1(t)$ with $k = \left(\int_T^b 1/(p(t)x^2(t))\,dt\right)^{-1}$. Then

$$\frac{x_0(t)}{x_1(t)} = \frac{x(t)}{x_1(t)} - k = \left(\int_T^t \frac{1}{p(s)x^2(s)}\,ds\right)^{-1} - k \to 0 \text{ as } t \to b-.$$

This implies that $x_0(t)$ is a principal solution of Eq. (h-s) at b and hence $x_1(t)$ is a nonprincipal solution of Eq. (h-s) at b since it is linearly independent of $x_0(t)$.

(b) Note that $x(t)$ is a nonprincipal solution of Eq. (h-s) at b means that $\int_t^b 1/(p(s)x^2(s))\,ds < \infty$ for any $t \in [T,b)$. Thus

$$x_0(t) = x(t) \int_t^b \frac{1}{p(s)x^2(s)}\,ds = x(t) \int_T^b \frac{1}{p(s)x^2(s)}\,ds - x(t) \int_T^t \frac{1}{p(s)x^2(s)}\,ds$$

is also a solution of Eq. (h-s), and

$$\frac{x_0(t)}{x(t)} = \int_t^b \frac{1}{p(s)x^2(s)}\,ds \to 0 \quad \text{as } t \to b-.$$

For any solution $\tilde{x}(t)$ of Eq. (h-s) linearly independent of $x_0(t)$, $\tilde{x}(t) = c_1 x_0(t) + c_2 x(t)$ with $c_2 \neq 0$. Then

$$\frac{x_0(t)}{\tilde{x}(t)} = \frac{x_0(t)}{c_1 x_0(t) + c_2 x(t)} \to 0 \quad \text{as } t \to b-.$$

Therefore, $x_0(t)$ is a principal solution of Eq. (h-s) at b.

$\square$

Now, we present some oscillation criteria for the equation

(6.4.5) $$x'' + q(t)x = 0$$

on the interval $[0, \infty)$, where $q \in C([0, \infty), \mathbb{R})$. As an immediate consequence of Corollary 6.3.1 we have

THEOREM 6.4.3. (a) *Assume* $q(t) \leq 0$ *for sufficiently large* t. *Then Eq.* (6.4.5) *is nonoscillatory.*

(b) *Assume* $q(t) \geq m > 0$ *for sufficiently large* t. *Then Eq.* (6.4.5) *is oscillatory.*

COROLLARY 6.4.1. (a) *Assume* $\limsup_{t\to\infty} q(t) < 0$. *Then Eq.* (6.4.5) *is nonoscillatory.*

(b) *Assume* $\liminf_{t\to\infty} q(t) > 0$. *Then Eq.* (6.4.5) *is oscillatory.*

PROOF. This follows from Theorem 6.4.3 easily by the definitions of the upper and lower limits. $\square$

Then we discuss the cases when either $q(t) \geq 0$ and can be arbitrarily close to zero or $q(t)$ is oscillatory on $[0, \infty)$.

THEOREM 6.4.4 (Fite–Wintner [16, 47]). *Assume*

(6.4.6) $$\int_0^\infty q(t)\,dt = \infty.$$

Then Eq. (6.4.5) *is oscillatory.*

PROOF. Assume the contrary, i.e., Eq. (6.4.5) is nonoscillatory on $[0, \infty)$. Then for any nontrivial solution $x(t)$ of Eq. (6.4.5), there exists a $T \geq 0$ such that $x(t) \neq 0$ on $[T, \infty)$. Define

(6.4.7) $$y(t) = \frac{x'(t)}{x(t)} \quad \text{for } t \in [T, \infty).$$

Then for $t \in [T, \infty)$, $y(t)$ satisfies the Riccati equation

$$(6.4.8) \qquad y'(t) \;=\; \frac{x''(t)x(t) - (x'(t))^2}{x^2(t)}$$

$$= \frac{-q(t)x^2(t) - (x'(t))^2}{x^2(t)} = -q(t) - y^2(t).$$

Replacing t by s in the above and integrating it from T to t we have that for $t \in [T, \infty)$

$$(6.4.9) \qquad y(t) = y(T) - \int_T^t q(s)\,ds - \int_T^t y^2(s)\,ds$$

and hence $y(t) \to -\infty$ as $t \to \infty$ since $\int_0^\infty q(t)\,dt = \infty$. This means that we can make T sufficiently large such that $y(t) < 0$, i.e., $x'(t)/x(t) < 0$ for $t \in [T, \infty)$. It follows that $x'(t) < 0$ and hence $x(t)$ is decreasing on $[T, \infty)$. Note from Theorem 6.4.1 and Definition 6.4.2 that $x(t)$ can be chosen as a nonprincipal solution at ∞. Then as $t \in [T, \infty)$ and $t \to \infty$,

$$\infty \leftarrow \int_T^t ds \leq \int_T^t \frac{x^2(T)}{x^2(s)}\,ds = x^2(T) \int_T^t \frac{1}{x^2(s)}\,ds < x^2(T) \int_T^\infty \frac{1}{x^2(s)}\,ds < \infty.$$

We have reached a contradiction. $\square$

REMARK 6.4.3. We point out that in Theorem 6.4.4, in order for Eq. (6.4.5) to be oscillatory, $q(t)$ is allowed to change sign on $[0, \infty)$, as long as $Q(t) := \int_0^t q(t)\,dt$ grows unbounded and approaches ∞ as $t \to \infty$. However, when $Q(t)$ remains bounded above, even if $q(t) > 0$ on $[0, \infty)$, the oscillation property of Eq. (6.4.5) may be lost.

EXAMPLE 6.4.1. Consider the Euler equation

$$(6.4.10) \qquad\qquad x'' + \frac{\mu}{t^2}x = 0 \quad \text{on } [1, \infty),$$

where $\mu > 0$ is a parameter. Here we replace the interval $[0, \infty)$ in Theorem 6.4.4 by $[1, \infty)$ to avoid the singularity of the function $q(t) = \mu/t^2$ at 0. We note that $\int_1^\infty (\mu/t^2)\,dt = \mu < \infty$ for any $\mu > 0$. Hence condition (6.4.6) is not satisfied.

It is easy to see that Eq. (6.4.10) is solvable since it can be changed to the equation with constant coefficients

$$(6.4.11) \qquad\qquad \frac{d^2x}{ds^2} - \frac{dx}{ds} + \mu x = 0$$

by the independent variable substitution $t = e^s$. Clearly, the roots of the characteristic equation for Eq. (6.4.11) are $r_{1,2} = (1 \pm \sqrt{\beta})/2$ with $\beta = 1 - 4\mu$. By solving Eq. (6.4.11) and using the relation $s = \ln t$, we obtain the general solution of Eq. (6.4.10)

$$x = \begin{cases} c_1 t^{r_1} + c_2 t^{r_2}, & \mu < 1/4; \\ t^{1/2}(c_1 + c_2 \ln t), & \mu = 1/4; \\ t^{1/2}\Big(c_1 \cos\big(\ln((\sqrt{-\beta}/2)t)\big) + c_2 \sin\big(\ln((\sqrt{-\beta}/2)t)\big)\Big), & \mu > 1/4. \end{cases}$$

This means that Eq. (6.4.10) is nonoscillatory for $\mu \in (0, 1/4]$ and oscillatory for $\mu \in (1/4, \infty)$.

The next result gives an improvement of Theorem 6.4.4.

THEOREM 6.4.5 (Wintner [47]). *Let* $Q(t) := \int_0^t q(t) \, dt$. *Assume*

$$(6.4.12) \qquad \lim_{t \to \infty} \frac{1}{t} \int_0^t Q(s) \, ds = \infty.$$

Then Eq. (6.4.5) is oscillatory.

PROOF. It is easy to check that condition (6.4.12) is equivalent to

$$(6.4.13) \qquad \lim_{t \to \infty} \frac{1}{t} \int_T^t \int_T^s q(\tau) \, d\tau ds = \infty$$

for any $T \in [0, \infty)$.

Assume the contrary, i.e., Eq. (6.4.5) is nonoscillatory on $[0, \infty)$. Then for any nontrivial solution $x(t)$ of Eq. (6.4.5), there exists a $T \geq 0$ such that $x(t) \neq 0$ on $[T, \infty)$. Without loss of generality, assume $x(t) > 0$ on $[T, \infty)$. Define $y(t)$ by (6.4.7). Then as shown in the proof of Theorem 6.4.4, (6.4.9) holds and hence

$$y(s) \leq y(T) - \int_T^s q(\tau) \, d\tau, \quad s \in [T, \infty).$$

Integrating the above inequality from T to t and using (6.4.7), we obtain that for $t \geq T$

$$\ln x(t) \leq \ln x(T) + y(T)(t - T) - \int_T^t \int_T^s q(\tau) \, d\tau ds.$$

Thus by (6.4.13),

$$\frac{1}{t} \ln x(t) \leq \frac{\ln x(T) + y(T)(t - T)}{t} - \frac{1}{t} \int_T^t \int_T^s q(\tau) \, d\tau ds \to -\infty.$$

As a result, we have that $\ln x(t) \to -\infty$ and hence $x(t) \to 0$ as $t \to \infty$. This implies that there exists a $T_1 \in [T, \infty)$ such that $x(T_1) = \max\{x(t) : t \in [T, \infty)\}$. Note from Theorem 6.4.1 and Definition 6.4.2 that $x(t)$ can be chosen as a nonprincipal solution at ∞. Then as $t \in [T_1, \infty)$ and $t \to \infty$,

$$\infty \leftarrow \int_{T_1}^t ds \leq \int_{T_1}^t \frac{x^2(T_1)}{x^2(s)} \, ds = x^2(T_1) \int_{T_1}^t \frac{1}{x^2(s)} \, ds < x^2(T_1) \int_{T_1}^\infty \frac{1}{x^2(s)} \, ds < \infty.$$

We have reached a contradiction. □

We observe that (6.4.12) implies (6.4.6) but the converse is not true in general. Hence Theorem 6.4.5 covers Theorem 6.4.4 as a special case. Consider that

$$\frac{1}{t} \int_0^t \int_0^s q(\tau) \, d\tau ds = \frac{1}{t} \int_0^t (t - s) q(s) \, ds.$$

Then the result below can be regarded as a further extension of Theorem 6.4.5.

THEOREM 6.4.6 (Kamenev [**29**]). *Assume that for some $\alpha > 1$,*

$$(6.4.14) \qquad \limsup_{t \to \infty} \frac{1}{t^\alpha} \int_0^t (t - s)^\alpha q(s) \, ds = \infty.$$

Then Eq. (6.4.5) is oscillatory.

PROOF. Assume the contrary, i.e., Eq. (6.4.5) is nonoscillatory on $[0, \infty)$. Then for any nontrivial solution $x(t)$ of Eq. (6.4.5), there exists a $T \geq 0$ such that $x(t) \neq 0$ on $[T, \infty)$. Define $y(t)$ by (6.4.7). Then as shown in the proof of Theorem 6.4.4, (6.4.8) holds for $t \in [T, \infty)$ and hence

$$q(s) = y'(s) - y^2(s), \quad s \in [T, \infty).$$

By multiplying the above equation by $(t - s)^\alpha$, integrating with respect to s from T to t, and using the integration by parts formula, we have

$$\int_T^t (t - s)^\alpha q(s) \, ds = \int_T^t (t - s)^\alpha [y'(s) - y^2(s)] \, ds$$

$$= -(t - T)^\alpha y(T) + \int_T^t [\alpha(t - s)^{\alpha-1} y(s) - (t - s)^\alpha y^2(s)] \, ds$$

$$= -(t - T)^\alpha y(T) + \int_T^t (t - s)^\alpha [\alpha(t - s)^{-1} y(s) - y^2(s)] \, ds.$$

By completing the square for y on the right-hand side, we have

$$\int_T^t (t - s)^\alpha q(s) \, ds$$

$$= -(t - T)^\alpha y(T) - \int_T^t \left((t - s)^\alpha [y(s) - \frac{\alpha}{2}(t - s)^{-1}]^2 - \frac{\alpha^2}{4}(t - s)^{\alpha-2} \right) ds$$

$$\leq -(t - T)^\alpha y(T) + \frac{\alpha^2}{4} \int_T^t (t - s)^{\alpha-2} \, ds$$

$$= -(t - T)^\alpha y(T) + \frac{\alpha^2}{4(\alpha - 1)}(t - T)^{\alpha-1}.$$

Dividing the above inequality by t^α and taking the upper limits as $t \to \infty$ we find that

$$\limsup_{t \to \infty} \frac{1}{t^\alpha} \int_T^t (t - s)^\alpha q(s) \, ds < \infty.$$

It follows that

$$\limsup_{t\to\infty} \frac{1}{t^\alpha} \int_0^t (t-s)^\alpha q(s)\, ds$$

$$\leq \quad \limsup_{t\to\infty} \frac{1}{t^\alpha} \int_T^t (t-s)^\alpha q(s)\, ds + \limsup_{t\to\infty} \frac{1}{t^\alpha} \int_0^T (t-s)^\alpha q(s)\, ds$$

$$\leq \quad \limsup_{t\to\infty} \frac{1}{t^\alpha} \int_T^t (t-s)^\alpha q(s)\, ds + \int_0^T |q(s)|\, ds < \infty.$$

This contradicts assumption (6.4.14).   □

REMARK 6.4.4. (i) Unlike Theorems 6.4.4 and 6.4.5, Theorem 6.4.6 uses only an upper limit, instead of a limit, of an integral involving the function $q(t)$ to determine oscillation.

(ii) We point out that (6.4.14) with $\alpha = 1$, i.e.,

$$\limsup_{t\to\infty} \frac{1}{t} \int_0^T (t-s)q(s)\, ds = \infty$$

alone is not sufficient for Eq. (6.4.5) to be oscillatory, see Hartman [24] for an example. The reader may check which step in the proof of Theorem 6.4.6 fails to work for the case when $\alpha = 1$.

The result below shows that the oscillation of Eq. (6.4.5) may also be caused by a severe oscillatory behavior of the function q.

THEOREM 6.4.7 (Hartman [24]). *Let* $Q(t) := \int_0^t q(t)\, dt$. *Assume*

$$(6.4.15) \qquad -\infty < \liminf_{t\to\infty} \frac{1}{t} \int_0^t Q(s)\, ds < \limsup_{t\to\infty} \frac{1}{t} \int_0^t Q(s)\, ds \leq \infty.$$

Then Eq. (6.4.5) is oscillatory.

The proof of Theorem 6.4.7 involves some knowledge from the functional analysis such as the Cauchy–Schwarz inequality and Hölder's inequality, which are beyond the scope of this book. The interested reader is referred to Hartman [24] for the proof. We point out that Theorem 6.4.7 fails to apply when

$$\liminf_{t\to\infty} \frac{1}{t} \int_0^t Q(s)\, ds = -\infty.$$

All the oscillation criteria given by the above theorems employ integrals involving the function q and hence require the information of q on the entire half-line $[0, \infty)$. It is difficult to apply them to the cases where q has a "bad" behavior on a big part of $[0, \infty)$, for example, when $\int_0^\infty q(t)\, dt = -\infty$. However, from the Sturm separation theorem, we see that oscillation is only an interval property, i.e., if there exists a sequence of subintervals $[a_i, b_i]_{i=1}^\infty$ of $[0, \infty)$ with $a_i \to \infty$ such that for each $i \in \mathbb{N}$, there exists a solution of Eq. (6.4.5) which has at least two zeros in $[a_i, b_i]$, then every solution of Eq. (6.4.5) is oscillatory, no matter how "bad" the function q is on the remaining parts of $[0, \infty)$. In the following theorem, we derive an interval criterion for oscillation, i.e., a criterion given by the behavior of q only on a sequence of subintervals of $[0, \infty)$. This result can be applied to some extreme cases such as $\int_0^\infty q(t)\, dt = -\infty$ and $\int_0^\infty q(t)\, dt = q^* \in (-\infty, \infty)$.

THEOREM 6.4.8 (Interval Oscillation Criterion [31]). *Assume that for each $T \geq 0$, there exist $a, b, c \in \mathbb{R}$ with $T \leq a < c < b$ such that for an $\alpha > 1$*

$$(6.4.16) \qquad \frac{1}{(c-a)^\alpha} \int_a^c (s-a)^\alpha q(s)\, ds + \frac{1}{(b-c)^\alpha} \int_c^b (b-s)^\alpha q(s)\, ds$$
$$> \frac{\alpha^2}{4(\alpha-1)} \left(\frac{1}{c-a} + \frac{1}{b-c} \right).$$

Then Eq. (6.4.5) is oscillatory.

PROOF. Assume the contrary, i.e., Eq. (6.4.5) is nonoscillatory on $[0, \infty)$. Then for any nontrivial solution $x(t)$ of Eq. (6.4.5), there exists a $T \geq 0$ such that $x(t) \neq 0$ on $[T, \infty)$. In particular, for $a, b, c \in \mathbb{R}$ with $T \leq a < c < b$ such that (6.4.16) holds, $x(t) \neq 0$ on $[a, b]$. Define $y(t)$ by (6.4.7). Then as shown in the proof of Theorem 6.4.4, (6.4.8) holds for $t \in [a, b]$ and hence

$$(6.4.17) \qquad q(s) = y'(s) - y^2(s), \quad s \in [a, b].$$

By the same argument as in the proof of Theorem 6.4.6 with T replaced by c, we have that for $t \in (c, b]$

$$\int_c^t (t-s)^\alpha q(s)\, ds \leq -(t-c)^\alpha y(c) + \frac{\alpha^2}{4(\alpha-1)}(t-c)^{\alpha-1}.$$

Letting $t = b$ and dividing both sides by $(b-c)^\alpha$ we obtain that

$$(6.4.18) \qquad \frac{1}{(b-c)^\alpha} \int_c^b (b-s)^\alpha q(s)\, ds \leq -y(c) + \frac{\alpha^2}{4(\alpha-1)(b-c)}.$$

On the other hand, we may multiply both sides of (6.4.17) by $(s-t)^\alpha$ and integrate it with respect to s from t to c for $t \in [a, c)$. Then by a similar argument to the above,

$$\int_t^c (s-t)^\alpha q(s)\, ds \leq (c-t)^\alpha y(c) + \frac{\alpha^2}{4(\alpha-1)}(c-t)^{\alpha-1}.$$

Letting $t = a$ and dividing both sides by $(c-a)^\alpha$ we obtain that

$$(6.4.19) \qquad \frac{1}{(c-a)^\alpha} \int_a^c (s-a)^\alpha q(s)\, ds \leq y(c) + \frac{\alpha^2}{4(\alpha-1)(c-a)}.$$

The combination of (6.4.18) and (6.4.19) leads to

$$\frac{1}{(c-a)^\alpha} \int_a^c (s-a)^\alpha q(s)\, ds + \frac{1}{(b-c)^\alpha} \int_c^b (b-s)^\alpha q(s)\, ds$$
$$\leq \frac{\alpha^2}{4(\alpha-1)} \left(\frac{1}{c-a} + \frac{1}{b-c} \right).$$

This contradicts assumption (6.4.16).   □

The result below is an immediate consequence of Theorem 6.4.8.

COROLLARY 6.4.2. *Assume that for each $T \geq 0$, there exist $a, c \in \mathbb{R}$ with $T \leq a < c$ such that for an $\alpha > 1$*

$$(6.4.20) \qquad \int_a^c (s-a)^\alpha [q(s) + q(2c - s)] \, ds > \frac{\alpha^2}{2(\alpha - 1)} (c - a)^{\alpha - 1}.$$

Then Eq. (6.4.5) is oscillatory.

PROOF. Let $b = 2c - a$. Then $b - c = c - a$, and by the substitution $s = 2c - \tau$,

$$\int_c^b (b-s)^\alpha q(s) \, ds = \int_a^c (\tau - a)^\alpha q(2c - \tau) \, d\tau.$$

Therefore, (6.4.20) implies (6.4.16) and the conclusion follows from Theorem 6.4.8. $\qquad \square$

The next result, which follows from Theorem 6.4.8 and Corollary 6.4.2, gives improvements of the Kamenev criterion by Theorem 6.4.6 for many cases.

COROLLARY 6.4.3. *Assume that for some $\alpha > 1$ and each $r \geq 0$, either*

(a) *the following two inequalities hold:*

$$(6.4.21) \qquad \limsup_{t \to \infty} \frac{1}{t^{\alpha - 1}} \int_r^t (s - r)^\alpha q(s) \, ds > \frac{\alpha^2}{4(\alpha - 1)}$$

and

$$(6.4.22) \qquad \limsup_{t \to \infty} \frac{1}{t^{\alpha - 1}} \int_r^t (t - s)^\alpha q(s) \, ds > \frac{\alpha^2}{4(\alpha - 1)}; \qquad or$$

(b) *the following inequality holds:*

$$\limsup_{t \to \infty} \frac{1}{t^{\alpha - 1}} \int_r^t (s - r)^\alpha [q(s) + q(2t - s)] \, ds > \frac{\alpha^2}{2(\alpha - 1)}.$$

Then Eq. (6.4.5) is oscillatory.

We leave the proof as an exercise, see Exercise 6.12.

EXAMPLE 6.4.2. Let

$$q(t) = \begin{cases} 5(t - 3n), & 3n \leq t \leq 3n + 1 \\ 5(-t + 3n + 2), & 3n + 1 < t \leq 3n + 2 \\ q_1(t), & 3n + 2 < t < 3n + 3 \end{cases}, \quad n \in \mathbb{N}_0;$$

where $q_1(t)$ makes $q(t)$ continuous on $[3n+2, 3n+3]$ and $\int_{3n+2}^{3n+3} q_1(t) \, dt = -n$. For any $T \geq 0$, there exists an $n \in \mathbb{N}_0$ such that $3n \geq T$. Let $a = 3n$ and $c = 3n + 1$. It is easy to verify that (6.4.20) holds for $\alpha = 2$ and hence Eq. (6.4.5) is oscillatory by Corollary 6.4.2. However, in this equation we have $\int_0^\infty q(t) \, dt = -\infty$.

EXAMPLE 6.4.3. Again, we consider the Euler equation

$$(6.4.23) \qquad\qquad x'' + \frac{\mu}{t^2} x = 0,$$

where $\mu > 0$ is a constant. From Example 6.4.1 we see that Eq. (6.4.23) is oscillatory if $\mu > 1/4$. However, none of the Theorems 6.4.3–6.4.7 can reveal this fact. Now, with Corollary 6.4.3, the verification of oscillation for Eq. (6.4.23) is trivial. Note that for $\alpha > 1$

$$\lim_{t \to \infty} \frac{1}{t^{\alpha-1}} \int_r^t (s-r)^\alpha \frac{\mu}{s^2} \, ds = \frac{\mu}{\alpha-1} \lim_{t \to \infty} \frac{(t-r)^\alpha}{t^\alpha} = \frac{\mu}{\alpha-1}.$$

For any $\mu > 1/4$, there exists an $\alpha > 1$ such that $\frac{\mu}{\alpha-1} > \frac{\alpha^2}{4(\alpha-1)}$. This means that (6.4.21) holds. Similarly, (6.4.22) holds for the same α. Applying Corollary 6.4.3, Part (a) we find that Eq. (6.4.23) is oscillatory for $\mu > 1/4$.

REMARK 6.4.5. The above discussion on interval oscillation criterion and its corollaries are excerpted from Kong [**31**], where an interval oscillation criterion of a more general form is given.

6.5. Boundary Value Problems

A boundary value problem (BVP) consists of an equation and certain boundary condition (BC). Here, we consider the BVPs consisting of the second-order linear equation

$$(\text{nh-s}) \qquad\qquad (p(t)x')' + q(t)x = h(t) \qquad \text{on } [a, b],$$

where

$$p, q, h \in C([a, b], \mathbb{R}) \text{ with } -\infty < a < b < \infty \text{ such that } p(t) > 0;$$

and either the separated BC

$$(\text{S}) \qquad\qquad \begin{cases} B_1(x) := \alpha_1 x(a) + \alpha_2 (px')(a) = 0 \\ B_2(x) := \beta_1 x(b) + \beta_2 (px')(b) = 0, \end{cases}$$

where $\alpha_i, \beta_i \in \mathbb{R}$ satisfying $(\alpha_1, \alpha_2) \neq (0, 0)$ and $(\beta_1, \beta_2) \neq (0, 0)$; or the periodic BC

$$(\text{P}) \qquad\qquad x(a) = x(b), \quad (px')(a) = (px')(b).$$

The separated BC (S) includes the following special cases: When $(\alpha_1, \alpha_2) = (1, 0)$ and $(\beta_1, \beta_2) = (1, 0)$, then BC (S) becomes the Dirichlet BC

$$x(a) = 0, \quad x(b) = 0;$$

and when $(\alpha_1, \alpha_2) = (0, 1)$ and $(\beta_1, \beta_2) = (0, 1)$, then BC (S) becomes the Neumann BC

$$(px')(a) = 0, \quad (px')(b) = 0.$$

REMARK 6.5.1. (i) We point out that the BCs (S) and (P) are different
from ICs in the sense that they are defined at the two end-points
of the interval $[a, b]$ rather than at an initial point. Unlike an IVP,
a BVP may or may not have a solution, and if it does, the solution
may or may not be unique; even if the functions in the equation is
sufficiently smooth.

(ii) BVPs (nh-s), (S) and (nh-s), (P) may have solutions for some right-
hand functions h, but have no solutions for some other right-hand
functions h.

In the following two subsections, we will discuss the two BVPs (nh-s),
(S) and (nh-s), (P) separately.

6.5.1. BVPs with Separated BCs. First, we present conditions which
guarantee that BVP (nh-s), (S) has a unique solution for each continuous
function h on $[a, b]$ using two linearly independent solutions of the homogeneous linear equation

(h-s) $$(p(t)x')' + q(t)x = 0.$$

We observe that $x \equiv 0$ is always a solution of BVP (h-s), (S).

THEOREM 6.5.1. *Let $x_1(t)$ and $x_2(t)$ be any linearly independent solutions of Eq. (h-s). Then the following statements are equivalent:*

(a) BVP (nh-s), (S) has a unique solution for every $h \in C([a, b], \mathbb{R})$,

(b) the determinant $\Delta(x_1, x_2) := \det \begin{bmatrix} B_1(x_1) & B_1(x_2) \\ B_2(x_1) & B_2(x_2) \end{bmatrix} \neq 0$,

(c) BVP (h-s), (S) has only the zero solution.

PROOF. (i) Let $x^*(t)$ be a solution of Eq. (nh-s). By Lemma 2.3.1, the
general solution of Eq. (nh-s) is given by

$$x(t) = c_1 x_1(t) + c_2 x_2(t) + x^*(t).$$

Since BC (S) is homogeneous linear in x, for $i = 1, 2$,

$$B_i(x) = 0 \iff c_1 B_i(x_1) + c_2 B_i(x_2) + B_i(x^*) = 0$$
$$\iff c_1 B_i(x_1) + c_2 B_i(x_2) = -B_i(x^*).$$

By linear algebra, BVP (nh-s), (S) has a unique solution for every
$h \in C([a, b], \mathbb{R})$ if and only if c_1 and c_2 are uniquely solvable from the
above systems of equations, and hence if and only if the coefficient
determinant of the system satisfy $\Delta(x_1, x_2) \neq 0$. This justifies the
equivalence between (a) and (b).

(ii) The general solution of Eq. (h-s) is given by

$$x(t) = c_1 x_1(t) + c_2 x_2(t).$$

Since BC (S) is homogeneous linear in x, for $i = 1, 2$,

$$B_i(x) = 0 \iff c_1 B_i(x_1) + c_2 B_i(x_2) = 0$$
$$\iff c_1 B_i(x_1) + c_2 B_i(x_2) = 0.$$

By linear algebra, BVP (nh-s), (S) has only the zero solution if and only if $c_1 = c_2 = 0$ is the only solution of the above systems of equations, and hence if and only if $\Delta(x_1, x_2) \neq 0$. This justifies the equivalence between (b) and (c).

$\square$

REMARK 6.5.2. It is seen that the determinant $\Delta(x_1, x_2)$ defined in Theorem 6.5.1 depends on the choice of the linearly independent solutions $x_1(t)$ and $x_2(t)$ of Eq. (h-s). However, by the equivalence relations given in Theorem 6.5.1 we find that if $\Delta(x_1, x_2) \neq 0$ for one pair of linearly independent solutions x_1 and x_2, then $\Delta(x_1, x_2) \neq 0$ for any pair of linearly independent solutions x_1 and x_2.

The next lemma plays a key role in deriving a formula for the unique solution of BVP (nh-s), (S) when it has a unique solution.

LEMMA 6.5.1. *BVP* (nh-s), (S) *has a unique solution for every* $h \in C([a, b], \mathbb{R}) \iff Eq.$ (h-s) *has two linearly independent solutions* $x_1(t)$ *and* $x_2(t)$ *satisfying*

$$(6.5.1) \qquad \begin{cases} B_1(x_1) = \alpha_1 x_1(a) + \alpha_2(px_1')(a) = 0, \\ B_2(x_2) = \beta_1 x_2(b) + \beta_2(px_2')(b) = 0. \end{cases}$$

PROOF. (i) Assume (6.5.1) holds with two linearly independent solutions $x_1(t)$ and $x_2(t)$ of Eq. (h-s). We claim that $B_1(x_2) \neq 0$ and $B_2(x_1) \neq 0$, and hence $\Delta(x_1, x_2) \neq 0$. In this case, by Theorem 6.5.1, BVP (nh-s), (S) has a unique solution for every $h \in C([a, b], \mathbb{R})$.

Otherwise, without loss of generality we may assume

$$\begin{cases} B_1(x_1) = \alpha_1 x_1(a) + \alpha_2(px_1')(a) = 0, \\ B_1(x_2) = \alpha_1 x_2(a) + \alpha_2(px_2')(a) = 0. \end{cases}$$

This means that both the vectors $(x_1(a), (px_1)(a))$ and $(x_2(a), (px_2)(a))$ are perpendicular to the vector (α_1, α_2) and hence they are linearly dependent. Thus, the Wronskian of x_1 and x_2 satisfies $W[x_1, x_2](a) = 0$. By Corollary 2.2.1, x_1 and x_2 are linearly dependent on $[a, b]$, contradicting the assumption.

(ii) Assume (6.5.1) does not hold with any two linearly independent solutions $x_1(t)$ and $x_2(t)$ of Eq. (h-s). We note that there are always solutions $x_1(t)$ and $x_2(t)$ of Eq. (h-s) satisfying $B_1(x_1) = 0$ and $B_2(x_2) = 0$. In fact, the solution $x_1(t)$ with the IC $x_1(a) = \alpha_2$, $(px_1')(a) = -\alpha_1$ satisfies $B_1(x_1) = 0$ and the solution $x_2(t)$ with the IC $x_2(b) = \beta_2$, $(px_2')(b) = -\beta_1$ satisfies $B_2(x_2) = 0$. Therefore, the assumption implies that x_1 and x_2 are linearly dependent and hence $x_1(t)$ satisfying both $B_1(x_1) = 0$ and $B_2(x_1) = 0$. For any solution $x_3(t)$ linearly independent of $x_1(t)$, we see that

$$\Delta(x_1, x_3) = \det \begin{bmatrix} B_1(x_1) & B_1(x_3) \\ B_2(x_1) & B_2(x_3) \end{bmatrix} = 0.$$

By Theorem 6.5.1, BVP (nh-s), (S) does not have a unique solution for every $h \in C([a,b], \mathbb{R})$.

$\square$

We are ready to derive a formula for the unique solution of BVP (nh-s), (S) under the assumption of Lemma 6.5.1.

THEOREM 6.5.2. *Assume Eq.* (h-s) *has two linearly independent solutions* $x_1(t)$ *and* $x_2(t)$ *satisfying* (6.5.1). *Define a function* $G : [a,b] \times [a,b] \to \mathbb{R}$ *by*

$$(6.5.2) \qquad G(t,s) = \begin{cases} \frac{1}{c} x_1(s) x_2(t), & a \leq s \leq t \leq b, \\ \frac{1}{c} x_1(t) x_2(s), & a \leq t \leq s \leq b; \end{cases}$$

where c *is a constant satisfying* $c = x_1(t)(px_2')(t) - x_2(t)(px_1')(t)$ *for any* $t \in [a,b]$. *Then*

$$(6.5.3) \qquad \tilde{x}(t) = \int_a^b G(t,s) h(s)\, ds$$

is the unique solution of BVP (nh-s), (S) *for every* $h \in C([a,b], \mathbb{R})$. *In this case, the function* $G(t,s)$ *is called the Green's function of BVP* (nh-s), (S).

PROOF. By the variation of parameters formula (6.1.8), the general solution of Eq. (h-s) is

$$\begin{aligned} x(t) &= c_1 x_1(t) + c_2 x_2(t) + \frac{1}{c} \int_a^t [-x_1(t) x_2(s) + x_2(t) x_1(s)] h(s)\, ds \\ &= x_1(t) \left[c_1 - \frac{1}{c} \int_a^t x_2(s) h(s)\, ds \right] + x_2(t) \left[c_2 + \frac{1}{c} \int_a^t x_1(s) h(s)\, ds \right]. \end{aligned}$$

Let $c_1 = \frac{1}{c} \int_a^b x_2(s) h(s)\, ds$ and $c_2 = 0$, we get a particular solution

$$(6.5.4) \qquad \tilde{x}(t) = \frac{1}{c} \left[x_2(t) \int_a^t x_1(s) h(s)\, ds + x_1(t) \int_t^b x_2(s) h(s)\, ds \right].$$

We show that $\tilde{x}(t)$ is the unique solution of BVP (nh-s), (S). In fact, by differentiating (6.5.4) we have

$$\begin{aligned} (6.5.5) \quad \tilde{x}'(t) &= \frac{1}{c} \Big[x_2'(t) \int_a^t x_1(s) h(s)\, ds + x_2(t) x_1(t) h(t) \\ &\qquad + x_1'(t) \int_t^b x_2(s) h(s)\, ds - x_1(t) x_2(t) h(t) \Big] \\ &= \frac{1}{c} \left[x_2'(t) \int_a^t x_1(s) h(s)\, ds + x_1'(t) \int_t^b x_2(s) h(s)\, ds \right]. \end{aligned}$$

From (6.5.4) and (6.5.5) we see that

$$\tilde{x}(a) = \left(\frac{1}{c} \int_a^b x_2(s) h(s)\, ds \right) x_1(a) \quad \text{and} \quad \tilde{x}'(a) = \left(\frac{1}{c} \int_a^b x_2(s) h(s)\, ds \right) x_1'(a).$$

Thus,

$$\alpha_1 \tilde{x}(a) + \alpha_2 (p\tilde{x}')(a) = \left(\frac{1}{c} \int_a^b x_2(s)h(s)\,ds \right) (\alpha_1 x_1(a) + \alpha_2 (px_1')(a)).$$

Since

$$B_1(x_1) = \alpha_1 x_1(a) + \alpha_2 (px_1')(a) = 0,$$

we have

$$B_1(\tilde{x}) = \alpha_1 \tilde{x}(a) + \alpha_2 (p\tilde{x}')(a) = 0.$$

Similarly, we can show that $B_2(\tilde{x}) = 0$. Therefore, $\tilde{x}(t)$ is a solution of BVP (nh-s), (S). With the Green's function $G(t,s)$ as defined by (6.5.2), the expression (6.5.4) for $\tilde{x}$ reduces to (6.5.3). This solution is unique due to Lemma 6.5.1.   □

REMARK 6.5.3. When the assumption of Theorem 6.5.2 does not hold, BVP (nh-s), (S) may have solutions for some right-hand functions h satisfying certain conditions. In this case, the solutions are not unique and cannot be obtained by the formula in Theorem 6.5.2.

We demonstrate the applications of Lemma 6.5.1 and Theorem 6.5.2 with simple examples.

EXAMPLE 6.5.1. We consider the BVP consisting of Eq. (nh-s) with $p(t) \equiv 1$ and $q(t) \equiv 0$ on $[a,b]$ and the Dirichlet BC, i.e., the problem

$$(6.5.6) \qquad\qquad x'' = h(t), \quad x(a) = 0, \ x(b) = 0,$$

where $h \in C([a,b], \mathbb{R})$. To solve the problem, we need to find two nontrivial solutions $x_1(t)$ and $x_2(t)$ of the corresponding homogeneous linear equation $x'' = 0$ satisfying the conditions $x_1(a) = 0$ and $x_2(b) = 0$, respectively. It is easy to see that we can choose $x_1 = t - a$ and $x_2 = b - t$. Clearly, $x_1(t)$ and $x_2(t)$ are linearly independent. Note that

$$c = \det \begin{bmatrix} t-a & b-t \\ 1 & -1 \end{bmatrix} = a - b.$$

Then the Green's function of BVP (6.5.6) is

$$G(t,s) = \begin{cases} \dfrac{1}{a-b}(b-t)(s-a), & a \le s \le t \le b, \\[2mm] \dfrac{1}{a-b}(t-a)(b-s), & a \le t \le s \le b. \end{cases}$$

Therefore, the unique solution of BVP (6.5.4) is

$$\begin{aligned}
\tilde{x}(t) &= \int_a^b G(t,s)h(s)\,ds \\
&= \frac{1}{a-b}\left[(b-t)\int_a^t (s-a)h(s)\,ds + (t-a)\int_t^b (b-s)h(s)\,ds \right].
\end{aligned}$$

EXAMPLE 6.5.2. We consider the BVP consisting of Eq. (nh-s) with $p(t) \equiv 1$ and $q(t) \equiv 0$ on $[a, b]$ and the Neumann BC, i.e., the problem

$$(6.5.7) \qquad x'' = h(t), \quad x'(a) = 0, \ x'(b) = 0,$$

where $h \in C([a, b], \mathbb{R})$. To solve the problem, we need to find two non-trivial solutions $x_1(t)$ and $x_2(t)$ of the corresponding homogeneous linear equation $x'' = 0$ satisfying the conditions $x_1'(a) = 0$ and $x_2'(b) = 0$, respectively. However, up to constant multiples, the only possibility for the solutions are $x_1(t) = x_2(t) = 1$. Hence $x_1(t)$ and $x_2(t)$ are linearly dependent. By Lemma 6.5.1, BVP (6.5.7) does not have a unique solution for every $h \in C([a, b], \mathbb{R})$.

On the other hand, BVP (6.5.7) has solutions for the function h satisfying a certain condition. In fact, the general solution of the equation $x'' = h(t)$ is

$$x = c_1 + c_2 t + \int_a^t (t - s) h(s) \, ds$$

with

$$x' = c_2 + \int_a^t h(s) \, ds.$$

For the solution to satisfy the condition $x'(a) = 0$, we have $c_2 = 0$. Then for the solution to satisfy the condition $x'(b) = 0$, the function h must satisfy the condition

$$\int_a^b h(s) \, ds = 0.$$

Under this condition, $x = c_1 + \int_a^t (t - s) h(s) \, ds$ is a solution of BVP (6.5.7) for any $c_1 \in \mathbb{R}$.

6.5.2. BVPs with Periodic BCs. To present the results on BVP (nh-s), (P), we need the two linearly independent solutions $u(t)$ and $v(t)$ of the homogeneous linear equation

$$(\text{h-s}) \qquad (p(t)x')' + q(t)x = 0$$

satisfying the ICs

$$(6.5.8) \qquad u(a) = 1, \ (pu')(a) = 0; \quad v(a) = 0, \ (pv')(a) = 1.$$

Let

$$\Delta = 2 - u(b) - (pv')(b).$$

Then we have the following theorem.

THEOREM 6.5.3. *BVP* (nh-s), (P) *has a unique solution for every* $h \in C([a, b], \mathbb{R}) \iff \Delta \neq 0$. *In this case, we define a function* $G : [a, b] \times [a, b] \to \mathbb{R}$ *by*

$$(6.5.9) \quad G(t, s) = \frac{v(b)}{\Delta} u(t)u(s) - \frac{(pu')(b)}{\Delta} v(t)v(s)$$

$$+ \begin{cases} \frac{(pv')(b)-1}{\Delta} u(t)v(s) - \frac{u(b)-1}{\Delta} u(s)v(t), & a \leq s \leq t \leq b, \\ \frac{(pv')(b)-1}{\Delta} u(s)v(t) - \frac{u(b)-1}{\Delta} u(t)v(s), & a \leq t \leq s \leq b. \end{cases}$$

Then

$$(6.5.10) \qquad \tilde{x}(t) = \int_a^b G(t,s)h(s)\, ds$$

is the unique solution of BVP (nh-s), (P) for every $h \in C([a,b], \mathbb{R})$*. The function* $G(t,s)$ *is called the Green's function of BVP (nh-s), (P).*

PROOF. Note that by Lemma 6.1.3, the Wronskian of the solutions u and v satisfy

$$(6.5.11) \quad W(t) \;=\; u(t)(pv')(t) - v(t)(pu')(t)$$
$$\equiv\; u(a)(pv')(a) - v(a)(pu')(a) = 1 \;\text{ on } [a,b].$$

From the variation of parameters formula (6.1.8), the general solution of (nh-s) is given by

$$(6.5.12) \qquad x(t) = c_1 u(t) + c_2 v(t) + \int_a^t [-u(t)v(s) + v(t)u(s)]h(s)\, ds$$

for $c_1, c_2 \in \mathbb{R}$. Hence

$$(6.5.13) \qquad x'(t) = c_1 u'(t) + c_2 v'(t) + \int_a^t [-u'(t)v(s) + v'(t)u(s)]h(s)\, ds.$$

Then $x(t)$ satisfies BC (P) $\iff$ c_1 and c_2 satisfy the system of equations
(6.5.14)

$$[u(b) - 1]c_1 + v(b)c_2 \;=\; \int_a^b [u(b)v(s) - v(b)u(s)]h(s)\, ds,$$
$$(pu')(b)c_1 + [(pv')(b) - 1]c_2 \;=\; \int_a^b [(pu')(b)v(s) - (pv')(b)u(s)]h(s)\, ds.$$

By linear algebra, the system has a unique solution c_1, c_2 for every $h \in C([a,b], \mathbb{R})$

$$\iff \det \begin{bmatrix} u(b) - 1 & v(b) \\ (pu')(b) & (pv')(b) - 1 \end{bmatrix} = 2 - u(b) - (pv')(b) \neq 0 \iff \Delta \neq 0.$$

This proves the first part of the theorem. To find the unique solution of BVP (nh-s), (P), we solve (6.5.14) for c_1, c_2 to obtain

$$\begin{bmatrix} c_1 \\ c_2 \end{bmatrix} = \begin{bmatrix} u(b) - 1 & v(b) \\ (pu')(b) & (pv')(b) - 1 \end{bmatrix}^{-1} \begin{bmatrix} \int_a^b [u(b)v(s) - v(b)u(s)]h(s)\, ds \\ \int_a^b [(pu')(b)v(s) - (pv')(b)u(s)]h(s)\, ds \end{bmatrix}$$

$$= \frac{1}{\Delta} \begin{bmatrix} (pv')(b) - 1 & -v(b) \\ -(pu')(b) & u(b) - 1 \end{bmatrix} \begin{bmatrix} \int_a^b [u(b)v(s) - v(b)u(s)]h(s)\, ds \\ \int_a^b [(pu')(b)v(s) - (pv')(b)u(s)]h(s)\, ds \end{bmatrix}.$$

Considering (6.5.11) we have that

$$(6.5.15)\; c_1 \;=\; \frac{1}{\Delta}\Big\{ ((pv')(b) - 1) \int_a^b [u(b)v(s) - v(b)u(s)]h(s)\, ds$$

$$- v(b) \int_a^b [(pu')(b)v(s) - (pv')(b)u(s)]h(s)\, ds \Big\}$$

$$= \frac{1}{\Delta}\Big[(1 - u(b)) \int_a^b v(s)h(s)\, ds + v(b) \int_a^b u(s)h(s)\, ds \Big].$$

Similarly,
(6.5.16)
$$c_2 = \frac{1}{\Delta} \left[-(pu')(b) \int_a^b v(s)h(s)\,ds + ((pv')(b) - 1) \int_a^b u(s)h(s)\,ds \right].$$

By substituting (6.5.15) and (6.5.16) into (6.5.13) and simplifying the expression, we see that the function $\tilde{x}(t)$ defined in (6.5.10) with the function $G(t, s)$ given in (6.5.9) is the unique solution of BVP (nh-s), (P) for every $h \in C([a, b], \mathbb{R})$. $\qquad \square$

REMARK 6.5.4. (i) In addition to the assumption that $p(t) > 0$ on $[a, b]$, if $q(t) \le 0$ and $q(t) \not\equiv 0$ on $[a, b]$, then BVP (nh-s), (P) has a unique solution for every $h \in C([a, b], \mathbb{R})$.

In fact, by integrating Eq. (h-s) and using the IC, we see that

(6.5.17)
$$u(t) = 1 - \int_a^t \left(\frac{1}{p(\tau)} \int_a^\tau q(s)u(s)\,ds \right) d\tau.$$

We claim that $u(t) > 0$ on $(a, b]$. Otherwise, since $u(a) = 1 > 0$, there exists a $t^* \in [a, b]$ such that $u(t) > 0$ on (a, t^*) and $u(t^*) = 0$. This contradicts (6.5.17) with $t = t^*$. Thus by (6.5.17), $u(b) > 1$. In the same way, we have

$$v(t) = \int_a^t \frac{1}{p(\tau)} \left(1 - \int_a^\tau q(s)v(s)\,ds \right) d\tau,$$

and hence $v(t) > 0$ on $(a, b]$. Since

$$(pv')(t) = 1 - \int_a^t q(s)v(s)\,ds,$$

we see that $(pv')(b) > 1$. As a result, $\Delta = 2 - u(b) - (pv')(b) < 0$. Then the conclusion follows from Theorem 6.5.3.

(ii) When $p(t) > 0$ and $q(t) \equiv 0$ on $[a, b]$, we have that $u(t) = 1$ and $v(t) = t$. It follows that $\Delta = 0$. By Theorem 6.5.3, BVP (nh-s), (P) does not have a unique solution for every $h \in C([a, b], \mathbb{R})$. By integrating Eq. (nh-s) we see that the general solution $x(t)$ satisfies

$$x = c_1 + c_2 \int_a^t \frac{1}{p(s)}\,ds + \int_a^t \left(\frac{1}{p(\tau)} \int_a^\tau h(s)\,ds \right) d\tau$$

and

$$p(t)x' = c_2 + \int_a^t h(s)\,ds.$$

By applying BC (P) and doing simple computations we find that BVP (nh-s), (P) has a solution for some $h \in C([a, b], \mathbb{R}) \iff \int_a^b h(t)\,dt = 0$. In this case, there are infinitely many solutions of BVP (nh-s), (P) in the form of

$$x = c_1 - \frac{\int_a^b (\int_a^\tau h(s)\,ds)/p(\tau)\,d\tau}{\int_a^b 1/p(s)\,ds} \int_a^t \frac{1}{p(s)}\,ds + \int_a^t \left(\frac{1}{p(\tau)} \int_a^\tau h(s)\,ds \right) d\tau,$$

where $c_1 \in \mathbb{R}$.

Now, we apply Theorem 6.5.3 to the cases where both $p(t)$ and $q(t)$ are nonzero constant functions. More specifically, consider the equation

(nh$\pm$) $x'' \pm r^2 x = h(t)$,

where $r > 0$. Let $\omega = b - a$. Then we have the following result.

COROLLARY 6.5.1. (a) BVP (nh-), (P) has a unique solution for every $h \in C([a,b], \mathbb{R})$, and the Green's function is

(6.5.18) $G(t,s) = -\dfrac{1}{2r(e^{r\omega} - 1)} \begin{cases} e^{r(t-s)} + e^{r(\omega-t+s)}, & a \le s \le t \le b, \\ e^{r(s-t)} + e^{r(\omega-s+t)}, & a \le t \le s \le b. \end{cases}$

(b) BVP (nh+), (P) has a unique solution for every $h \in C([a,b], \mathbb{R})$ $\iff r \ne 2k\pi/\omega$ with $k \in \mathbb{N}$. In this case, the Green's function is

(6.5.19)

$G(t,s) = \dfrac{1}{2r(1 - \cos(r\omega))} \begin{cases} \sin(r(t-s)) + \sin(r(\omega - t + s)), & a \le s \le t \le b, \\ \sin(r(s-t)) + \sin(r(\omega - s + t)), & a \le t \le s \le b. \end{cases}$

PROOF. (a) For the solutions $u(t)$ and $v(t)$ of the equation $x'' - r^2 x = 0$ satisfying (6.5.8), we have

$$u = \text{ch}\,(r(t-a)),\ u' = r\text{sh}\,(r(t-a));\quad v = \frac{1}{r}\text{sh}\,(r(t-a)),\ v' = \text{ch}\,(r(t-a));$$

where $\text{ch}\,t$ and $\text{sh}\,t$ are the hyperbolic functions. Thus $\Delta = 2 - 2\text{ch}\,(r\omega) < 0$. By Theorem 6.5.3, BVP (nh-), (P) has a unique solution for every $h \in C([a,b], \mathbb{R})$. The Green's function in (6.5.18) can be obtained by substituting $u(t)$ and $v(t)$ into (6.5.9) and applying identities for hyperbolic functions. We leave the detail as an exercise, see Exercise 6.18.

(b) For the solutions $u(t)$ and $v(t)$ of the equation $x'' + r^2 x = 0$ satisfying (6.5.8), we have

$$u = \cos(r(t-a)),\ u' = -r\sin(r(t-a));\quad v = \frac{1}{r}\sin(r(t-a)),\ v' = \cos(r(t-a)).$$

Thus $\Delta = 2 - 2\cos(r\omega) \ne 0 \iff r \ne 2k\pi/\omega$ with $k \in \mathbb{N}$. Then the first part is confirmed by Theorem 6.5.3. The Green's function in (6.5.19) can be obtained by substituting $u(t)$ and $v(t)$ into (6.5.9) and applying identities of trigonometric functions. We leave the detail as an exercise, see Exercise 6.19.

$\square$

We point out that the Green's function for BVPs with separated BCs given in Theorem 6.5.2 is derived based on Hartmann [22, Section XI, 2]; and the Green's function for BVPs with periodic BCs given in Theorem 6.5.3 originates from Atici and Guseinov [1].

Finally, we comment that Green's functions for linear BVPs have been widely used to study the existence of solutions of nonlinear BVPs. For instance, by Theorem 6.5.2 [Theorem 6.5.3], a function $x(t)$ is a solution of the BVP consisting of the nonlinear equation

$$(p(t)x')' + q(t)x = f(t, x)$$

and BC (S) [BC (P)] if and only if it satisfies the integral equation

$$x(t) = \int_a^b G(t, s) f(s, x(s)) \, ds,$$

where $G(t, s)$ is the Green's function of the BVP (nh-s), (S) [BVP (nh-s), (P)]. This means that $x(t)$ is a "fixed point" of the operator T defined by

$$(Tx)(t) := \int_a^b G(t, s) f(s, x(s)) \, ds$$

in some space. Therefore, the existence of solutions of the nonlinear BVP can be solved by applying certain fixed point theorems. However, this is beyond the scope of this book.

6.6. Sturm–Liouville Problems

Many linear BVPs in partial differential equations lead, by separation of variables, to certain Sturm–Liouville eigenvalue problems, or simply Sturm–Liouville problems (SLPs). An SLP consists of the Sturm–Liouville equation

(S-L) $-(p(t)x')' + q(t)x = \lambda w(t)x$ on $[a, b]$

and a "self-adjoint" BC, for example, the separated BC

(S) $\begin{cases} \cos \alpha \, x(a) - \sin \alpha \, (px')(a) = 0, & \alpha \in [0, \pi), \\ \cos \beta \, x(b) - \sin \beta \, (px')(b) = 0, & \beta \in (0, \pi]; \end{cases}$

or the periodic BC

(P) $x(a) = x(b), \quad (px')(a) = (px')(b).$

Here we assume that $p, q, w \in C([a, b], \mathbb{R})$ with $-\infty < a < b < \infty$ such that $p(t) > 0$ and $w(t) > 0$ on $[a, b]$; and $\lambda \in \mathbb{C}$ is an eigenvalue parameter. We note that BC (S) here is a normalization of the BC (S) in Sect. 6.5.

As in Sect. 6.5 we see that $x \equiv 0$ is always a solution of SLP (S-L), (S) [SLP (S-L), (P)]. However, SLP (S-L), (S) [SLP (S-L), (P)] may or may not have a nontrivial solution for a given value of the parameter λ.

DEFINITION 6.6.1. $\lambda = \lambda^* \in \mathbb{C}$ is said to be an eigenvalue of SLP (S-L), (S) [SLP (S-L), (P)] with an associated eigenfunction $u(t)$ if $x = u(t)$ is a nontrivial solution of SLP (S-L), (S) [SLP (S-L), (P)] with $\lambda = \lambda^*$.

It is easy to see that if $u(t)$ is an eigenfunction associated with an eigenvalue λ^*, then any nonzero constant multiple of $u(t)$ is also an eigenfunction associated with the same eigenvalue λ^*. Moreover, the set of eigenfunctions associated with λ^* form a vector space.

DEFINITION 6.6.2. Let λ^* be an eigenvalue of SLP (S-L), (S) [SLP (S-L), (P)]. Then λ^* is said to be geometrically simple if the eigenfunction space associated with λ^* has dimension 1, i.e., if $u(t)$ and $v(t)$ are both eigenfunctions associated with λ^*, then $u(t)$ and $v(t)$ are linearly dependent; λ^* is said to be geometrically double if the eigenfunction space associated with λ^* has dimension 2, i.e., every nontrivial solution of Eq. (S-L) with $\lambda = \lambda^*$ is an eigenfunction associated with λ^*.

Although the eigenvalue parameter λ is allowed to be complex-valued, it can be shown that all eigenvalues of SLPs with "self-adjoint" BCs turn out to be real-valued. In particular, we have the following result.

LEMMA 6.6.1. *All eigenvalues of SLP (S-L), (S) [SLP (S-L), (P)] are real.*

PROOF. Let λ be an eigenvalue of SLP (S-L), (S) with an associated eigenfunction $u(t)$. Then

$$(6.6.1) \qquad -(p(t)u'(t))' + q(t)u(t) = \lambda w(t)u(t) \quad \text{on } [a, b].$$

Taking the complex conjugate on both sides of (6.6.1) and noting that p, q, w are real-valued, we have

$$(6.6.2) \qquad -(p(t)\bar{u}'(t))' + q(t)\bar{u}(t) = \bar{\lambda} w(t)\bar{u}(t) \quad \text{on } [a, b].$$

Multiplying both sides of (6.6.1) and (6.6.2) by $\bar{u}(t)$ and $u(t)$, respectively, and then subtracting them, we obtain that

$$(6.6.3) \quad \bar{u}(t)(p(t)u'(t))' - u(t)(p(t)\bar{u}'(t))' = (\bar{\lambda} - \lambda)w(t)u(t)\bar{u}(t) \quad \text{on } [a, b].$$

By integration by parts we have

$$\int_a^b [\bar{u}(t)(p(t)u'(t))' - u(t)(p(t)\bar{u}'(t))'] \, dt = [\bar{u}(t)(p(t)u'(t)) - u(t)(p(t)\bar{u}'(t))]_a^b = 0$$

since both $u(t)$ and $\bar{u}(t)$ satisfy BC (S). Thus, integrating both sides of (6.6.3) from a to b leads to

$$(\bar{\lambda} - \lambda) \int_a^b w(t)u(t)\bar{u}(t) \, dt = (\bar{\lambda} - \lambda) \int_a^b w(t)|u(t)|^2 \, dt = 0.$$

Note that $w(t) > 0$ and $u(t) \not\equiv 0$ on $[a, b]$. It follows that $\bar{\lambda} - \lambda = 0$ and hence λ is real.

The proof for SLP (S-L), (P) is essentially the same and hence is omitted. $\square$

To study the existence and properties of eigenvalues of SLPs, we first consider three special SLPs with $p(t) \equiv w(t) \equiv 1$ and $q(t) \equiv 0$.

EXAMPLE 6.6.1. (a) Consider the SLP with the Dirichlet BC

$$-x'' = \lambda x, \quad x(0) = 0, x(1) = 0.$$

It is easy to see that the general solution for different values of λ is as follows:

$$x = c_1 e^{\sqrt{-\lambda}t} + c_2 e^{-\sqrt{-\lambda}t}, \quad \lambda < 0,$$
$$x = c_1 + c_2 t, \quad \lambda = 0,$$
$$x = c_1 \cos \sqrt{\lambda}t + c_2 \sin \sqrt{\lambda}t, \quad \lambda > 0.$$

For the case when $\lambda < 0$, by applying the BC to the general solution we obtain a system for the constants c_1 and c_2

$$\begin{cases} c_1 + c_2 = 0, \\ c_1 e^{\sqrt{-\lambda}} + c_2 e^{-\sqrt{-\lambda}} = 0. \end{cases}$$

Since the coefficient determinant

$$\Delta = \begin{vmatrix} 1 & 1 \\ e^{\sqrt{-\lambda}} & e^{-\sqrt{-\lambda}} \end{vmatrix} \neq 0,$$

we see that the only solution is $c_1 = c_2 = 0$. Therefore, there is no nontrivial solution of the SLP with this λ. This means that $\lambda < 0$ cannot be an eigenvalue.

For the case when $\lambda = 0$, by applying the BC to the general solution we obtain a system for the constants c_1 and c_2

$$\begin{cases} c_1 = 0, \\ c_1 + c_2 = 0. \end{cases}$$

Obviously, the only solution is $c_1 = c_2 = 0$. Therefore, there is no nontrivial solution of the SLP with this λ. This means that $\lambda = 0$ cannot be an eigenvalue.

For the case when $\lambda > 0$, by applying the BC to the general solution we obtain a system for the constants c_1 and c_2

$$\begin{cases} c_1 = 0, \\ c_2 \sin \sqrt{\lambda} = 0. \end{cases}$$

Note that the system has nonzero solutions $(0, c_2)$ for any $c_2 \neq 0$ $\iff \sin \sqrt{\lambda} = 0 \iff \lambda_n = ((n+1)\pi)^2$ for $n \in \mathbb{N}_0$. Therefore, for each $n \in \mathbb{N}_0$, $\lambda_n = ((n+1)\pi)^2$ is an eigenvalue with an associated eigenfunction $u_n(t) = \sin((n+1)\pi t)$, and no other values of λ are eigenvalues. Note that λ_n is geometrically simple and $u_n(t)$ has exactly n zeros in $(0, 1)$.

(b) Consider the SLP with the Neumann BC

$$-x'' = \lambda x, \quad x'(0) = 0, x'(1) = 0.$$

Similar to Part (a), we find that there is no nontrivial solution of the problem for $\lambda < 0$. By a further computation we see that for each $n \in \mathbb{N}_0$, $\lambda_n = (n\pi)^2$ is an eigenvalue with an associated eigenfunction $u_n(t) = \cos(n\pi t)$, and no other values of λ are eigenvalues. Note that λ_n is geometrically simple and $u_n(t)$ has exactly n zeros in $(0, 1)$.

(c) Consider the SLP with the periodic BC

$$-x'' = \lambda x, \quad x(0) = x(1), \ x'(0) = x'(1).$$

Similar to Part (a), we find that there is no nontrivial solution of the problem for $\lambda < 0$. By a further computation we see that $\lambda_0 = 0$ is a geometrically simple eigenvalue with an associated eigenfunction $u_0 = 1$; and for each $n \in \mathbb{N}$, $\lambda_{2n-1} = \lambda_{2n} = (2n\pi)^2$ are geometrically double eigenvalues. Hence both $\sin(2n\pi t)$ and $\cos(2n\pi t)$ are associated eigenfunctions. Note that the eigenfunctions do not satisfy the same zero-counting properties as in Parts (a) and (b).

We observe from Example 6.6.1 that both SLPs with separated BCs and SLPs with periodic BCs may have a sequence of real eigenvalues. To investigate the general case, we discuss the two types of SLPs (S-L), (S) and (S-L), (P) separately.

6.6.1. SLPs with Separated BCs. The results in Example 6.6.1, Parts (a) and (b) can be extended to the general SLPs with separated BCs as stated in the theorem below.

THEOREM 6.6.1. *SLP (S-L), (S) has a countably infinite number of eigenvalues λ_n, $n \in \mathbb{N}_0$, which are all real, bounded below and unbounded above, and can be arranged to satisfy*

(6.6.4) $\lambda_0 < \lambda_1 < \cdots < \lambda_n < \cdots, \quad \text{and } \lambda_n \to \infty \text{ as } n \to \infty.$

Moreover, for $n \in \mathbb{N}_0$, the eigenvalue λ_n is geometrically simple and the eigenfunctions associated with λ_n have exactly n zeros in (a, b).

To prove this theorem, we need to establish two more lemmas on the properties of the eigenvalues and eigenfunctions of SLP (S-L), (S) and on the behavior of the Prüfer transformations of solutions of Eq. (S-L).

LEMMA 6.6.2. *All eigenvalues of SLP (S-L), (S) are geometrically simple.*

PROOF. By definition, we need to show that if $u(t)$ and $v(t)$ are both eigenfunctions associated with an eigenfunction λ, then $u(t)$ and $v(t)$ are linearly dependent. Let $W(t)$ be the Wronskian of $u(t)$ and $v(t)$. Then

$$W(t) = u(t)(pv')(t) - v(t)(pu')(t).$$

Since both $u(t)$ and $v(t)$ satisfy BC (S), we have

$$W(a) = u(a)(pv')(a) - v(a)(pu')(a) = 0.$$

By Corollary 2.2.1, $u(t)$ and $v(t)$ are linearly dependent. $\square$

LEMMA 6.6.3. *Let $x(t, \lambda)$ be the solution of Eq. (S-L) satisfying the IC*

(6.6.5) $$x(a, \lambda) = \sin \alpha, \quad (px')(a, \lambda) = \cos \alpha$$

and $\theta(t, \lambda)$ its Prüfer angle. Then for each $t \in (a, b]$,

  (a) $\theta(t, \lambda)$ *is strictly increasing in* λ,
  (b) $\theta(t, \lambda) \to \infty$ *as* $\lambda \to \infty$,
  (c) $\theta(t, \lambda) \to 0$ *as* $\lambda \to -\infty$.

PROOF. (a) Let $\lambda_1, \lambda_2 \in \mathbb{R}$ such that $\lambda_1 < \lambda_2$. Then $x(t, \lambda_1)$ and $x(t, \lambda_2)$ are solutions of the equations

(6.6.6) $$-(p(t)x')' + q(t)x = \lambda_1 w(t)x$$

and

(6.6.7) $$-(p(t)x')' + q(t)x = \lambda_2 w(t)x,$$

respectively. By Definition 6.3.1, Eq. (6.6.7) is a strict-Sturm-majorant of Eq. (6.6.6) on $[a, b]$ with $c = a$. Note that for $i = 1, 2$, $\theta(t, \lambda_i)$ is the Prüfer angle of $x(t, \lambda_i)$ satisfying $\theta(a, \lambda_i) = \alpha$. Then by Theorem 6.3.1 and Remark 6.3.1, $\theta(t, \lambda_1) < \theta(t, \lambda_2)$ for $t \in (a, b]$. This shows that $\theta(t, \lambda)$ is strictly increasing in λ for $t \in (a, b]$.

To prove Parts (b) and (c), we make use of the independent variable substitution $s = s(t) := \int_a^t 1/p(\tau)\, d\tau$. By Remark 6.1.1, Part (ii), the function $y(s, \lambda) := x(t(s), \lambda)$ with $t(s)$ the inverse function of $s(t)$ is the solution of the IVP consisting of the equation

(6.6.8) $$\frac{d^2 y}{ds^2} + [-p(t(s))q(t(s)) + \lambda p(t(s))w(t(s))]y = 0 \quad \text{on } [0, \bar{b}]$$

and the IC

(6.6.9) $$y(0, \lambda) = \sin \alpha, \quad \frac{dy}{ds}(0, \lambda) = \cos \alpha,$$

where $\bar{b} = \int_a^b 1/p(\tau)\, d\tau$. Moreover, $\theta(t(s), \lambda)$ is the Prüfer angle of $y(s, \lambda)$. We will compare the solution $y(s, \lambda)$ of IVP (6.6.8), (6.6.9) with the solution $z(s, m)$ of the IVP

(6.6.10) $$\frac{d^2 z}{ds^2} + mz = 0, \quad z(0, m) = \sin \alpha, \quad \frac{dz}{ds}(0, m) = \cos \alpha,$$

where $m \in \mathbb{R}$ is a parameter.

(b) By solving Eq. (6.6.10) for $m > 0$ we see that for each $n \in \mathbb{N}$ and $s \in (0, \bar{b}]$, there exists a sufficiently large $m > 0$ such that $z(\cdot, m)$ has at least $n + 1$ zeros in $(0, s]$. We observe that for a fixed $m > 0$, Eq. (6.6.8) is a Sturm-majorant of Eq. (6.6.10) when λ is sufficiently large. Thus by Theorem 6.3.2, for each $n \in \mathbb{N}$, there exists a $\lambda_n > 0$ such that $y(\cdot, \lambda)$ has at least n zeros in $(0, s]$ for $\lambda \geq \lambda_n$. This shows that $\theta(t(s)) \geq n\pi$ for $\lambda \geq \lambda_n$. Therefore, we have $\theta(t, \lambda) \to \infty$ as $\lambda \to \infty$ for $t \in (a, b]$.

(c) From (6.6.5) we see that $\theta(a, \lambda) = \alpha \in [0, \pi)$ for any $\lambda \in \mathbb{R}$. By Theorem 6.2.1, $\theta(t, \lambda) > 0$ for $t \in (a, b]$ and $\lambda \in \mathbb{R}$.

For $m < 0$, by solving the IVP (6.6.10) we obtain that

(6.6.11)
$$z(s, m) = \frac{1}{2}\left[\left(\sin\alpha + \frac{\cos\alpha}{\sqrt{-m}}\right)e^{\sqrt{-m}s} + \left(\sin\alpha - \frac{\cos\alpha}{\sqrt{-m}}\right)e^{-\sqrt{-m}s}\right]$$

and

(6.6.12)
$$\frac{dz}{ds}(s, m) = \frac{\sqrt{-m}}{2}\left[\left(\sin\alpha + \frac{\cos\alpha}{\sqrt{-m}}\right)e^{\sqrt{-m}s} - \left(\sin\alpha - \frac{\cos\alpha}{\sqrt{-m}}\right)e^{-\sqrt{-m}s}\right].$$

Let $\phi(s, m)$ be the Prüfer angle of $z(s, m)$. Then $\phi(0, m) = \alpha \in [0, \pi)$ for any $m < 0$. It follows that $\phi(s, m) > 0$ for any $s \in (0, \bar{b}]$ and $m < 0$. From (6.6.11) and (6.6.12) we find that for any $\alpha \in [0, \pi)$ and $s \in (0, \bar{b}]$, $z(s, m) > 0$ and $dz/ds(s, m) > 0$ when $m < 0$ is sufficiently small. Hence for $s \in (0, \bar{b}]$, $0 < \phi(s, m) < \pi/2$ when $m < 0$ is sufficiently small. On the other hand, by the definition of the Prüfer angle and (6.6.11), (6.6.12) we find that for $s \in (0, \bar{b}]$

$$\tan\phi(s, m) = \left(\frac{z}{dz/ds}\right)(s, m) \to 0 \quad \text{as } m \to -\infty.$$

Hence $\phi(s, m) \to 0$ as $m \to -\infty$. We observe that for a fixed $m < 0$, Eq. (6.6.10) is a Sturm-majorant of Eq. (6.6.8) when $\lambda < 0$ is sufficiently small. Thus by Theorem 6.3.1, we have $\theta(t, \lambda) \to 0$ as $\lambda \to -\infty$ for $t \in (a, b]$.

□

Now we are ready to prove Theorem 6.6.1.

PROOF OF THEOREM 6.6.1. It is known from Lemmas 6.6.1 and 6.6.2 that all eigenvalues of SLP (S-L), (S) are real and geometrically simple.

Let $x(t, \lambda)$ be the solution of Eq. (S-L) satisfying IC (6.6.5). Then $x(t, \lambda)$ satisfies the first BC in (S) for any $\lambda \in \mathbb{R}$. Let $\theta(t, \lambda)$ be the Prüfer angle of $x(t, \lambda)$. Then by Lemma 6.6.3, for each $n \in \mathbb{N}_0$, there exists a unique $\lambda_n \in \mathbb{R}$ such that $\theta(b, \lambda_n) = n\pi + \beta$. This means that $x(t, \lambda_n)$ satisfies the second BC in (S), and hence λ_n is an eigenvalue of SLP (S-L), (S) with $u_n(t) := x(t, \lambda_n)$ as an associated eigenfunction. Clearly, the sequence $\{\lambda_n\}_{n=0}^{\infty}$ satisfies (6.6.4), and for each $n \in \mathbb{N}_0$, $u_n(t)$ has exactly n zeros in (a, b) due to Theorem 6.2.1.

□

We comment that the idea for the proof of Theorem 6.6.1 is from [22, Section XI, 4] which utilizes the Prüfer angle analysis. A modified proof is given in [45, Chapter 13] for the same result but under a more general assumption on the coefficient functions p, q, and w. Theorem 6.6.1 can also be proved using the Fourier transformation or the spectral theory of self-adjoint operators on Hilbert spaces. However, these methods are beyond the scope of this book.

The next theorem shows the relations among all eigenfunctions of SLP (S-L), (S).

THEOREM 6.6.2. *Let λ_n be the n-th eigenvalue of SLP (S-L), (S) with an associated eigenfunction u_n for $n \in \mathbb{N}_0$. Then for any $m, n \in \mathbb{N}_0$ such that $m \neq n$, u_m and u_n are orthogonal to each other in the sense that*

$$(6.6.13) \qquad \int_a^b w(t)u_m(t)u_n(t)\,dt = 0.$$

PROOF. From the assumption we see that for $t \in [a, b]$

$$(6.6.14) \qquad -(p(t)u_m{}'(t))' + q(t)u_m(t) = \lambda_m w(t)u_m(t)$$

and

$$(6.6.15) \qquad -(p(t)u_n{}'(t))' + q(t)u_n(t) = \lambda_n w(t)u_n(t).$$

By multiplying both sides of (6.6.14) and (6.6.15) by $u_n(t)$ and $u_m(t)$, respectively, and then subtracting them, we obtain that

$$(6.6.16) \quad -u_n(t)(p(t)u_m'(t))' + u_m(t)(p(t)u_n{}'(t))' = (\lambda_m - \lambda_n)w(t)u_m(t)u_n(t).$$

By integration by parts we have

$$\int_a^b [-u_n(t)(p(t)u_m'(t))' - u_m(t)(p(t)u_n{}'(t))']\,dt$$
$$= [-u_n(t)(p(t)u_m'(t)) - u_m(t)(p(t)u_n{}'(t))]_a^b = 0$$

since both $u_m(t)$ and $u_n(t)$ satisfy BC (S). Thus, integrating both sides of (6.6.16) from a to b leads to

$$(\lambda_m - \lambda_n) \int_a^b w(t)u_m(t)u_n(t)\,dt$$
$$= \int_a^b [-u_n(t)(p(t)u_m'(t))' - u_m(t)(p(t)u_n{}'(t))']\,dt = 0.$$

Therefore, (6.6.13) holds since $\lambda_m \neq \lambda_n$. □

The last theorem will tell us how the first eigenvalue λ_0 of SLP (S-L), (S) changes when the functions q and w in the equation and the angles α and β in the BC change. Here, by saying that λ_0 is strictly increasing in q we mean that when the function q is replaced by q_1 and q_2, respectively, where $q_1, q_2 \in C([a, b], \mathbb{R})$ such that $q_1(t) \leq q_2(t)$, and $q_1(t) \not\equiv q_2(t)$ on $[a, b]$, then the corresponding first eigenvalues λ_0^1 and λ_0^2 satisfy $\lambda_0^1 < \lambda_0^2$. The monotone property of λ_0 in w is similarly defined.

THEOREM 6.6.3. *Let λ_0 be the first eigenvalue of SLP (S-L), (S). Then*

(a) λ_0 is strictly increasing in q;

(b) λ_0 is strictly decreasing in w when $\lambda_0 > 0$, and is strictly increasing in w when $\lambda_0 < 0$;

(c) λ_0 is strictly decreasing in the BC angle α, and is strictly increasing in the BC angle β.

PROOF. We only prove Part (a). The proofs for Parts (b) and (c) are left as exercises, see Exercise 6.27.

Let $q_1, q_2 \in C([a, b], \mathbb{R})$ such that $q_1(t) \leq q_2(t)$, and $q_1(t) \not\equiv q_2(t)$ on $[a, b]$. Let λ_0^1 and λ_0^2 be the first eigenvalues of SLP (S-L), (S), where q is replaced by q_1 and q_2, with associated eigenfunctions u_1 and u_2, respectively. Since u_1 and u_2 do not have zeros in (a, b), we may assume that $u_1(t) > 0$ and $u_2(t) > 0$ on (a, b). Then for $t \in [a, b]$

$$(6.6.17) \qquad -(p(t)u_1{}'(t))' + q_1(t)u_1(t) = \lambda_0^1 w(t)u_1(t)$$

and

$$(6.6.18) \qquad -(p(t)u_2{}'(t))' + q_2(t)u_2(t) = \lambda_0^2 w(t)u_2(t).$$

By multiplying both sides of (6.6.17) and (6.6.18) by $u_2(t)$ and $u_1(t)$, respectively, and then subtracting them, we obtain that

$$(6.6.19) \quad -u_2(t)(p(t)u_1{}'(t))' + u_1(t)(p(t)u_2{}'(t))' + (q_1(t) - q_2(t))u_1(t)u_2(t)$$
$$= (\lambda_0^1 - \lambda_0^2)w(t)u_1(t)u_2(t).$$

By integration by parts we have

$$\int_a^b [-u_2(t)(p(t)u_1{}'(t))' + u_1(t)(p(t)u_2{}'(t))'] \, dt$$
$$= [-u_2(t)(p(t)u_1'(t)) + u_1(t)(p(t)u_2{}'(t))]_a^b = 0$$

since both $u_1(t)$ and $u_2(t)$ satisfy BC (S). Thus, integrating both sides of (6.6.19) from a to b leads to

$$(\lambda_0^1 - \lambda_0^2) \int_a^b w(t)u_1(t)u_2(t) \, dt = \int_a^b [(q_1(t) - q_2(t))u_1(t)u_2(t)] \, dt < 0.$$

This shows that $\lambda_0^1 < \lambda_0^2$ and hence justifies the conclusion. $\qquad \square$

REMARK 6.6.1. (i) It is also true that the first eigenvalue λ_0 is strictly increasing in p. However, it cannot be proved this way. I would like the reader to figure out the reason.

(ii) The monotone properties of the first eigenvalue λ_0 given in Theorem 6.6.3 and Part (a) of this remark can be extended to the other eigenvalues λ_n, $n \in \mathbb{N}$. However, the proofs involve the notion of "Frechét derivatives" of functions, see [32, Theorem 4.2] or [49, Theorem 3.6.1] for the detail.

The corollary below follows from Theorem 6.6.3.

COROLLARY 6.6.1. Let λ_0 be the first eigenvalue of SLP (S-L), (S).

(a) Assume $q(t) \equiv 0$ on $[a, b]$ and $\alpha = \beta = \pi/2$. Then $\lambda_0 = 0$.

(b) Assume $q(t) \geq 0$ on $[a, b]$, $\alpha \leq \pi/2$, and $\beta \geq \pi/2$. Moreover, the conditions in Part (a) are not satisfied. Then $\lambda_0 > 0$.

(c) Assume $q(t) \leq 0$ on $[a, b]$, $\alpha \geq \pi/2$, and $\beta \leq \pi/2$. Moreover, the conditions in Part (a) are not satisfied. Then $\lambda_0 < 0$.

PROOF. Note that when $q(t) \equiv 0$ on $[a, b]$ and $\alpha = \beta = \pi/2$, $\lambda = 0$ is an eigenvalue of SLP (S-L), (S) with $u(t) \equiv 1$ as an associated eigenfunction. Since $u(t)$ does not have zeros in (a, b), $\lambda = 0$ is the first eigenvalue. Therefore, Part (a) holds.

Parts (b) and (c) are immediate consequences of Part (a) and Theorem 6.6.3. □

COROLLARY 6.6.2. *To emphasize the dependence on the BC angles α and β, let $\lambda_n := \lambda_n(\alpha, \beta)$ be the n-th eigenvalue of SLP (S-L), (S) for $n \in \mathbb{N}_0$. Then we have*

$$\lambda_n(\alpha, \beta) \leq \lambda_n(0, \pi) \quad \text{for all } \alpha \in [0, \pi) \text{ and } \beta \in (0, \pi],$$

and the equal sign holds only if $(\alpha, \beta) = (0, \pi)$. In other words, among the n-th eigenvalues of all SLPs consisting of Eq. (S-L) and separated BCs, the one with Dirichlet BC is the largest.

PROOF. This follows from Theorem 6.6.3, Part (c) and Remark 6.6.1, Part (ii) directly. □

We point out that Theorem 6.6.3 and Corollaries 6.6.1 and 6.6.2 are extracted from Kong and Zettl [32] with a simplified proof.

6.6.2. SLPs with Periodic BCs. The results on the SLPs with periodic BCs can be obtained by using the abstract spectral theory for self-adjoint operators. In this book, we employ an elementary approach. More specifically, we will study the SLP with periodic BC (S-L), (P) based on the results in Subsection 6.6.1 for the SLP with separated BC (S-L), (S).

Let $\{\mu_n\}_{n=0}^{\infty}$ be the eigenvalues of the SLP consisting of the Eq. (S-L) and the Dirichlet BC

(D) $$x(a) = 0, \quad x(b) = 0;$$

and $\{\nu_n\}_{n=0}^{\infty}$ be the eigenvalues of the SLP consisting of the Eq. (S-L) and the Neumann BC

(N) $$(px')(a) = 0, \quad (px')(b) = 0.$$

Then the results in Example 6.6.1, Part (c) can be extended to the general SLPs with periodic BCs as stated in the theorem below. Here, $\{A, B\}$ means both A and B.

THEOREM 6.6.4. *SLP (S-L), (P) has a countably infinite number of eigenvalues λ_n, $n \in \mathbb{N}_0$, which are all real, bounded below and unbounded above, and can be arranged to satisfy the coupling relations with μ_n and ν_n for $n \in \mathbb{N}_0$:*

$$
\begin{aligned}
(6.6.20) \quad \nu_0 &\leq \lambda_0 < \{\mu_0, \nu_1\} < \lambda_1 \leq \{\mu_1, \nu_2\} \leq \lambda_2 \\
&< \{\mu_2, \nu_3\} < \lambda_3 \leq \{\mu_3, \nu_4\} \leq \lambda_4 < \cdots \\
&< \{\mu_{2n}, \nu_{2n+1}\} < \lambda_{2n+1} \leq \{\mu_{2n+1}, \nu_{2n+2}\} \leq \lambda_{2n+2} < \cdots.
\end{aligned}
$$

Furthermore, we have the following:

(a) λ_0 *is geometrically simple; and for $n \geq 1$, λ_n may be geometrically simple or double, and λ_n is geometrically double $\iff \lambda_n = \mu_i = \nu_j$ for some $i, j \in \mathbb{N}_0$.*

(b) *The eigenfunctions associated with λ_0 has no zeros in $[a, b]$, and for $n \in \mathbb{N}_0$, the eigenfunctions associated λ_{2n+1} and λ_{2n+2} have exactly $2n + 2$ zeros in the half-open interval $[a, b)$.*

REMARK 6.6.2. From Theorems 6.6.4 and 6.6.1 we see that the eigenvalues λ_n, $n \in \mathbb{N}_0$, of SLP (S-L), (P) satisfy

$$\lambda_0 < \lambda_1 \leq \lambda_2 < \lambda_3 \leq \lambda_4 < \cdots < \lambda_{2n+1} \leq \lambda_{2n+2} < \cdots, \quad \text{and } \lambda_n \to \infty \text{ as } n \to \infty.$$

To prove Theorem 6.6.4, we change SLP (S-L), (P) to a problem for a first-order system. Let $y = \begin{bmatrix} x \\ px' \end{bmatrix}$. Then SLP (S-L), (P) becomes the problem consisting of the vector-valued equation

$$(6.6.21) \qquad y' = A(t, \lambda)y \quad \text{with } A = \begin{bmatrix} 0 & 1/p(t) \\ q(t) - \lambda w(t) & 0 \end{bmatrix}$$

and the BC

$$(6.6.22) \qquad\qquad\qquad y(a) = y(b).$$

It is easy to see that $\lambda \in \mathbb{R}$ is an eigenvalue of SLP (S-L), (P) with an eigenfunction $x(t, \lambda) \iff \lambda \in \mathbb{R}$ is an eigenvalue of the problem (6.6.21), (6.6.22) with the eigenfunction $y(t, \lambda) := \begin{bmatrix} x \\ px' \end{bmatrix}(t, \lambda)$.

Let $u(t, \lambda)$ and $v(t, \lambda)$ be the solutions of Eq. (S-L) satisfying

$$u(a, \lambda) = 1, \quad (pu')(a, \lambda) = 0$$

and

$$v(a, \lambda) = 0, \quad (pv')(a, \lambda) = 1.$$

Denote

$$Y(t, \lambda) = \begin{bmatrix} u & v \\ pu' & pv' \end{bmatrix}(t, \lambda)$$

and

$$\Phi(\lambda) = \begin{bmatrix} \phi_{11} & \phi_{12} \\ \phi_{21} & \phi_{22} \end{bmatrix}(\lambda) := Y(b, \lambda).$$

Clearly, $Y(t, \lambda)$ is the principal matrix solution of Eq. (6.6.21) at a. Note that $\operatorname{tr} A(t, \lambda) \equiv 0$ on $[a, b]$. By Theorem 2.2.3, we have that $\det Y(t, \lambda) \equiv \det Y(a, \lambda) = 1$ for all $t \in [a, b]$. In particular, $\det \Phi(\lambda) = 1$. We define

$$(6.6.23) \qquad D(\lambda) := \operatorname{tr} \Phi(\lambda) = \phi_{11}(\lambda) + \phi_{22}(\lambda).$$

Then we have the following result:

LEMMA 6.6.4. $\lambda \in \mathbb{R}$ *is an eigenvalue of SLP* (S-L), (P) $\iff D(\lambda) = 2$.

PROOF. By Lemma 2.2.3, Part (c), if $y(t, \lambda)$ is a nontrivial solution of Eq. (6.6.21), then $y(t, \lambda) = Y(t, \lambda)c$ for some $c \neq 0$. Thus, we see that

$\lambda \in \mathbb{R}$ is an eigenvalue of SLP (S-L), (P)

$\Longleftrightarrow$ there exists a nontrivial solution $y(t, \lambda)$ such that $y(a, \lambda) = y(b, \lambda)$

$\Longleftrightarrow$ $(\Phi(\lambda) - I)c = 0$ for some $c \neq 0$

$\Longleftrightarrow$ $\det(\Phi(\lambda) - I) = 0$

$\Longleftrightarrow$ $\det \Phi(\lambda) - D(\lambda) + 1 = 0$

$\Longleftrightarrow$ $D(\lambda) = 2$.

$\square$

The next two lemmas characterize the function $D(\lambda)$ at μ_n and ν_n.

LEMMA 6.6.5. (a) $\lambda = \mu_n$ for some $n \in \mathbb{N}_0 \iff \phi_{12}(\lambda) = 0$. Moreover, $v(t, \mu_n)$ is an eigenvector associated with μ_n.

(b) $\lambda = \nu_n$ for some $n \in \mathbb{N}_0 \iff \phi_{21}(\lambda) = 0$. Moreover, $u(t, \nu_n)$ is an eigenvector associated with ν_n.

PROOF. We only prove Part (a). The proof for Part (b) is similar and hence is omitted.

Let $\lambda = \mu_n$ for some $n \in \mathbb{N}_0$. Then SLP (S-L), (D) with $\lambda = \mu_n$ has an eigenfunction $x(t, \mu_n)$ associated with μ_n. Hence $x(a, \mu_n) = x(b, \mu_n) = 0$. Note that $v(a, \lambda) = 0$ for any $\lambda \in \mathbb{R}$. This means that $x(t, \mu_n)$ and $v(t, \mu_n)$ are linearly dependent on $[a, b]$. As a result, $\phi_{12}(\mu_n) = v(b, \mu_n) = x(b, \mu_n) = 0$.

Assume $\phi_{12}(\lambda) = 0$, i.e., $v(b, \lambda) = 0$, for some $\lambda \in \mathbb{R}$. Note that $v(a, \lambda) = 0$. This means that $v(t, \lambda)$ is an eigenfunction of SLP (S-L), (D) with eigenvalue λ. Therefore, $\lambda = \mu_n$ for some $n \in \mathbb{N}_0$. $\square$

LEMMA 6.6.6. For $n \in \mathbb{N}_0$ we have

(6.6.24) $\quad (-1)^{n+1} D(\mu_n) \geq 2 \quad and \quad (-1)^n D(\nu_n) \geq 2$.

Moreover,

(6.6.25) $\quad \nu_0 < \{\mu_0, \nu_1\} < \{\mu_1, \nu_2\} < \cdots < \{\mu_n, \nu_{n+1}\} < \cdots$.

PROOF. By Lemma 6.6.5 we have that $v(t, \mu_n)$ is an eigenfunction associated with μ_n and $\phi_{12}(\mu_n) = 0$. Since

$$\det \Phi(\mu_n) = \det \begin{bmatrix} \phi_{11} & 0 \\ \phi_{21} & \phi_{22} \end{bmatrix} (\mu_n) = 1,$$

we see that $\phi_{22}(\mu_n) = [\phi_{11}(\mu_n)]^{-1}$. Then

(6.6.26) $\quad D(\mu_n) = \phi_{11}(\mu_n) + \phi_{22}(\mu_n) = \phi_{11}(\mu_n) + [\phi_{11}(\mu_n)]^{-1}$.

It follows from Theorem 6.6.1 that $v(t, \mu_n)$ has exactly n zeros in (a, b). Note that $v(a, \mu_n) = v(b, \mu_n) = 0$, by Theorem 6.3.3, $u(t, \mu_n)$ has exactly $n + 1$ zeros in (a, b). Since $u(a, \mu_n) > 0$, we have $(-1)^{n+1} u(b, \mu_n) > 0$. This together with (6.6.26) shows that $(-1)^{n+1} D(\mu_n) \geq 2$. Similarly we can prove that $(-1)^n D(\nu_n) \geq 2$.

From Theorem 6.6.1 we have that

$$\mu_0 < \mu_1 < \cdots < \mu_n < \cdots$$

and

$$\nu_0 < \nu_1 < \cdots < \nu_n < \cdots.$$

Also from Corollary 6.6.2 we see that $\nu_0 < \mu_0$. Then (6.6.25) follows from (6.6.24) and the zero-counting properties for $u(t, \mu_n)$ and $v(t, \nu_n)$. $\qquad\square$

To establish (6.6.20), we need the following results for the function $D(\lambda)$:

LEMMA 6.6.7. $D(\lambda)$ is continuously differentiable on $\mathbb{R}$, and for any $\lambda \in \mathbb{R}$

$$(6.6.27) \quad 4\phi_{12}(\lambda)D'(\lambda) = -\int_a^b [2\phi_{12}(\lambda)u(s, \lambda) + (\phi_{22} - \phi_{11})(\lambda)v(s, \lambda)]^2 w(s)\, ds$$

$$-[4 - D^2(\lambda)]\int_a^b v^2(s, \lambda)w(s)\, ds$$

and

$$(6.6.28) \quad 4\phi_{21}(\lambda)D'(\lambda) = \int_a^b [2\phi_{21}(\lambda)v(s, \lambda) - (\phi_{22} - \phi_{11})(\lambda)u(s, \lambda)]^2 w(s)\, ds$$

$$+[4 - D^2(\lambda)]\int_a^b u^2(s, \lambda)w(s)\, ds.$$

PROOF. The continuous differentiability of $D(\lambda)$ follows from Theorem 1.5.3. To prove (6.6.27) and (6.6.28) we denote $Y_\lambda(t, \lambda) = \dfrac{\partial}{\partial \lambda}Y(t, \lambda)$. From (6.6.21) and Theorem 1.5.3 we have

$$Y_\lambda'(t, \lambda) = A(t, \lambda)Y_\lambda(t, \lambda) + A_\lambda(t, \lambda)Y(t, \lambda), \quad Y_\lambda(a, \lambda) = 0.$$

This together with the variation of parameters formula (2.3.2) implies that

$$Y_\lambda(t, \lambda) = \int_a^t Y(t, \lambda)Y^{-1}(s, \lambda)A_\lambda(s, \lambda)Y(s, \lambda)\, ds.$$

Below we abbreviate the notation by omitting the arguments. By (6.6.23),

$$
\begin{aligned}
D' &= \operatorname{tr}\Phi_\lambda = \operatorname{tr}\int_a^b \Phi Y^{-1}A_\lambda Y\, ds \\
&= \operatorname{tr}\int_a^b \begin{bmatrix} \phi_{11} & \phi_{12} \\ \phi_{21} & \phi_{22} \end{bmatrix}\begin{bmatrix} pv' & -v \\ -pu' & u \end{bmatrix}\begin{bmatrix} 0 & 0 \\ -w & 0 \end{bmatrix}\begin{bmatrix} u & v \\ pu' & pv' \end{bmatrix} ds \\
&= \operatorname{tr}\int_a^b \begin{bmatrix} \phi_{11}uv - \phi_{12}u^2 & * \\ * & \phi_{21}v^2 - \phi_{22}uv \end{bmatrix} w\, ds \\
&= \int_a^b [\phi_{21}v^2 - (\phi_{22} - \phi_{11})uv - \phi_{12}u^2]w\, ds.
\end{aligned}
$$

Note that

$$4 - D^2 = 4(\phi_{11}\phi_{22} - \phi_{12}\phi_{21}) - (\phi_{11} + \phi_{22})^2 = -(4\phi_{12}\phi_{21} + (\phi_{22} - \phi_{11})^2).$$

Thus,

$$
\begin{aligned}
4\phi_{12}D' &= \int_a^b [4\phi_{12}\phi_{21}v^2 - 4\phi_{12}(\phi_{22} - \phi_{11})uv - 4\phi_{12}^2 u^2]w\,ds \\
&= \int_a^b \left[-\left(2\phi_{12}u + (\phi_{22} - \phi_{11})v\right)^2 + \left(4\phi_{12}\phi_{21} + (\phi_{22} - \phi_{11})^2\right)v^2 \right]w\,ds \\
&= -\int_a^b [2\phi_{12}u + (\phi_{22} - \phi_{11})v]^2 w\,ds - (4 - D^2)\int_a^b v^2 w\,ds,
\end{aligned}
$$

i.e., (6.6.27) holds. The identity (6.6.28) can be proved similarly. □

Based on Lemma 6.6.7, we derive relations among $\{\lambda_n\}_{n=0}^\infty$, $\{\mu_n\}_{n=0}^\infty$, and $\{\nu_n\}_{n=0}^\infty$.

LEMMA 6.6.8. *For any $n \in \mathbb{N}_0$,*

(a) *λ_n is geometrically double $\Longleftrightarrow$ there exist $i, j \in \mathbb{N}_0$ such that $\lambda_n = \mu_i = \nu_j$;*

(b) *in general, there exist $i, j \in \mathbb{N}_0$ such that $\lambda_n \leq \{\mu_i, \nu_j\} \leq \lambda_{n+1}$.*

PROOF. (a) By definition, λ_n is a geometrically double eigenvalue means that for $\lambda = \lambda_n$, every nontrivial solution of Eq. (S-L) is an eigenfunction of the SLP (S-L), (P). This is equivalent to saying that

$$\Phi(\lambda_n) = Y(b, \lambda_n) = Y(a, \lambda_n) = I.$$

If λ_n is geometrically double, then $\phi_{12}(\lambda_n) = \phi_{21}(\lambda_n) = 0$. By Lemma 6.6.5, we have $\lambda_n = \mu_i = \nu_j$ for some $i, j \in \mathbb{N}_0$.

If $\lambda_n = \mu_i = \nu_j$ for some $i, j \in \mathbb{N}_0$, again by Lemma 6.6.5, we have $\phi_{12}(\lambda_n) = \phi_{21}(\lambda_n) = 0$. Note that $D(\lambda_n) = \phi_{11}(\lambda_n) + \phi_{22}(\lambda_n) = 2$. From (6.6.27) or (6.6.28) we see that $\phi_{11}(\lambda_n) - \phi_{22}(\lambda_n) = 0$. It follows that $\phi_{11}(\lambda_n) = \phi_{22}(\lambda_n) = 1$ and hence $\Phi(\lambda_n) = I$. This shows that λ_n is geometrically double.

(b) If one of λ_n and λ_{n+1} is geometrically double, then the conclusion follows from Part (a). Assume both λ_n and λ_{n+1} are geometrically simple. Note that $D(\lambda_n) = D(\lambda_{n+1}) = 2$ and $D'(\lambda)$ has opposite signs at λ_n and λ_{n+1}. It is easy to see from (6.6.27) or (6.6.28) that $D'(\lambda) \neq 0$ for $\lambda = \lambda_n, \lambda_{n+1}$. Thus $D'(\lambda_n)D'(\lambda_{n+1}) < 0$. Applying this to (6.6.27) and (6.6.28) we obtain that

$$\phi_{12}(\lambda_n)\,\phi_{12}(\lambda_{n+1}) < 0 \quad \text{and} \quad \phi_{21}(\lambda_n)\,\phi_{21}(\lambda_{n+1}) < 0.$$

By the continuity of the functions $\phi_{12}(\lambda)$ and $\phi_{21}(\lambda)$, each of them has a zero in $(\lambda_n, \lambda_{n+1})$. Then the conclusion follows from Lemma 6.6.5.

□

Now we are ready to prove Theorem 6.6.4.

PROOF OF THEOREM 6.6.4. By Lemma 6.6.6, Theorem 6.6.1, and the continuity of the function $D(\lambda)$, we see that the equation $D(\lambda) = 2$ has an infinite number of zeros λ_n, $n \in \mathbb{N}_0$ such that $\lambda_n \to \infty$ as $n \to \infty$. By Lemma 6.6.4, λ_n, $n \in \mathbb{N}_0$, are the eigenvalues of SLP (S-L), (S).

We claim that $\lambda_0 \geq \nu_0$. Otherwise, either (i) $\lambda_0 \leq \lambda_1 < \nu_0$, or (ii) $\lambda_0 < \nu_0 \leq \lambda_1$. Case (i) contradicts Lemma 6.6.8 since there is no $j \in \mathbb{N}_0$ such that $\lambda_0 \leq \nu_j \leq \lambda_1$. Assume Case (ii) holds. By Lemma 6.6.8, there exists an $i \in \mathbb{N}_0$ such that $\lambda_0 \leq \mu_i \leq \lambda_1$. From (6.6.25), we have $\nu_0 < \mu_0$ and hence $i = 0$. It implies that μ_0 and ν_0 have the same sign. However, this contradicts (6.6.24) with $n = 0$. Therefore, the coupling relations (6.6.20) follow from Lemmas 6.6.6 and 6.6.8.

The result on the geometric multiplicities of λ_n is given in Lemma 6.6.8, Part (a).

Finally, for any $n \in \mathbb{N}_0$, if an eigenfunction $x(t)$ associated with λ_n has zeros in $[a, b)$, then the number of zeros must be an even number. In fact, due to the periodic BC (P), either $x(a) = x(b) \neq 0$, hence $x(t)$ must have an even number of zeros in (a, b); or $x(a) = x(b) = 0$ and $x'(a) = x'(b) \neq 0$, hence $x(t)$ must have an even number of zeros in $[a, b)$. Then the zero-counting properties of the eigenfunctions associated with λ_n are obtained by applying the inequalities in (6.6.20) and the Sturm comparison theorem 6.3.2. We leave it as an exercise, see Exercise 6.30. $\qquad \square$

REMARK 6.6.3. From (6.6.20) we see that for $n \in \mathbb{N}_0$, $\mu_n \geq \lambda_n$. Combining this fact and Corollary 6.6.2, we conclude that for $n \in \mathbb{N}_0$, among all n-th eigenvalues of SLPs consisting of Eq. (S-L) and separated or periodic BCs, the Dirichlet eigenvalue μ_n is the largest.

The coupling relations in (6.6.20) are illustrated in Fig. 6.2 below.

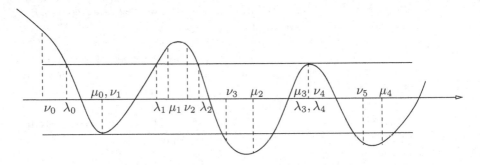

FIGURE 6.2. D-function

With essentially the same proof as that of Theorem 6.6.2, we have the following result.

THEOREM 6.6.5. *For any $m, n \in \mathbb{N}_0$, let λ_m and λ_n be eigenvalues of SLP (S-L), (P) with u_m and u_n any associated eigenfunctions, respectively. Assume $\lambda_m \neq \lambda_n$. Then u_m and u_n are orthogonal to each other in the sense of (6.6.13).*

We comment that the above results for SLP with periodic BC (S-L), (P) are partially from Coddington and Levenson [11, Section 8.3]. These results have been extended to SLPs with general self-adjoint coupled BCs, see Eastham, Kong, Wu, and Zettl [**14**] for the details.

Exercises

6.1. Consider the equation $(p(t)x')' + q(t)\, x = 0$ where $p, q \in C((a, b), \mathbb{R})$, $p(t) > 0$, and $-\infty \leq a < b \leq \infty$.

    (a) Show that every nontrivial solution has a finite number of zeros on $[c, d] \subset (a, b)$.

    (b) Assume $x_1(t)$ and $x_2(t)$ are nontrivial solutions satisfying $x_1'(c) = x_2'(c) = 0$ for some $c \in (a, b)$. Show that $x_1(t)$ and $x_2(t)$ have exactly the same zeros in (a, b).

6.2. Assume $a_1 \in C^1(a, b)$ and $a_2 \in C(a, b)$. Find a function $\phi(t) \in C^2(a, b)$ such that $x = \phi(t)y$ transforms the Eq. (nh) for x to an equation in the form of (6.1.5) for y.

6.3. (a) Show that every solution of the equation $x'' + \dfrac{2t}{t^2 + 1}\, x = 0$ has at most one zero on $[0, \pi]$.

    (b) What can you say about the number of zeros of any solution of the equation

$$x'' + \left(1 + \frac{2t}{t^2 + 1}\right) x = 0 \text{ on } [0, \infty), \text{ and why?}$$

6.4. Let $q_1, q_2 \in C[0, 1]$ be such that $q_1(t) \leq q_2(t)$, $0 \leq t \leq 1$. Suppose that the problems

$$x_1'' + q_1(t)x_1 = 0, \quad x_2'' + q_2(t)x_2 = 0, \quad x_1(0) = x_2(0) = 0, \quad x_1'(1) = 0,$$

have nontrivial solutions $x_1(t)$ and $x_2(t)$. Show that $x_2'(c) = 0$ for some $c \in (0, 1]$.

6.5. Let $q \in C[a, b]$. Suppose that the differential inequality $y'' + q(t)y \leq 0$ has a C^2-solution $y(t)$ satisfying $y(t) > 0$ on $[a, b]$. Show that every nontrivial solution of the differential equation $x'' + q(t)x = 0$ has at most one zero in the interval $[a, b]$.

6.6. Consider Bessel's equation

$$x'' + \frac{x'}{t} + \left(1 - \frac{\mu^2}{t^2}\right) x = 0$$

where $\mu \in \mathbb{R}$. Show that every solution of this equation has an infinite number of zeros t_n, $n = 1, 2, \ldots$, satisfying $\lim_{n \to \infty}(t_{n+1} - t_n) = \pi$.

6.7. Use Remark 6.1.1, Part (b) to extend the results of Corollary 6.3.1 for Eq. (6.3.4) to Eq. (h-s) when $\int_0^\infty 1/p(t)\,dt = \infty$.

6.8. Show that

    (a) if $\limsup_{t\to\infty} t^2 q(t) < 1/4$, then Eq. (6.4.5) is nonoscillatory;

    (b) if $\liminf_{t\to\infty} t^2 q(t) > 1/4$, then Eq. (6.4.5) is oscillatory.

6.9. Assume that $\int_a^b 1/p(t)\,dt = \infty$ and $\int_a^b q(t)\,dt = \infty$. Show that Eq. (h-s) is oscillatory on $[a, b)$.

6.10. Find a function $q \in C([0, \infty), \mathbb{R})$ which satisfies (6.4.12) but not (6.4.6).

6.11. In light of Exercise 6.9, extend Theorem 6.4.5 to Eq. (h-s).

6.12. Prove Corollary 6.4.3.

6.13. (a) Show that Eq. (6.4.5) with $q(t) = t\sin t$ is oscillatory using Theorem 6.4.7 and Corollary 6.4.3, respectively.

    (b) Show that Eq. (6.4.5) with $q(t) = t^2 \sin t$ is oscillatory using Corollary 6.4.3. Does Theorem 6.4.7 work for this case?

    (c) Show that Eq. (6.4.5) with $q(t) = k\left(\sin(\frac{2\pi}{3}t) - \frac{1}{2}\right)$ is oscillatory for $k > 32$ using Corollary 6.4.3. Note that in this case, $\int_0^\infty q(t)\,dt = -\infty$.

6.14. Consider the following BVPs:

    (a) $x'' + x = h(t)$, $x(0) = 0$, $x(\pi) = 0$;

    (b) $x'' + x = h(t)$, $x(0) = 0$, $x'(\pi) = 0$;

    (c) $x'' + x = h(t)$, $x'(0) = 0$, $x'(\pi/2) = 0$.

For each of them, determine if the BVP has a unique solution for any $h \in C[0, \pi]$; and if it does, find the unique solution using the Green's function.

6.15. Consider the BVP

$$x'' - 2x' + x = h(t), \quad x(0) = 0, \; 2x(1) - x'(1) = 0.$$

    (a) Show that this BVP does not have a Green's function.

    (b) Find a condition on the function h such that the BVP has a solution, and then find the solutions of the BVP for this case.

6.16. Determine if the BVP

$$x'' - x' = h(t), \quad x'(0) = 0, \; x'(1) - 2x(1) = 0$$

has a unique solution for any $h \in C[0, 1]$; and if it does, find the solution using the Green's function.

6.17. For any $h \in C[0, 1]$ and $a, b \in \mathbb{R}$, find the solution of the BVP

$$x'' = h(t), \quad x(0) = a, \; x(1) = b.$$

6.18. Derive Formula (6.5.18) for the Green's function of the BVP (nh-), (P) from (6.5.9).

6.19. Derive Formula (6.5.19) for the Green's function of the BVP (nh+), (P) from (6.5.9).

6.20. Find the solution of the BVP

$$x'' - x = e^t \quad x(0) = x(1), \ x'(0) = x'(1).$$

6.21. Find the solution of the BVP

$$x'' + x = \cos t \quad x(0) = x(\pi), \ x'(0) = x'(\pi).$$

6.22. Consider the BVP

$$x'' + x = h(t), \quad x(0) = x(2\pi), \ x'(0) = x'(2\pi).$$

(a) Show that this BVP does not have a Green's function.
(b) Find a condition on the function h such that the BVP has a solution, and then find the solutions of the BVP for this case.

6.23. Verify the details of Example 6.6.1, Parts (b) and (c).

6.24. Find the eigenvalues and eigenfunctions of the SLP

$$-(tx')' = \frac{\lambda}{x}, \quad x(1) = 0, \ x(e) = 0.$$

6.25. Note that the eigenvalue problem

$$t^2 x'' - \lambda t x' + \lambda x = 0, \quad x(1) = 0, \ x(e) = 0$$

is not an SLP. Show that it does not have any real eigenvalues.

6.26. For Parts (a) and (b) of Example 6.6.1, respectively, verify that the eigenfunctions are orthogonal as defined in Theorem 6.6.2.

6.27. Prove Theorem 6.6.3, Parts (b) and (c).

6.28. (a) Consider the SLP consisting of the equation

$$-(p(t)x')' + [q(t) + l\, w(t)]x = \lambda w(t)x$$

and BC (S), where $p, q, w \in C[a, b]$. For $n \in \mathbb{N}_0$ let $\lambda_n(l)$ be the n-th eigenvalue of this SLP. Show that $\lambda(l) \to \pm\infty$ as $l \to \pm\infty$.
(b) Extend the result in Part (a) to the SLP consisting of the equation

$$-(p(t)x')' + [q(t) + l\, r(t)]x = \lambda w(t)x$$

and BC (S), where $r \in C[a, b]$ such that $r(t) > 0$. Justify your answer.

6.29. For SLP (S-L), (S), denote the n-th eigenvalue by $\lambda_n = \lambda_n(\alpha)$, $n \in \mathbb{N}_0$, to emphasize the dependence of λ_n on α as α changes. By Theorem 6.6.3, Part (b), we know that $\lambda_0(\alpha)$ is strictly decreasing on $[0, \pi)$. From this we conclude that

$$\lim_{\alpha \to \pi-} \lambda_0(\alpha) = -\infty \quad \text{and} \quad \lim_{\alpha \to \pi-} \lambda_{n+1}(\alpha) = \lambda_n(0) \ \text{ for } n \in \mathbb{N}_0.$$

Explain why this is true. You may assume that $\lambda_n(\alpha)$ depends on α continuously.

State a parallel result for the behavior of λ_n, $n \in \mathbb{N}_0$, as β changes.

6.30. Use the Sturm comparison theorem 6.3.2 to prove Theorem 6.6.4, Part (b).

6.31. Let λ_0 be the first eigenvalue of SLP (S-L), (P). Show that $\lambda_0 \geq 0$ if $q(t) \geq 0$ on $[a, b]$.

6.32. Consider the SLP consisting of Eq. (S-L) and the anti-periodic BC

(AP) $x(a) = -x(b), \quad (px')(a) = -(px')(b).$

    (a) Are all eigenvalues of SLP (S-L), (AP) real, and why?

    (b) Show that $\lambda \in \mathbb{R}$ is an eigenvalue of SLP (S-L), (AP) $\iff$ $D(\lambda) = -2$.

    (c) Establish parallel results on the eigenvalues $\{\bar{\lambda}_n\}_{n=0}^{\infty}$ for SLP (S-L), (AP) to those in Theorem 6.6.4 for SLP (S-L), (P).

    (d) How do you compare $\bar{\lambda}_n$ with μ_n for $n \in \mathbb{N}_0$?

6.33. Calculate the eigenvalues and eigenfunctions of the SLP

$$-x'' = \lambda x, \quad x(0) = -x(1), \quad x'(0) = -x'(1);$$

and then verify that the results in Exercise 6.23, Part (c) are satisfied by this problem.

Answers and Hints

Chapter 1

1.1. Use the contradiction method to show that f is not Lipschitz in any neighborhood of 0.

1.4. Use the results in Exercise 1.3.

1.5. (a) Let $r(t) = \int_{t_0}^{t} h(s)u(s)\,ds$.

    (b) Note that $\int_{t_0}^{t} f(s)h(s)e^{\int_{s}^{t} h(\sigma)\,d\sigma}\,ds \le f(t)\int_{t_0}^{t} h(s)e^{\int_{s}^{t} h(\sigma)\,d\sigma}\,ds$.

1.6. Let $t = -\tau$ and $u(t) = v(\tau)$ and apply the original Gronwall inequality to $v(\tau)$.

1.7. (a) IVP has a unique solution.

    (b) IVP has a unique solution when $x_0 \ne 0$ and has at least one solution when $x_0 = 0$.

    (c) IVP has an infinite number of solutions. For example,

$$x = \begin{cases} \left[\frac{2}{3}(t - c)\right]^{3/2}, & t \ge c \\ 0, & t < c \end{cases}$$

    is a solution for any $c > 0$.

1.8. (b) Change the IVP to an integral equation.

1.9. (a) Solution exists and is unique.

    (b) Solution exists.

    (c) Solution does not exist.

1.10. Solution exists when $t_0 \ne k\pi + \pi/2$, solution exists and is unique when $t_0 \ne k\pi + \pi/2$ and $a_2 \ne 0$, and solution does not exist when $t_0 = k\pi + \pi/2$. Here $k \in \mathbb{Z}$.

1.11. Use the uniqueness and the contradiction method.

1.12. Consider the ICs satisfied by the function and use the uniqueness argument.

1.13. (a) Let $x_1(t)$ and $x_2(t)$ be two solutions of the IVP and consider the differential equation for $[x_1(t) - x_2(t)]^2$.

1.14. This is an extension of Example 1.3.1.

1.15. The maximal interval of existence of the solution through $(0,0)$ is $(-\infty, \infty)$, and the maximal interval of existence of the solution through $(\ln 2, -3)$ is $(0, \infty)$.

1.16. Use Corollary 1.4.1.

© Springer International Publishing Switzerland 2014
Q. Kong, *A Short Course in Ordinary Differential Equations*, Universitext,
DOI 10.1007/978-3-319-11239-8

1.18. Use the inequality that $\sqrt{a+b} \leq \sqrt{a}+\sqrt{b}$ for $a, b \geq 0$ and the Gronwall inequality.

1.19. (a) Use the Gronwall inequality.
(b) The solution exists on (a, b).

1.20. Note that $f(t, x) = [f(t, x) - f(t, 0)] + f(t, 0)$ and use the result in Exercise 1.19.

1.21. Use the contradiction method and integration after separating variables.

1.22. Show that $x(t)$ keeps the same sign as x_0 for all $t \in [t_0, \infty)$ by the contradiction method.

1.23. Change the two IVPs to integral equations and then subtract them.

1.24. (b) Use Theorem 1.5.1.

Chapter 2

2.1. Impossible. Use the Wronskian to show.

2.2. Use the Wronskian.

2.4. Use the Wronskian.

2.5. Use Theorem 2.2.3.

2.6. I. Use the uniqueness argument for the first equality and then let $s = -t$ in the first equality to obtain the second equality; or II. use the expression $X(t) = e^{At}$.

2.7. Note that $\langle \phi(t), \psi(t) \rangle = \phi^T(t)\psi(t)$ and show that $(\phi^T(t)\psi(t))' = 0$.

2.8. Let $y_i(t) = x_i(t) - x_{n+1}(t)$, $i = 1, \ldots, n$. Show that $y_i(t)$, $i = 1, \ldots, n$, are solutions of Eq. (H) and are linearly independent.

2.9. (b) $x(t) = c_1 e^t + c_2 e^{3t} + \int_{t_0}^t \dfrac{1}{2e^{3s}(1+e^{-s})}(-e^{t+2s} + e^{3t})\, ds$.

2.11. Write the equation for $y(t)$ as $y' = A(t)y + [B(t) - A(t)]y$, and use the variation of parameters formula.

2.12. (a) $\begin{bmatrix} e^{-t} & e^{-t} - e^{-2t} \\ 0 & e^{-2t} \end{bmatrix}$,

(b) $e^{2t} \begin{bmatrix} \cos(\sqrt{3}t) - \frac{1}{\sqrt{3}}\sin(\sqrt{3}t) & -\frac{1}{\sqrt{3}}\sin(\sqrt{3}t) \\ \frac{4}{\sqrt{3}}\sin(\sqrt{3}t) & \cos(\sqrt{3}t) + \frac{1}{\sqrt{3}}\sin(\sqrt{3}t) \end{bmatrix}$,

(e) $\begin{bmatrix} e^t & 0 & e^t - e^{2t} \\ 0 & e^{2t} & te^{2t} \\ 0 & 0 & e^{2t} \end{bmatrix}$,

(f) Treat the matrix as a diagonal block matrix.

(g) Compare it with Example 2.4.4.

2.15. $e^{At} = \dfrac{1}{3}\begin{bmatrix} e^{5t} + 2e^{-t} & e^{5t} - e^{-t} \\ 2e^{5t} - 2e^{-t} & 2e^{5t} + e^{-t} \end{bmatrix}$, $x(t) = \dfrac{1}{3}\begin{bmatrix} 2e^{5t} + e^{-t} \\ 4e^{5t} - e^{-t} \end{bmatrix}$.

2.16. (a) $e^{At} = e^t \begin{bmatrix} 1 & t \\ 0 & 1 \end{bmatrix}$.

(b) $x(t) = \begin{bmatrix} te^t - (e^t + e^{-t})/2 \\ e^t \end{bmatrix}$.

2.17. Use Lemma 2.5.1 and Theorem 2.2.2.

2.18. (a) Let $X(t)$ be a fundamental matrix solution and V its transition matrix. Show that $x(t) = X(t)c$ satisfies $x(t + \omega) = kx(t) \iff Vc = kc$.

   (b) Immediate from (a) with $k = 1$.

2.19. Use the uniqueness argument.

2.20. (a) Use the uniqueness argument.

   (b) Show $X(-t)$ is also a fundamental matrix solution of equation (H-p) and hence $X(-t) \equiv X(t)$.

2.21. (a) Use the variation of parameters formula in Theorem 2.3.1 with $X(t)$ the principal matrix solution at $t_0 = 0$, Exercise 2.20, and Theorem 2.5.1.

   (b) Use the result in (a).

2.22. (a) Directly from Theorem 2.5.2.

   (b) Use the proof of Theorem 2.5.2.

2.23. (a) Use Corollary 2.5.4.

   (b) Use Theorem 2.5.2.

2.24. (a) Change the second-order equation to a first-order system and then use Corollary 2.5.4.

   (b) Use the results of problem 22.

2.25. Assume $x(t)$ is an $n\omega$-periodic solution, then $x_1(t) := x(t+\omega)$ is an $n\omega$-periodic solution. Show that $x_1(t)$ and $x(t)$ are linearly independent.

Chapter 3

3.1. Use the results in Example 3.1.4.

3.2. (b) Use $X(t - t_0)X^{-1}(0)$ as the principal matrix solution in Theorem 3.2.1.

3.3. Use Theorem 2.5.1 to show $X(t)X^{-1}(t_0) = X(t - m\omega)X^{-1}(t_0 - m\omega)$ for $m \in \mathbb{Z}$.

3.4. See Remark 3.3.1, Part (i).

3.7. (a) Uniformly stable and asymptotically stable, (b) unstable, (c) uniformly stable and asymptotically stable.

3.8. Uniformly stable and asymptotically stable. Show that $\mu_2(e^{-t}I + A(t)) \le e^{-t} + k$.

3.9. (a) Use the Lozinskii measure μ_2.

3.10. (a) Uniformly stable and asymptotically stable for $\alpha < 0$, uniformly stable for $\alpha = 0$, unstable for $\alpha > 0$.

   (b) Uniformly stable and asymptotically stable for $\alpha < 0$, unstable for $\alpha > 0$, undetermined by linearization for $\alpha = 0$.

   (c) Uniformly stable and asymptotically stable for $\alpha < 0$, uniformly stable for $\alpha = 0$, undetermined by linearization for $\alpha > 0$.

3.11. (b) Unstable.

3.12. Use Theorem 3.4.4.

3.13. See the proof of Theorem 3.4.1, Part (a).

3.14. (a) Using Theorem 3.4.2.
      (b) Nothing.

3.15. Unstable.

3.17. (a) Use Theorems 3.2.1 and 3.4.2, Part (a).
      (b) Use Theorems 3.2.1 and 3.4.2, Part (b).

3.18. (a) Uniformly stable and asymptotically stable, (b) unstable, (c) uniformly stable and asymptotically stable, (d) uniformly stable and asymptotically stable.

3.19. Uniformly stable and asymptotically stable.

3.20. Uniformly stable and asymptotically stable for $k < 0$, uniformly stable for $k = 0$, and unstable for $k > 0$.

3.21. (a) Unstable,
      (b) Uniformly stable and asymptotically stable, consider $V(x,y) = ax^{2m} + by^{2n}$.

3.22. $(0,0)$ is uniformly stable and asymptotically stable, (a,b) is unstable.

3.23. Stable but not asymptotically stable, modify the Lyapunov function $V = x^2 + y^2$ by adding fourth-order terms.

3.24. (a) Uniformly stable, use $V = (x + z)^2 + y^2 + z^2$;
      (b) Unstable, use $V = -x^2 - y^2 + z^2$.

3.25. Use the mechanical energy function as the Lyapunov function.

3.26. Uniformly stable and asymptotically stable, use the mechanical energy function as the Lyapunov function.

3.27. Show $f(0) = 0$ and then show $V(x,y) = y^2/2 + \int_0^x (as^3 + f(s))\, ds$ is positive definite.

3.28. Let $V(t,x)$ be the mechanical energy function.

3.29. Let $V(t,x)$ be the mechanical energy function.

3.30. (a) Asymptotically stable, (b) unstable.

Chapter 4

4.3. (a) Yes since the function $\{(x_1(t), y_1(t)) : t \in \mathbb{R}\}$ is a shift of the solution $\{(x(t), y(t)) : t \in \mathbb{R}\}$.
     (b) No since the curves $\{(x_2(t), y_2(t)) : t \in \mathbb{R}\}$ and $\{(x(t), y(t)) : t \in \mathbb{R}\}$ are distinct and intersect.

4.5. Show that the equilibria and the equations for orbits of the two systems are the same.

4.6. (b) Use the result in problem 4 and note that $(0,0)$ is an equilibrium.
     (c) Use the result in problem 4 and note that all points on the circle $x^2 + y^2 = 1$ are equilibria.

4.7. (b) Use the system of differential equations which induces the dynamical system. Show $(0,0)$ is a saddle point, and then find the straight lines which contain orbits by solving the equation for orbits.

4.8. (a) $\phi_t\left(\begin{bmatrix} x_{01} \\ x_{02} \end{bmatrix}\right) = \begin{bmatrix} x_{01}e^{-t} \\ -(x_{01}^2/4)e^{-2t} + (x_{02} + x_{01}^2/4)e^{2t} \end{bmatrix}.$

(b) Show that $x_0 \in E$ implies that $\phi_t(x_0) \in E$ for any $t \in \mathbb{R}$.

4.10. Use the contradiction method.

4.11. (a) Use the contradiction method.

4.12. Show that $(0,0)$ is the only equilibrium which is an unstable spiral-point, the lines $x = \pm 1$ are orbits. By finding the directions of the vector field, show that any other orbits in between spiral away from $(0,0)$ and toward the lines $x = \pm 1$.

4.13. Use the contradiction method.

4.14. (a) $(0,0)$: stable improper node.

(b) $(1,-2)$: saddle-point.

(c) $(3,1)$: unstable spiral-point.

(d) $(1,0)$: center.

4.16. $(1,1)$: stable spiral-point.

4.18. Use Theorem 4.4.3 or change the system to its polar-coordinate form.

4.19. Change the second-order equation to a first-order system of the form as required in Theorem 4.4.3 and then use Theorem 4.4.3.

4.20. (a) $(0,0)$: stable improper node; (a,b): saddle-point.

(b) Use the Lyapunov function $V = x^2 + y^2$ and the Poincaré–Bendixson theorem.

4.21. Use Lemma 4.5.2.

4.22. (a) Use the generalized Poincaré–Bendixson theorem 4.5.3 and Exercise 4.13.

(b) Use the contradiction method and the Poincaré–Bendixson theorem.

4.23. (a) $(0,0)$: saddle point; $(1,0)$: center, using Theorem 4.4.3.

4.24. (a) Equilibrium: $(0,0)$; closed orbit: $r = 1$.

(b) Equilibria: $(0,0)$, $(1,0)$; homoclinic orbit on the unit circle; for all orbits Γ inside the unit circle, $A(\Gamma^-) = \{(0,0)\}$ and $\Omega(\Gamma^+) = \{(1,0)\}$; and for all orbits Γ outside the unit circle, $A(\Gamma^-) = \emptyset$ and $\Omega(\Gamma^+) = \{(1,0)\}$.

(c) Equilibria: $(0,0)$, $(\pm 1,0)$; for all orbits Γ inside the unit circle and above the x-axis, $A(\Gamma^-) = \{(0,0)\}$ and $\Omega(\Gamma^+) = \{(-1,0)\}$; for all orbits Γ inside the unit circle and below the x-axis, $A(\Gamma^-) = \{(0,0)\}$ and $\Omega(\Gamma^+) = \{(1,0)\}$; for all orbits outside the unit circle and not on the x-axis, $A(\Gamma^-) = \emptyset$ and $\Omega(\Gamma^+) = \{(-1,0)\}$.

4.25. Use the Lyapunov function $V = 2x^2 + y^2$ and the Poincaré–Bendixson theorem.

4.26. Use the method in Example 4.6.5, or make the independent variable change $t = -s$ and then apply the result in Example 4.6.5.

4.27. Use Theorem 4.6.1.

4.28. Use Theorem 4.6.2 with $r(x,y) = e^{-2x}$.

4.30. Use Theorem 4.6.2. For $k < 0$, let $r(x, y) = e^{ax}$ and determine the value of a. Yes, let $V(x, y) = x$.

4.31. (a) There is at most 1 closed orbit. Use the contradiction method and the Green's formula on the region bounded by the two closed orbits.

   (b) There are at most n closed orbits.

4.32. (a) Use the contradiction method.

   (b) Use the result in Part (a).

4.33. Use the polar-coordinate form of the system and the Poincaré–Bendixson theorem.

   (a) $r = 1$ is the only limit cycle which is orbitally asymptotically stable.

   (b) $r = 1$ is the only limit cycle which is orbitally unstable.

4.34. (b) Use the polar-coordinate form of the system and the Poincaré–Bendixson theorem.

4.35.   Case 1: $c \neq 0$. Let $\bar{a}(s) = 1 + (a - 1)s^2$, and $\bar{d}(s)$ and $\bar{q}(s)$ be defined in the same way. Let $\bar{c}(s) = cs$ and $\bar{b}(s) = [(\bar{a}\bar{d} - \bar{q})/\bar{c}](s)$.

   Case 2: $b \neq 0$. Similar to Case 1.

   Case 3: $b = c = 0$ and $a, d > 0$. Let $\bar{a}(s) = 1 + (a - 1)s$, $\bar{d}(s) = 1 + (d - 1)s$, and $\bar{c}(s) = \bar{b}(s) = 0$.

   Case 4: $b = c = 0$ and $a, d < 0$. Choose $\bar{a} \in C([0, 1], \mathbb{R})$ such that $\bar{a}(1) = a$, $\bar{a}(1/2) = 0$, and $\bar{a}(0) = 1$; $\bar{d}(s)$ is chosen in the same way. Let $\bar{c}(s) = -\bar{b}(s) = s(1 - s)$.

4.37. Use the contradiction method and Theorem 4.7.2. Note that there are orbits on both sides of $(0, 0)$ in the x-axis.

4.38. (a) $(1, 0)$: stable improper-note, 1; $(3, -1)$: saddle-point, -1.

   (b) Use the contradiction method and Theorem 4.7.2. Note that there are orbits on both sides of $(1, 0)$ in the x-axis.

4.39. Use the homotopy method and Lemma 4.7.2.

4.40. Use the homotopy method and Lemma 4.7.2.

4.41. Find a nonsingular matrix $T \in \mathbb{R}^{4 \times 4}$ such that $B := T^{-1}AT$ is in the form in Lemma 4.8.1.

4.42. Follow the approach in Example 4.8.2. There are an infinite number of center manifolds for Part (b).

4.44. (a) $\mathcal{H}(x_1, x_2) = (x_1, x_2 + x_1^2/5)^T$.

   (b) $\mathcal{H}(x_1, x_2, x_3) = (x_1, x_2 - x_3^2/3, x_3)^T$.

4.45. (a) Stable,   (b) asymptotically stable,   (c) unstable.

Chapter 5

5.1. (a) $\mu = 3$ is a bifurcation value. Subcritical pitchfork bifurcation for $\alpha = 3$ and transcritical bifurcation for $\alpha = 4$.

   (b) $\mu = 0$ is a bifurcation value. Saddle-node bifurcation for $\alpha = 1$ and transcritical bifurcation for $\alpha = 2$.

5.2. (a) Equilibria: $x_1 = 0$ and $x_\pm = 1 \pm \sqrt{1 - \mu}$. Transcritical bifurcation at $\mu = 0$ with $x = 0$ and saddle-node bifurcation at $\mu = 1$ with $x = 1$.

    (d) Saddle-node bifurcation at $\mu = -1/4$ with $x = 1/2$, transcritical bifurcation at $\mu = 0$ with $x = 0$ and at $\mu = 2$ with $x = -1$, subcritical pitchfork bifurcation at $\mu = 1$ with $x = 0$.

5.3. (a) See Examples 5.2.4 and 5.2.5.

    (b) Either do it directly using Theorem 5.2.2 or rewrite the equation to $x' = x^{p-1}(\mu x \pm x^{2m+1} + o(|x^{2m+1}| + |x\mu|)$ and then use the results in Example 5.2.5 and Part (a).

5.4. Equilibria: $(0,0)$, $(\mu/2, 0)$, $(0, 1)$, $(\mu - 1, 2 - \mu)$. Transcritical bifurcations at $\mu = 0$ with $(x_1, x_2) = (0,0)$ and at $\mu = 2$ with $(x_1, x_2) = (1, 0)$, transcritical bifurcation without stability change at $\mu = 1$.

5.7. (a) Transcritical bifurcation at $\mu = 0$ with all equilibria unstable.

    (b) Subcritical pitchfork bifurcation at $\mu = 0$.

5.8. (a) $(0,0)$ is an unstable spiral-point when $\mu > 0$, and is an stable spiral-point when $\mu \le 0$; a limit cycle appears as a circle surrounding $(0,0)$ with $r = \sqrt{\mu}$ when $\mu > 0$.

    (b) $(0,0)$ is an unstable spiral-point when $\mu \ge 0$, and is an stable spiral-point when $\mu < 0$; two limit cycles appear as circles surrounding $(0,0)$ when $\mu \in (-1/4, 0)$ with the inner one orbitally unstable and the outer one orbitally asymptotically stable.

5.9. Use Theorem 5.4.3.

5.10. Use the Lyapunov function $V = \dfrac{1}{2}x_2^2 + (1 - \cos x_1)$ for $\mu = 0$, and use Theorems 5.4.1 and 5.4.2.

Chapter 6

6.1. (a) Use the contradiction method and Lemma 6.1.2.

    (b) Use the Wronskian of x_1 and x_2.

6.3. (a) Compare the given equation with the equation $u'' + u = 0$.

6.4. Use Theorem 6.3.1.

6.5. Find a function $\tilde{q}(t)$ such that $y''(t) + \tilde{q}(t)y(t) = 0$ and then use Theorem 6.3.2.

6.6. Use the transformation given in Problem 2 to change the equation to an equation in the form of (6.1.5).

6.8. Use the substitutions $x = t^{1/2}y$ and $t = e^s$ to change Eq. (6.4.5) to the equation

$$\frac{d^2y}{ds^2} + t^2(q(t) - 1/(4t^2))y = 0 \quad \text{with } t = e^s$$

and then apply Corollary 6.4.1.

6.9. Let $y(t) = p(t)x'(t)/x(t)$.

6.12. For any $T \ge 0$, choose $a, b, c \in \mathbb{R}$ with $T \le a < c < b$ such that (6.4.16) holds.

6.13. (c) Let $a = 3n+1/4$ and $c = 3n+3/4$. Note that $q_1(t) := \sin(\frac{2\pi}{3}t) - \frac{1}{2}$
       satisfies $q_1(a) = 0$, $(q_1(c) - q_1(a))/(c - a) = 1$, and q_1 is concave
       down on (a, c).

6.14. (a) The BVP does not have a unique solution for every $h \in C[0, \pi]$.

       (b) $G(t, s) = \begin{cases} -\sin s \cos t, & 0 \le s \le t \le \pi \\ -\sin t \cos s, & 0 \le t \le s \le \pi. \end{cases}$

6.15. (b) $\int_0^1 sh(s)e^{-s} \, ds = 0$ and $x = c_2 te^t + e^t \int_0^t h(s)e^{-s}(-s+t) \, ds$ for any
       $c_2 \in \mathbb{R}$.

6.16. $G(t, s) = \begin{cases} e^{t-s} - e^{1-s}/2, & 0 \le s \le t \le 1, \\ 1 - e^{1-s}/2, & 0 \le t \le s \le 1. \end{cases}$

       Change the equation to the self-adjoint form (nh-s) and note that the
       right-hand side function becomes $e^{-t}h(t)$. (The Green's function is for
       the original problem.)

6.17. Find the solutions x_1 and x_2 of the BVPs

$$x'' = h(t), \quad x(0) = 0, \quad x(1) = 0$$

       and

$$x'' = 0, \quad x(0) = a, \quad x(1) = b;$$

       respectively.

6.24. Make the independent variable substitution $t = e^s$ to change the equa-
       tion to an equation with constant coefficients.

6.25. Use the same substitution as in the preceding problem.

6.27. (b) Show that

$$(\lambda_0^1 - \lambda_0^2) \int_a^b w_1(t)u_1(t)u_2(t) \, dt = \lambda_0^2 \int_a^b (w_2(t) - w_1(t))u_1(t)u_2(t) \, dt.$$

       (c) For the BC angle β, show that

$$(\lambda_0^1 - \lambda_0^2) \int_a^b w(t)u_1(t)u_2(t) \, dt = [u_1(pu_2') - u_2(pu_1')](b)$$

       and then use the BC (S). Similarly for BC angle α.

6.28. (a) Change the equation to $-(p(t)x')' + q(t)x = (\lambda - l)w(t)x$.

       (b) Consider the equation

$$-(p(t)x')' + [q(t) + (lr_*/w^*)w(t)]x = \lambda w(t)x,$$

       where $r_* = \min\{r(t) : t \in [a, b]\}$ and $w^* = \max\{w(t) : t \in [a, b]\}$,
       and use Theorem 6.6.3, Part (a).

6.29. Note that $\alpha = 0$ and $\alpha = \pi$ both represent the Dirichlet BC at a and
       use the monotone property of $\lambda_0(\alpha)$ to show that $\lim_{\alpha \to \pi_-} = -\infty$, and
       then use the zero-counting properties of the eigenfunctions to show the
       second equality.

6.31. Use Theorem 6.6.3, Part (a) and (6.6.20).

6.32. (a) Yes, with a similar proof to Lemma 6.6.1.

   (b) See the proof of Lemma 6.6.4.

   (c) The inequalities parallel to (6.6.20) are:

$$\nu_0 < \lambda_0 \; \leq \; \{\mu_0, \nu_1\} \leq \lambda_1 < \{\mu_1, \nu_2\} < \lambda_2$$
$$\leq \; \{\mu_2, \nu_3\} \leq \lambda_3 < \{\mu_3, \nu_4\} < \lambda_4 \leq \cdots$$
$$\leq \; \{\mu_{2n}, \nu_{2n+1}\} \leq \lambda_{2n+1} < \{\mu_{2n+1}, \nu_{2n+2}\} < \lambda_{2n+2} \leq \cdots.$$

   For $n \in \mathbb{N}_0$, the eigenfunctions associated with λ_{2n} and λ_{2n+1} have exactly $2n + 1$ zeros in the half-open interval $[a, b)$.

   (d) $\bar{\lambda}_n \leq \mu_n$ for all $n \in \mathbb{N}_0$.

Bibliography

[1] F.M. Atici, G.Sh. Guseinov, On the existence of positive solutions for nonlinear differential equations with periodic boundary conditions. J. Comp. Appl. Math. **132**, 341–356 (2001)

[2] E.A. Barbashin, N.N. Krasovskii, On the stability of motion as a whole. Dokl. Akad. Nauk SSSR **86**, 453–456 (1952). (in Russian)

[3] G. Birkhoff, G-C. Rota, *Ordinary Differential Equations* (Wlley, New York, 1978)

[4] R. Bronson, G.B. Costa, *Schaum's Outline of Differential Equations*, 3rd ed. (McGraw-Hill, New York, 2006)

[5] F. Brauer, J.A. Nohel, *The Qualitative Theory of Ordinary Differential Equations* (W. A. Benjamin, New York, 1969)

[6] J. Carr, *Applications of Center ManifoldTheory* (Springer, New York, 1981)

[7] N.G. Chetaev, A theorem on instability. Dokl. Akad. Nauk SSSR **1**, 529–531 (1934). (in Russian)

[8] N.G. Chetaev, *Stability of Motion* (Moscow, Nauka, 1965). (in Russian)

[9] C. Chicone, *Ordinary Differential Equations with Applications*, 2nd ed. (Springer, New York, 2006)

[10] S.N. Chow, J.K. Hale, *Methods of Bifurcation Theory* (Springer, New York, 1982)

[11] E.A. Coddington, N. Levinson, *Theory of Ordinary Differential Equations* (McGraw Hill, New York, 1955)

[12] W.A. Coppel, *Stability and Asymptotic Behavior of Differential Equations* (D. C. Heath and Company, Boston, 1965)

[13] G. Dahlquist, Stability and error bounds in the numerical integration of ordinary differential equations. Kungl. Tekn. Högsk. Handl. Stockholm **130**, 87 pp. (1959)

[14] M. Eastham, Q. Kong, H. Wu, A. Zettl, Inequalities among eigenvalues of Sturm-Liouville problems. J. Inequalities Appl. **3**, 25–43 (1999)

[15] A.F. Filippov, An elementary proof of Jordan's theorem. Uspekhi Mat. Nauk **5:5**(39), 173176 (1950)

[16] W.B. Fite, Concerning the zeros of the solutions of certain differential equations. Trans. Am. Math. Soc. **19**, 341–352 (1918)

© Springer International Publishing Switzerland 2014
Q. Kong, *A Short Course in Ordinary Differential Equations*, Universitext,
DOI 10.1007/978-3-319-11239-8

[17] E.D. Gaughan, *Introduction to Analysis*, 5th edn. (Brooks/Cole, Pacific Grove, 1997)

[18] R. Grimshaw, *Nonlinear Ordinary Differential Equations* (CBC Press, New York, 1993)

[19] D. Grobman, Homeomorphisms of systems of differential equations. Dokl. Akad. Nauk SSSR **128**, 880–881 (1959)

[20] J. Guckenheimer, P. Holmes, *Nonlinear Oscillations, Dynamical Systems and Bifurcations of Vector Fields* (Springer, New York, 1983)

[21] J.K Hale, H. Kocak, *Dynamics and Bifurcations* (Springer, New York, 1991)

[22] P. Hartman, *Ordinary Differential Equations*, 2nd edn. (Birkha"user, Boston, 1982)

[23] P. Hartman, A lemma in the theory of structural stability of differential equations. Proc. Am. Math. Soc. **11**, 610–620 (1960)

[24] P. Hartman, On non-oscillatory linear differential equations of second order. Am. J. Math. **74**, 389–400 (1952)

[25] M. Hirsch, S. Smale, R.L. Devaney, *Differential Equations, Dynamical Systems, and Linear Algebra* (Academic Press, New York, 1974)

[26] M. Hirsch, C. Pugh, M. Shub, Invariant manifolds, *Lecture notes in Mathematics*, vol. 583 (Springer, New York, 1977)

[27] K. Hoffman, R. Kunze, *Linear Algebra*, 2nd edn. (Prentice-Hall, Englewood Cliffs, 1971)

[28] G. Iooss, D.D. Joseph, *Elementary Stability and Bifurcation Theory* (Springer, New York, 1989)

[29] I. Kamenev, Some specifically nonlinear oscillation theorems. Mat. Zametki **10**, 128–134 (1971) (in Russian)

[30] A. Kelley, The stable, center-stable, center, center-unstable, unstable manifolds. J. Differ. Equat. **3**, 546–570 (1967)

[31] Q. Kong, Interval criteria for oscillation of second order linear ordinary differential equations. J. Math. Anal. Appl. **229**, 258–270 (1999)

[32] Q. Kong, A. Zettl, Eigenvalues of regular Sturm-Liouville problems. J. Differ. Equat. **131**, 1–19 (1996)

[33] N.N. Krasovskii, *Problems of the Theory of Stability of Motion* (Russian), 1959; English translation: (Stanford University Press, Stanford, CA, 1963)

[34] Y. A. Kuznetsov, *Elements of Bifurcation Theory*, Applied Mathematical Sciences, Vol. 112, 2nd edn. (Springer, New York, 1998)

[35] J.P. LaSalle, Some extensions of Liapunov's second method. IRE Trans. Circuit Theory **CT-7**, 520–527 (1960)

[36] S.M. Lozinskii, Error estimates for the numerical integration of ordinary differential equations (Russian), I. Izv. Vyss. Ucebn. Zaved. Matematike **6**, 149–159 (1958)

[37] J.H. Liu, *A First Course in the Qualitative Theory of Differential Equations* (Printice-Hall, Upper Saddle River, 2003)

[38] Q. Lu, L. Peng, Z. Yang, *Ordinary Differential Equations and Dynamical Systems* (Beihang University Press, Beijing, 2010) (in Chinese)

[39] Z.E. Ma, Y.C. Zhou, *Qualitative and Stability Methods for Ordinary Differential Equations* (Science Press, Beijing, 2001) (in Chinese)

[40] J. Marsden, M. McCracken, *The Hopf-Bifurcation and Its Applications* (Springer, New York, 1976)

[41] L. Perko, *Differential Equations and Dynamical Systems*, 2nd edn. (Springer, New York, 1996)

[42] W. Reid, *Sturmian Theorey for Ordinary Differential Equations* (Springer, New York/Heidelberg/Berlin, 1980)

[43] M. Vidyasagar, *Nonlinear Systems Analysis* (Prentice-Hall, Englewood Cliffs, 1978)

[44] W. Walter, translated by Russel Thompson, *Ordinary Differential Equations* (Springer, New York, 1998)

[45] J. Weidmann, *Spectral Theory of Ordinary Differential Operators*, Lecture notes in Mathematics, vol. 1258 (Springer, Berlin, Heidelberg, 1980)

[46] S. Wiggins, *Introduction to Applies Nonlinear Dynamical Systems and Chaos*, 2nd edn. (Springer, New York, 2003)

[47] A. Wintner, A criterion of oscillatory stability. Quart. Appl. Math. **7**, 115–117 (1949)

[48] T. Yoshizawa, *Stability Theory by Liapunov's Second Method* (Gakujutsutosho Printing, Tokyo, 1966)

[49] A. Zettl, *Sturm-Liouville Theory*, Mathematical Surveys and Monographs, vol. 121 (Amer. Math. Soc., Providence, 2005)

[50] J.Y. Zhang, *Geometric Theory and Bifurcation Problems for Ordinary Differential Equations* (Beijing University Press, Beijing, 1981) (in Chinese)

[51] Z. Zhang, T. Ding, W. Huang, *Qualitative Theory of Differential Equations* (Science Press, Beijing, 1985)

Index

Printed in the United States
Bookmasters